AF294052

# THE FRONTIERS COLLECTION

*Series Editors:*
A.C. Elitzur L. Mersini-Houghton M.A. Schlosshauer M.P. Silverman
J.A. Tuszynski R. Vaas H.D. Zeh

The books in this collection are devoted to challenging and open problems at the forefront of modern science, including related philosophical debates. In contrast to typical research mono-graphs, however, they strive to present their topics in a manner accessible also to scientifically literate non-specialists wishing to gain insight into the deeper implications and fascinating questions involved. Taken as a whole, the series reflects the need for a fundamental and interdisciplinary approach to modern science. Furthermore, it is intended to encourage active scientists in all areas to ponder over important and perhaps controversial issues beyond their own speciality. Extending from quantum physics and relativity to entropy, consciousness and complex systems – the Frontiers Collection will inspire readers to push back the frontiers of their own knowledge.

## *Other Recent Titles*

**Weak Links**
The Universal Key to the Stability of Networks and Complex Systems
By P. Csermely

**Entanglement, Information, and the Interpretation of Quantum Mechanics**
By G. Jaeger

**Particle Metaphysics**
A Critical Account of Subatomic Reality
By B. Falkenburg

**The Physical Basis of the Direction of Time**
By H.D. Zeh

**Mindful Universe**
Quantum Mechanics and the Participating Observer
By H. Stapp

**Decoherence and the Quantum-To-Classical Transition**
By M.A. Schlosshauer

**The Nonlinear Universe**
Chaos, Emergence, Life
By A. Scott

**Symmetry Rules**
How Science and Nature Are Founded on Symmetry
By J. Rosen

**Quantum Superposition**
Counterintuitive Consequences of Coherence, Entanglement, and Interference
By M.P. Silverman

For all volumes see back matter of the book

# Hildegard Meyer-Ortmanns · Stefan Thurner

## Editors

# PRINCIPLES OF EVOLUTION

## From the Planck Epoch to Complex Multicellular Life

 Springer

*Editors*

Hildegard Meyer-Ortmanns
Jacobs University Bremen
School of Engineering & Science
Campus Ring 8
28759 Bremen
Germany
h.ortmanns@jacobs-university.de

Stefan Thurner
Medizinische Universität Wien
Inst. für Wissenschaft Komplexer
Systeme
Währinger Gürtel 18–20
1090 Wien
Austria
thurner@univie.ac.at

*Series Editors:*

Avshalom C. Elitzur
Bar-Ilan University, Unit of Interdisciplinary Studies, 52900 Ramat-Gan, Israel
email: avshalom.elitzur@weizmann.ac.il

Laura Mersini-Houghton
Dept. Physics, University of North Carolina, Chapel Hill, NC 27599-3255, USA
email: mersini@physics.unc.edu

Maximilian A. Schlosshauer
Niels Bohr Institute, Blegdamsvej 17, 2100 Copenhagen, Denmark
email: schlosshauer@nbi.dk

Mark P. Silverman
Trinity College, Dept. Physics, Hartford CT 06106, USA
email: mark.silverman@trincoll.edu

Jack A. Tuszynski
University of Alberta, Dept. Physics, Edmonton AB T6G 1Z2, Canada
email:jtus@phys.ualberta.ca

Rüdiger Vaas
University of Giessen, Center for Philosophy and Foundations of Science, 35394 Giessen, Germany
email: ruediger.vaas@t-online.de

H. Dieter Zeh
Gaiberger Straße 38, 69151 Waldhilsbach, Germany
email: zeh@uni-heidelberg.de

ISSN 1612-3018
ISBN 978-3-642-26802-1          ISBN 978-3-642-18137-5 (eBook)
DOI 10.1007/978-3-642-18137-5
Springer Heidelberg Dordrecht London New York



*Cover design:* KuenkelLopka GmbH, Heidelberg

Printed on acid-free paper

Springer is part of Springer Science+Business Media (www.springer.com)

*To Carlo (H.M.-O.)*
*To Franziska (S.T.)*

# Preface

The year 2009 was celebrated worldwide as Darwin year, 200 years after Charles Darwin's birth and 150 years after the appearance of his seminal book *On the Origin of Species*. Countless contributions in both the scientific and popular media revived the ongoing controversy between Darwinism and modern evolution theory on one hand and creationism – devoid of any scientific basis – on the other. The discussions were often polemic and varied in their degree of objectiveness. This inspired us to suggest a comprehensive survey of recent achievements in understanding evolution from a purely scientific point of view.

The contributions are from internationally known experts from various disciplines, writing about evolutionary theory from the perspective of their own fields, ranging from mathematics, physics, and cosmology, to biochemistry and cell biology. The concentration on natural sciences means that only some aspects of cultural evolution are covered. Seemingly simple questions are posed, concerning the origin of life, the origin of the universe, the concept of self-organization without the need for external interference, the probability of life coming into existence "by chance", and the role of contingency as compared to necessity; but instead of simple and premature answers being proposed, the questions are refined and specified until an answer can be arrived at using tools from the natural sciences. Scientific challenges correspond to reducing a complex problem by breaking it up into parts, finding common underlying mechanisms in the plethora of phenomena, or reproducing evolutionary processes in lab experiments or computer simulations; deeper insight and involvement in these challenges usually go along with fascination, amazement, and, finally, respect for the outcome of evolution in its full variety. We hope to contribute to all these aspects with our compilation of progress in the theory of evolution.

The book is addressed to readers with a background in life sciences and interest in mathematical modeling, or with a mathematical background in modeling and an interest in biological applications.

We thank the authors for their contributions and are indebted to the staff of Springer-Verlag for their support during the preparation of this book.

<table>
<tr><td>Bremen, Germany</td><td align="right">Hildegard Meyer-Ortmanns</td></tr>
<tr><td>Vienna, Austria</td><td align="right">Stefan Thurner</td></tr>
<tr><td>August 2010</td><td></td></tr>
</table>

# Contents

# Contributors


**Matthias Bartelmann** Institut für Theoretische Astrophysik, Zentrum für Astronomie, Universität Heidelberg, D-69120 Heidelberg, Germany mbartelmann@ita.uni-heidelberg.de

**Philippe Bastiaens** Department of Systemic Cell Biology, Max Planck Institute of Molecular Physiology, D-44227 Dortmund, Germany; Department of Chemical Biology, Technische Universität, D-44227 Dortmund, Germany, philippe.bastiaens@mpi-dortmund.mpg.de

**Richard A. Blythe** SUPA, School of Physics, University of Edinburgh, Edinburgh EH9 3JZ, UK, r.a.blythe@ed.ac.uk

**Leif Dehmelt** Department of Systemic Cell Biology, Max Planck Institute of Molecular Physiology, D-44227 Dortmund, Germany; Department of Chemical Biology, Technische Universität, D-44227 Dortmund, Germany, leif.dehmelt@mpi-dortmund.mpg.de

**Harold Fellermann** Department of Physics and Chemistry, FLinT Center for Living Technology, University of Southern Denmark, DK-5230 Odense M, Denmark, harold@ifk.sdu.dk

**Christian Hilbe** Faculty of Mathematics, Universität Wien, 1090 Wien, Austria, christian.hilbe@univie.ac.at

**Erwin Frey** Arnold Sommerfeld Center for Theoretical Physics and Center for NanoScience, Ludwig-Maximilians-Universität München, D-80333 München, Germany, frey@lmu.de

**Sanjay Jain** Department of Physics and Astrophysics, University of Delhi, Delhi 110 007, India, jain_physics@yahoo.co.in

**Byungnam Kahng** Department of Physics and Astronomy, Seoul National University, Seoul, Republic of Korea, bkahng@snu.ac.kr; kahng@phya.snu.ac.kr

**Kunihiko Kaneko** Department of Pure and Applied Sciences, University of Tokyo, Tokyo 153-8902, Japan, kaneko@complex.c.u-tokyo.ac.jp

**Claus Kiefer** Institute for Theoretical Physics, University of Cologne, D-50937 Köln, Germany, kiefer@thp.uni-koeln.de

**Pureun Kim** Department of Physics, Seoul National University, Seoul 151-747, Republic of Korea, lucky625@snu.ac.kr

**Sandeep Krishna** National Centre for Biological Sciences, Bangalore 560065, India, sandeep@ncbs.res.in

**Deokjae Lee** Department of Physics, Seoul National University, Seoul 151-747, Republic of Korea, lee.deokjae@gmail.com

**Hildegard Meyer-Ortmanns** School of Engineering and Science, Jacobs University Bremen, D-28725 Bremen, Germany, h.ortmanns@jacobs-university.de

**Alexander S. Mikhailov** Department of Physical Chemistry, Fritz Haber Institute of the Max Planck Society, D-14195 Berlin, Germany, mikhailov@fhi-berlin.mpg.de

**Tobias Reichenbach** Howard Hughes Medical Institute and Laboratory of Sensory Neuroscience, The Rockefeller University New York, NY 10065-6399, USA; Arnold Sommerfeld Center for Theoretical Physics and Center for NanoScience, Ludwig-Maximilians-Universität München, D-80333 München, Germany, tobias.reichenbach@rockefeller.edu

**Peter Schuster** Institut für Theoretische Chemie der Universität Wien, Währingerstraße 17, A-1090 Wien, Austria, pks@tbi.univie.ac.at

**Gunter M. Schütz** Institut für Festkörperforschung, Forschungszentrum Jülich GmbH, D-52425 Jülich, Germany, g.schuetz@fz-juelich.de

**Petra Schwille** Biophysics/BIOTEC, Technische Universität Dresden, D-01307 Dresden, Germany; Max Planck Institute for Molecular Cell Biology and Genetics, D-01307 Dresden, Germany, schwille@biotec.tu-dresden.de

**Karl Sigmund** Faculty of Mathematics, University of Vienna, A-1090 Wien, Austria, karl.sigmund@univie.ac.at

**Stefan Thurner** Science of Complex Systems, Medical University of Vienna, A-1090 Wien, Austria; Santa Fe Institute, Santa Fe, NM 87501, USA; IIASA, A-2361 Laxenburg, Austria, stefan.thurner@meduniwien.ac.at

# Chapter 1
# Introduction

**Hildegard Meyer-Ortmanns**

**Abstract** This introductory section serves as a summary of overarching concepts, universal features, recurrent puzzles, a common language, and striking parallels between units of life, ranging from simple to complex organisms. We start with a short chronology of evolution, since nowhere else in this book is the historical path of evolution followed. An underlying common methodology of all articles here is reductionism. The section on reductionism illustrates the great success of this approach with examples from physics and biology, including also a hint on its abuse when it is pushed to extremes. Guided by the success of reductionism, one may wonder whether there is a universal theory of evolution. Such a universal theory does not exist, but striving for universal laws makes sense when it is based on striking similarities between seemingly very different realizations of systems whose dynamics is governed by the very same mechanism. In this case recurrent behavior goes beyond a superficial analogy. Again we illustrate the concept of universality with examples from physics, but also indicate limitations in view of universal equations. Beyond universal laws, universal principles of organization may be at work. One such overarching principle is self-organization. In addition to its known successful application in various disciplines such as physics, chemistry, and cell biology, it leads to a challenge for future research on how far one can further stretch this concept to explain all complex outcome of evolution as self-organized. Common to the various examples of self-organization in later chapters is the emergence of a complex structure out of less structured or even random initial conditions. The very choice of initial conditions is often the art of the game. In the spirit of reductionism, the initial conditions should involve as little structure as possible to let the complex structure emerge from the very rules of evolutionary processes. This demand may lead to "chicken and egg"-like dilemmas. Such dilemmas appear in many facets in and outside natural science. They are intimately related to questions of origin, in particular the origin of life. Questions of life's origin go along with an estimate of the

H. Meyer-Ortmanns (✉)
School of Engineering and Science, Jacobs University Bremen
D-28725 Bremen, Germany
e-mail: h.ortmanns@jacobs-university.de

H. Meyer-Ortmanns, S. Thurner (eds.), *Principles of Evolution*,
The Frontiers Collection, DOI 10.1007/978-3-642-18137-5_1,
© Springer-Verlag Berlin Heidelberg 2011

date when first forms of life appeared. Therefore another evocative topic concerns the very probability of life coming into existence in the course of evolution. This relates to the tension between contingency and necessity, stochastic fluctuations and deterministic rules. The question arises as to whether, if we were able to rewind the tape of evolution and replay it again, contingency would lead to minor differences or even changes in the gross features of the evolutionary outcome. Rewinding the whole tape is science fiction, but rerunning short sequences of this tape is reality. We collect a few such attempts from contemporary lab experiments under controlled initial conditions or related computer simulations. Without mathematical modeling, seemingly natural extrapolations lead to premature or even false conclusions on the evolutionary potential. Therefore we disentangle the reduction of complexity from misleading oversimplifications and conclude with an appeal for mathematical modeling also in biology. Finally we summarize all chapters of this book to embed their content in the context of this book.

## 1.1  A Short Chronology of Evolution

The contributions to this book refer to essentially three different epochs of the universe: the very early state about 14 billion years ago, which was the time when the conditions were fixed for the later structure formation of the universe; a second epoch, about 4 billion years ago, after our planet had formed and early forms of life were created, and a third epoch, when life organized in cells and societies as of today. In view of these three periods we present a very coarse-grained view in the following. The reason for the coarse-grained selection is that in the first and second epochs we cannot follow the historical path, owing to lack of data. Most detailed is the history of evolution from earlier to current life forms, because it is based on the fossil record, but the fossil record gives no hint about pathways of chemical evolution and the bridge to biological evolution [1]. Owing to lack of data from the remote past, the historical route to life is very speculative and will remain so in the future. Therefore in this book, no statements will be found on how life "really" emerged, but only how it may have emerged, and what the possible pathway could have been, and the proposed options are certainly not unique; they kindle controversial discussions on the pros and cons of "genetic first" models, whose representatives argue in favor of an RNA-world to start with, or "metabolism first" models, with (auto)catalytic reaction networks first, or "compartment first" models, with a lipid world as starting point. The partition into these three options for what came first reflects the three necessary ingredients of a primitive cell: a molecular realization of coding and storing information in the form of RNA and DNA, metabolism for providing the fuel for the "machinery" to run, and compartmental structures with a division mechanism needed for the heredity of information.

Going backwards in time from the prebiotic to the chemical world, governed by the laws of chemistry and physics, and further back to the world governed merely by the laws of physics due to its elementary constituents, we end up at the big bang, about 14 billion years ago. "Big bang" stands for an extrapolated event prior to some finite time in the past when the universe originated from an extremely hot and dense state of matter, as predicted by so-called Friedmann models (for a more detailed discussion, refer to Chap. 7). One may be tempted to extrapolate classical notions of space and time back to time "zero", but such a naive extrapolation leads to singularities in the formulas, the classical description becomes meaningless. In standard chronologies one finds instead of "time zero" an extremely tiny number assigned to the big bang, namely $10^{-44}$ s. What sense does it make to assign such an incredibly small number to an event that initiated an evolution lasting over 14 billion years? $10^{-44}$ s is the time light takes to travel a distance on the order of the Planck length, which is $\sim 10^{-35}$ m. The Planck length in turn is defined in terms of three fundamental constants of physics, the velocity of light (in vacuum) $c$, Newton's gravitational constant $G$, and the (reduced) Planck action constant $\hbar$, according to $l_{\mathrm{P}} = \sqrt{\hbar G/c^3}$. The mass that can be constructed in terms of these quantities by dimensional arguments is $m_{\mathrm{P}} = \sqrt{\hbar c/G} \approx 10^{19}$ GeV$/c^2 \approx 10^{-8}$ kg. This number sounds less exotic than the Planck length: it is not extremely large, orders of magnitude $(10^{-7})$ smaller than the mass of a bacterium (since we deal in this book with bacteria in another context, the comparison is natural), neither is it extremely small, it is $10^{19}$ times larger than the mass of a proton, and the energy required to create such a particle out of the vacuum is far too high to be achieved in labs via particle accelerators.

So the question remains why these fundamental constants and units matter at all in view of the big bang. Backwards in time the universe got denser and hotter. When it reached a density on the order of the Planck mass per Planck length, quantum mechanical and gravitational effects became important at the same time. The reason is the following: The Compton wavelength that can be uniquely assigned to a particle is the length scale within which quantum properties become relevant. On the other hand, also the Schwarzschild radius can be uniquely assigned to the same particle: if the mass of this particle is confined to a sphere with a radius as small as its Schwarzschild radius, it ends up as a black hole, and by then at the latest gravity can no longer be ignored. So one may ask for which mass density both lengths are (roughly) the same. Here it turns out that if the Planck mass is concentrated over the tiny region with an extension on the order of the Planck length, both length scales are almost the same, that is, the particle's Compton wavelength is of the same order as its Schwarzschild radius. (The restriction "of the same order" and "almost the same" refers to prefactors on the order of one, while the Planck mass and Planck length are known to high accuracy due to the corresponding accuracy for the fundamental constants that are used for their very definition.) Now, if quantum mechanical and gravitational effects play a role at the same time, which happens (probably earlier, but certainly) no later than when the universe has reached the Planck density, classical notions of space and time as used in Einstein's general relativity become meaningless: a unified theory of quantum gravity is needed. Quantum gravity, however, is one of the most challenging branches of theoretical physics and one easily encounters paradoxes when it is applied to cosmology. To date there is no proposal of a candidate that is convincing in all desired theoretical properties, but proposals have been made, and we shall learn about one of them in Chap. 8.

Let us return to the time that light takes to travel the Planck length, that is, $10^{-44}$ s. This number marks the smallest time scale to which an extrapolation of the classical notion of time is meaningful. The big bang then summarizes events outside (not to say "before") the classical realm. Chapter 8 deals with this so-called Planck epoch. For a certain proposal of quantum gravity it is demonstrated how the very notion of time fades away on a formal level so that time becomes an emergent quantity at later stages of the universe.

Less speculative than the description of the earliest epoch is the standard cosmological model, which is a so-called Friedmann model, specified for a certain (single) set of parameters. It describes cosmic evolution from a few minutes after the big bang until today (see Chap. 7). This model also predicts when thermal radiation could freely propagate. That was possible when the universe was about 400,000 years old, and it then became transparent within about 40,000 years. This thermal radiation is the famous cosmic microwave background (CMB), whose tiny anisotropies are interpreted as an imprint of earlier density fluctuations in the universe. As pointed out in Chap. 7, it is essentially Newtonian hydrodynamics that governs the structure formation of the universe from that time on, once the CMB data are appropriately implemented as initial conditions.

What happened between the big bang and the decoupling of the cosmic microwave radiation falls in the realm of theoretical particle physics within the framework of the Standard Model and its possible extensions (see, e.g., [2–4]). Accordingly, the originally unified forces became separated into the four fundamental interactions of today: the strong, weak, electromagnetic, and gravitational interactions; moreover, an inflationary expansion of the universe is thought to have occurred, which amplified initial quantum fluctuations that finally led to structure formation in the classical universe. More matter than antimatter remained. Nucleons (protons and neutrons) were made out of the initially free quarks and gluons (when the temperature had cooled down to the order of $10^{12}$ K, where K stands for Kelvin). "Plasma"-time started when electromagnetism became the dominant force and photons collided with electrons so often that the electrons could not bind to atomic nuclei. However, when about 40,000 years later the universe had further cooled down to $\sim 3,000$ K, the radiation lost energy and the electrons now could be captured by atomic nuclei. Atoms were formed while photons moved freely through the universe as CMB radiation.

From then on gravity determined what happened on cosmic scales with the cosmic gas: quasars and galaxies arose along with black holes in their centers. Within the galaxies smaller clouds contracted to form stars, and within the stars heavier elements than the primordial ones formed. Not finally, but in particular, our blue planet formed from the accretion disc revolving around the sun. This was about 4.6 billion years before the present time and about 9 billion years after the big bang.

Formation of the earth is usually taken as the starting point of another chronology, namely the chronology of life on earth [5]. Although life may have developed (possibly in other forms) at other places in the universe as well, we do not know any historical facts, so that a chronology would not make sense, but there is a relatively young field of research, called astrobiology, that studies possible forms of life under very different and – from our point of view – very extreme conditions.

For the chronology of evolution of life on earth "time zero" is defined as "today" and years are counted backwards from now. The first main epoch is the Hadean Eon, which extends from 4.6 to 3.8 billion years ago, during which the first precursor of life occurred, which we shall not discuss along a hypothetical historical path but along a possible path. The second epoch, the Archean Eon, includes the split between bacteria and archaea (about 3.5 billion years ago), and in particular the new use of photosynthesis by cyanobacteria. During the Proterozoic Eon, 2.5–0.5 billion years ago, eukaryotic cells appeared, sexual reproduction occurred, and first simple and later more complex multicellular organisms developed. About 500 million years ago the phyla of animals started. The Phanerozoic Eon extends from 542 million years ago until today. Now the resolution of time proceeds in terms of the appearances of sponges, worms, fish and proto-amphibians, land plants, insects, reptiles, mammals, birds, flowers, and finally, about 200,000 years ago, *Homo sapiens* in a form resembling us today. With respect to the phase of today, we discuss cells in vivo (Chap. 9) and similarities in organization between cells, swarms, and societies (Chaps. 13, 14, 15, and 16).

Usually chronologies of the evolution of life select major phenotypic changes, such as a change in the natural habitat from sea to land, or the occurrence of new phyla, moving along some milestones of the phylogenetic tree. The selection of steps could be based on different criteria as well, such as the way in which information is transmitted between generations. These criteria underlie the so-called major transitions in evolution according to Maynard Smith and Szathmáry [6]. To these belong the origin of chromosomes, the origin of eukaryotes, the origin of sex, the origin of multicellular organisms, the origin of social groups such as ants, bees, and termites, and – although only indicated by the authors – the development of natural neural networks.

Finally we would like to point out the ambiguity in what is called a major event or a sudden event in evolution. A drastic change over a short time period may be an effect of the overall gross time scale, on which the event appears as almost instantaneous. It may be also an artifact of current ignorance of the intermediate mechanisms. What looks like a big and rather unlikely mutation may later turn out to be a cascade of combined small-step changes.

## 1.2 Scientific Reductionism

Throughout this book we find reductionism in the methodology. Phenomena on a given scale are explained in terms of interacting units on a smaller scale or a scale on a lower level in a hierarchical system. This reductionism includes the occurrence of emergent phenomena. (A typical example of an emergent phenomenon is the full segregation say of black and white balls into two separated areas as result of local migration rules, although the local rules would not suggest a complete separation.)

Reductionism is a very successful concept in physics, particularly when it is related to symmetries. The special-relativity principle of Poincaré, Einstein, and Lorentz, for example, postulates that the laws of physics should be the same for all observers who move with a constant velocity with respect to each other. Including not only classical mechanics, but also electrodynamics and Maxwell's equations, this implies that transformations between such moving inertial systems ("attached" to the observers) should leave the velocity of light invariant. This, in turn, restricts the allowed transformations between different inertial systems to Lorentz transformations. And the implied postulate therefore reads that the laws of physics should be covariant (form-invariant) under Lorentz transformations, resulting in strong constraints on the models that are compatible with this postulate. (An extension of this postulate to reference systems that are accelerating with respect to each other led Einstein to his construction of general relativity.)

Another most powerful principle with far-reaching consequences is that of local gauge invariance, which goes back to Hermann Weyl at the beginning of the last century [7]. According to this postulate, physical laws for the fundamental interactions should be invariant under local gauge transformations. Without going into more detail here of what this actually means, we can state that the implementation of this postulate for different gauge groups led to the construction of the theories of the four fundamental interactions (i.e., gauge theories of the strong, weak, electromagnetic, and gravitational forces) [8]. The strong and the electroweak gauge theories together make up the so-called Standard Model, which is experimentally confirmed to high accuracy. Remarkably, the construction of these gauge theories from the postulate is almost unique. The degree of uniqueness can only be appreciated by those who know about the zoo of elementary particles and the plethora of experimental results on collisions between these particles, which become ordered, understandable, and predictable in terms of these theories. All other forces, ranging from van der Waals forces to intermolecular forces in biology, in principle derive from the fundamental ones (in principle, not in practice).

Another success of reductionism is known from statistical mechanics, when macroscopic observables such as the temperature or magnetization of a system are expressed in terms of microscopic variables such as the kinetic energy or spin of its atomic constituents. Such a mathematical representation provides not only a deeper understanding, but also possible control of the system via its microscopic parameters; one should admit, however, that this is only possible for relatively simple systems (as compared to living matter), and in general bridging the gap between microscopic and macroscopic phenomena in terms of a mathematical derivation is quite hard.

Because of parallels in current developments in systems biology, we mention one further successful manifestation of reductionism: the renormalization group approach [9, 10]. In contrast to the previously mentioned bridging of scales in one step, the renormalization group is an iterative procedure. It aims at a derivation from first principles along with a reduction of complexity: when appropriate new variables are introduced, their interactions should be analytically derived on the new scale upon integration of effects from the smaller scale (where the scale can refer to configuration or momentum space). The process is then iterated a number of times. This concept is indeed very successful in explaining critical phenomena, and if the new variables and the new interactions on the coarse scale are of the same type as those on the fine scale, it corresponds to a special case (and is summarized under the keyword of scale invariance); in general, variables and laws on the coarse scale may be quite different from those on the fine scale. In the simplest realization, however, the new variables may be chosen just as the normalized sums over a selected range of the former variables whenever this normalized sum may be considered as representative variable for the new scale. Similarly, if a whole chain of enzymes always acts together in the same way, it seems to be justified to gather the collective action of enzymes into one new effective enzyme variable. In general, however, the new variables will be intricate functions of the old ones, and the new variables should match the criterion that they should lead to simple interaction laws. It is already quite hard to derive phenomena of nuclear physics in terms of quarks and gluons, that is, in terms of the variables of the underlying theory of strong interactions, that is quantum chromodynamics. To bridge larger gaps in scales in an analytic derivation will be even harder.

Nowadays certain directions of systems biology proceed in a similar spirit, when specific behavior of metabolic networks is traced back to the underlying level of genes (see, e.g., [11]). Multiscale modeling of natural hierarchically organized materials such as bones aims at a similar form of reductionism: to explain malfunction of bone on a macroscopic level (manifest in the form of diseases) in terms of failure of signaling pathways on a cellular level; this opens the possibility of curing the cause rather than the symptoms. Multiscale modeling again is an iterative procedure, integrating features from scale to scale.

So far our examples from physics and systems biology argue in favor of reductionism when it leads to a deeper understanding, allows us to control systems via a few essential parameters, or reduces the complexity. In the light of present and future computer simulations one may wonder why one should choose an iterative procedure such as the renormalization group approach or multiscale modeling at all, why not choose the brute force method and simulate a little "nanomachine" like a virus in terms of its atomic constituents to push the principle of reductionism even further? This is actually what people do with success. More precisely, molecular dynamics simulations of the complete satellite tobacco mosaic virus (STMV) have been performed [12] with 1 million atoms over a time interval of 50 ns. The virus consists of a capsid composed of 60 identical copies of a single protein, and a $1,058$ kb RNA genome. It is modeled via 949 nucleotides out of the complete genome, arranged into 30 double-stranded helical segments of 9 base pairs each. An important outcome of these simulations was that the capsid is unstable when there is no RNA inside, and the implications for assembly and infection mechanisms were discussed. Results of this kind provide useful insights for identifying the relevant input for a certain behavior. In common with all molecular dynamics simulations, the endeavor of starting from the atomic decomposition of macromolecules and simulating certain aspects of their evolution in time covers a short time window out of the total time evolution. The feasible time window in the simulations should be long enough to follow the real kinetics of the process in question. For the time window of 50 ns in the quoted numerical experiments of [12], advanced computer algorithms were already needed along with parallel computing with sophisticated communication between the processors running in parallel; otherwise, for example on a typical laptop of today, the simulations would take decades of CPU-time.

For a moment let us be optimistic and extrapolate the power of future computers so that the accessible time windows become larger and larger and the evolution of macromolecules can be followed over longer and longer periods. What would be missing in such an approach, even if it produces reliable results on the larger scale, is some insight into how the results come about. What the computer simulations can definitely never provide is an understanding in terms of new simple variables, since emergent phenomena on a gross scale may require new (e.g., collective) variables with new (effective) rules of interactions, in terms of which the laws on the larger scale are simple and transparent again. In the framework of the renormalization group, computers are not able to propose on their own the effective new variables for the next iteration step. Simple laws and simple (collective) variables are essential for what is called understanding. If reductionism is pushed to its extreme, the price is a loss of transparency. A representation of DNA in its atomic constituents yields constraints on its bending, coiling, and other physical properties, and is therefore of use, but it does not reveal its emergent role as a carrier of information. Basic laws from physics still provide useful constraints on the hardware of a cell, but the "software" needs other disciplines like biochemistry and cell biology. Therefore living systems need intrinsically an interdisciplinary approach. Reductionism in the various disciplines is quite useful, but should not be overstretched, and such an approach is followed throughout the articles of this book.

## 1.3 Universal Features, Universal Processes, and Striving for Universal Laws

Again, let us start from the physics perspective. Here the universal features we shall describe are not abstractions, or oversimplifications, or mere metaphors for universality, but striking quantitative experimental facts: substances that are obviously very different, such as binary liquids and ferromagnets, show the same critical exponents. Critical exponents characterize the behavior of these systems close to a critical point. In a liquid–vapor transition, a critical point terminates the liquid–vapor coexistence curve, in a binary mixture such as isobutyric acid (2-methylpropanoic acid) plus water it marks the temperature at which phase separation can first take place as the temperature is lowered, and in a spin system it marks the transition from a paramagnetic to a ferromagnetic state. Now, the remarkable experimental observation is that these systems show common features in how they approach the critical point. These features are characterized by critical exponents: in particular, and common to these different systems, the typical length over which units of the system are correlated diverges at the critical point according to some inverse power of the distance from the critical point, and it is this power that is called the critical exponent of the correlation length. The correlation length then dominates all other length scales of the system. The systems show collective behavior near the transition, and the behavior depends only on a few generic characteristics, such as the space dimension and the type of interaction (long- or short-ranged); microscopic details do not affect the exponents. The collective behavior, parametrized in terms of sets of critical exponents, is the same for various systems at the critical point, and those systems which share identical sets of critical exponents are gathered into so-called universality classes. Universality between liquid–gas and ferromagnetic systems can be explained in terms of a transformation that maps one system into the other. In general, universality can be understood in the framework of the renormalization group as formulated by Wilson [9] and Kadanoff [10]. No doubt, on the atomic level isobutyric acid and water look quite different from an Ising ferromagnet, but obviously these microscopic details are irrelevant in the region of the phase transition, and the behavior can be explained in terms of a subset of variables, the so-called relevant degrees of freedom.

In a similar spirit, in relation to the theory of evolution, one may search for classes of population dynamics that lead to quantitatively the same critical behavior of phase transitions in terms of bursts and extinctions of populations, which may be independent of the concrete realization of the population's individuals (as chemical or animal species, for example).

Behind the shared universal behavior of members of the same universality class are shared universal mechanisms, characterizing the transition from one phase of the system to another. In common with the following notions of universality is the feature that the universal behavior (here at critical phenomena) refers to selected parameters (out of the critical region), subsets of variables (those termed as relevant variables), and underlying processes (such as the emergence of long-range correlations).

So let us next discuss universal *processes*, universal in the sense that they are observed on different scales, some of them ranging from the microscopic scale of elementary particles to macroscopic scales. To these belong creation and annihilation or birth and death events, replication, mutation, competition, selection, fragmentation, composition, recombination, diffusion, drift, and migration. All processes can be specified in terms of rates, which are taken from experimental measurements or postulated on the basis of a theoretical understanding of the involved interactions. On the level of elementary particle physics the rates are determined by the well understood fundamental interactions. They allow us to predict cross sections at the large collider experiments. For chemical species the equations can be cast in the form of chemical rate equations (see Chap. 2). If our variables stand for normalized frequencies of individuals of a given population, we obtain equations of population genetics, and depending on the rules of interactions, equations of (evolutionary) game theory result, some of which are discussed in Chaps. 13 and 14.

For a moment let us assume that the above list of elementary processes is complete, in the sense that we can decompose an arbitrary concatenation of evolutionary processes on an arbitrary scale in terms of these elementary ones. Say, we consider the evolution of a population of several species of bacteria, fluctuating in number owing to birth and death processes, undergoing mutations during replication, diffusing and migrating on a plate, and interacting only with their nearest neighbors. One may be tempted to jump to conclusions and wonder whether there is just one corresponding universal "chemical" master equation, *the* master equation of the theory of evolution, independent of the subset of processes being considered. The answer is negative. The very form of the equation depends on the choice of included processes. As pointed out in Chap. 13, there is a hierarchy also in the mathematical complexity of description. In the context of population dynamics it can be summarized as follows: As long as stochastic effects due to mutations in the species or fluctuations in the population size can be neglected, the equations will take the form of deterministic rate equations, for example in the form of (nonlinear) replicator equations. When the included processes lead to fluctuations, in particular also in the population size, the population size itself becomes a random variable, the equations take the form of (chemical) master equations and no longer determine the size, but only the probability of finding the population at a given time to have a certain size. All this description holds as long as no spatial organization (that is, information on who interacts with whom) and no diffusion or migration processes are taken into account. This is obviously a special case of a generic situation. An inclusion of the influence of mobility and connectivity, in addition to the effects of finite size and noise of other origin, finally leads to equations that take the form of stochastic partial differential equations. For further details we refer to Chap. 13. Currently, the effect of spatio-temporal correlations on population dynamics is an active field of research. Therefore, in general, an extension towards inclusion of further processes on top of a given set is more than adding further components to a given set; it usually leads to an interplay of the former set of reactions with the new set and therefore to a rich spectrum of possible behavior that is reflected in an increasing complexity of the mathematical description.

Still one may search for universal laws in the plethora of phenomena. What we call universal laws in the context of the theory of evolution are characterized by a set of equations that allow very different realizations in natural systems. An example is the motif of three mutually repressing species. The interaction here is mutual repression, species $A$ represses $B$, $B$ represses $C$, and $C$ represses $A$: a game that is played on many scales, ranging from the genetic level [13] to bacteria [14] and human players of the rock–paper–scissors game. Common to the different realizations are terms in the set of equations, which describe mutual repression, and the mechanism of how various species can coexist and diversity be maintained. Another motif of this kind that allows for many interpretations is the SIR model for populations, whose individuals can be susceptible (S), infected (I), or recovered (R) [15].

In summary, universal laws refer to recurrent features on different scales, providing the underlying mechanism in common. A universal "theory" that aims at an explanation of complex behavior in all its facets in one theoretical framework is neither realistic nor a topic of this book. (If we nevertheless use the notion of "theory of evolution", "theory" stands for a conglomerate of mathematical models with varying degree of predictive power.)

## 1.4 The Concept of Self-Organization

Self-organization is a key concept for explaining complex structures emerging out of a less complex or an even random start on the basis of local rules. The local rules can be coded in the form of differential equations. It has a long tradition in chemistry, physics, and morphogenesis, and in the meantime has entered life sciences, including the fundamental unit of life, the cell. The intimate relation of self-organization to the science of evolution is the chance of explaining the emergence of complex structures without the need for a creator, an architect, a designer who guides this evolution from an elevated vantage point with a blueprint in his mind.

As a first example let us consider Turing patterns, named after Alan Turing, who is known as a mathematician, logician, cryptanalyst, and computer scientist but also studied the chemical basis of morphogenesis. He predicted oscillations in chemical reactions, such as Belousov–Zhabotinsky reactions, as early as the 1950s [16]. These patterns are neither random nor regular but complex in their structure. Turing patterns in animal coats of zebras, cheetahs, jaguars, leopards, and snails result from instabilities in the diffusion of morphogenetic chemicals in the animal's skin during the embryonic stage of development; for reviews see [17, 18]. Since the proposed mechanism cannot be directly tested at the embryonic stage, a very different set of experiments was performed, experiments with vibrating plates in which the density fluctuations of the air were visualized by holographic interferometry. Varying the shape and size of the plates enabled the patterns of visualized amplitude fluctuations to be tuned between various patterns of spots and stripes, strongly resembling patterns of animal coats [19]. What is the reason behind this amazing possibility of mimicking and visualizing pattern formation? The answer lies in the same mathematical structure of the associated differential equations. As indicated in [19], the reaction–diffusion equations for two substances (which act as positive and negative inhibitors, not further specified in their chemical composition, but finally determining the melanin concentration on the skin) can be approximately mapped to an equation that describes the amplitude fluctuations of a vibrating membrane, here chosen as an elastic plate whose initial and boundary conditions can be easily varied. So the experiments serve to analyze the important role of size and shape of animal's skin on pattern formation on this skin. In principle, therefore, a detailed understanding of the mechanisms behind Turing patterns in animal coats is possible (even if these patterns sometimes look like they are designed by talented artists if we think of butterflies, for example).

As we shall see in Chap. 7, it is "just" the combination of physical laws of gravitation and hydrodynamics that is able to predict the formation of cosmic structures such as galaxies in the universe if the initial conditions for the equations are chosen compatible with the CMB data. So the structure is coded in the initial conditions along with the dynamical laws of evolution. A hint to how the structure of the CMB data itself might have arisen is given in Chap. 8.

From the predictive power of self-organization on cosmic scales and of pattern formation in animal coats down to scales of daily life one may be tempted again to jump to conclusions. Can ultimately all evolution, in particular the emergence of life, be understood as self-organized? Self-organized in the sense that we know the laws and their mutual interplay and can thus reproduce or predict the very formation of living entities from appropriately chosen initial conditions? In order to estimate how far we are still off from answering this question, it is instructive to consider the basic unit of life: a living cell. First of all it should be noticed that a single cell (whether prokaryotic or eukaryotic) is a structure of much higher degree of complexity than mass densities in the universe or pigment densities in the skin, and the inherent hierarchical structure of the cell refers to a functional hierarchy, which contains a subset of processes that are able to control other processes on a different scale. Therefore it is very instructive to focus on self-organization within a single cell (Chap. 9). In their contribution, Dehmelt and Bastiaens carefully distinguish between different organizational principles within a cell: apart from self-organization there are "key-regulators" and "blueprints", similarly to the distribution of work between an architect, his blueprint of the final building, and the workers who build the house accordingly. Key-regulators are steering units – without key-regulators no performance would be possible – whereas steered units, larger in number and less complex, may be more easily substituted when they fail so that the performance goes on. Still, aster and spindle formation in the cell are meanwhile understood as self-organized processes via microtubules and molecular motors [20], but "who" constructed the molecular motors? If self-organization within the cell is only one among other organizational principles, do we need architects and designers in general in the same way as the cell needs key-regulators and blueprints, coded in its DNA, to function well? Can a system produce its own key-regulator if the system is not a society with its self-elected representative or ruler, but a precursor of a cell? The answer is open.

A challenge for future research will certainly be to construct dynamical systems in which spontaneous symmetry breaking or nonlinear interactions lead to the generation of a dynamical hierarchy, a hierarchy organized by the system itself and leading to asymmetric relationships between steering and steered units. There is some chance for such a description if we take self-organized dynamical hierarchy of evolved turbulence in a viscous fluid as a metaphor of how it may occur. In an abstract way let us summarize what is needed. If we describe a hierarchical system as consisting of several "horizontal" levels, labeled along a "vertical" axis. It seems to be quite generic that the different levels are characterized by different time scales (accompanied by different interaction strengths) that are characteristic for the intrinsic dynamics on this level. So the time scales may be used as possible labels. The interactions between units within the levels on one hand and between the levels on the other hand are then quite different: the upper level will enter the dynamics on the lower level as control parameters; their slow change may induce bifurcations or phase transitions on the lower level, but there is no direct interaction with the fast variables, so the slow dynamics will control the fast dynamics and not vice versa. While the fast dynamics is able to adapt to slow changes of the control parameters, the slow dynamics is not, it cannot resolve the short time scale, it merely sees the average values of the fast variables. A reflection of this asymmetry may be found in steering and steered processes: slow processes are able to steer fast ones, fast ones influence slow ones in a different way. The whole scenario of a dynamical hierarchy is actually realized in a self-organized way via the Navier–Stokes equations, due to nonlinear interactions of the fluid molecules. Note that our previous examples of galaxy formation and aster and spindle formation in the cell are self-organized processes within one level of the formerly mentioned type of hierarchy. (The functional hierarchy should be distinguished from the hierarchy of spatial scales involved in the structure formation of the universe.) What is missing for the cell is a set of dynamical equations that predicts self-organization in the "vertical" direction, that is, the emergence of key-regulators on top of the regulated units, in analogy to the Navier–Stokes equations, which predict the formation of slowly moving and large vortices that enslave the fast-moving and small vortices. The reader will find an extensive discussion of dynamical hierarchies in the book by Mikhailov and Calenbuhr [35] and a summary in Chap. 16.

## 1.5 Chicken-and-Egg Problems in Many Facets

No egg without a chicken and no chicken without an egg. How can we break up this cycle and find out which was first (if one of them was first at all)? Obviously the chicken-and-egg problem is intimately related to evolution, and in a metaphoric sense it is a ubiquitous dilemma of the kind "Which came first if $A$ cannot exist without $B$ and $B$ cannot exist without $A$?" Although in this book we study such dilemmas in a scientific context, it should be mentioned that even ancient philosophers such as Aristotle, and Plutarch were aware of this puzzle in terms of bird-and-egg or hen-and-egg problems and also recognized their generic nature, which finally leads to the question of the origin of the universe. When time is considered as cyclic, as in Buddhism, there is no "first", and this may be a better way to bear the march of time, but even if our universe ends up in what is called a big crunch and restarts with a new big bang, this would not answer our question when we only want to focus on the segment from $A$ to $B$ or, alternatively and not necessarily equivalently, from $B$ to $A$, in order to understand the intermediate evolutionary steps of this segment.

Not only in natural science and philosophy, but also in the social sciences, chicken-and-egg dilemmas are very familiar, and here they can be related to vicious circles. An advertisement for a job usually requires experience and experience requires already having a job, or a place of residence requires employment and employment requires a place of residence. The energy costs to break up these cycles and overcome the barriers in practice act as selection criteria. Chicken-and-egg dilemmas are also known in medical applications. In medical treatment of diseases the "dilemma" occurs when symptoms can be the cause and consequence at the same time, although there a successful treatment of the symptoms would fortunately also remove the cause.

Back to chicken-and-egg problems in the scientific context. Well known from mathematics and physics are differential or integral equations posed as self-consistency problems. If the left hand side stands for the required solution, the trouble occurs when the right hand side requires the very same solution as input.

Here a way out may be an approximative ansatz inserted in the right hand side, leading to an approximate solution on the left hand side. Iterating this procedure, the difference between the succeeding approximative solutions may converge to zero, so that the right solution can be self-consistently determined.

More generally, what is recurrent in scientific versions of the chicken-and-egg problem is the struggle for how much information or structure should be encoded in the initial conditions or, more generally, in the chosen initial state of the system, and how much structure should come out as result of the dynamic laws that evolve the initial to the final state. From the reductionist's point of view, one would like to put in as little structure as possible, so that the complexity of the final state is larger than that of the initial one (whatever the complexity precisely refers to). In general, the specification cannot be reduced to assigning initial values to a certain function or a functional that evolves in time, in particular not if we deal with experimental realizations. The very selection of what should evolve poses a problem and often is the art of the game.

In the context of this book, the question of whether the prebiotic world was an RNA-world, a "metabolism-first" world, or a lipid world (as mentioned in Sect. 1.1) certainly comes closest to the chicken-and-egg problem in the literal sense. This is an ongoing debate among experts. A fundamental distinction in biology is between nucleic acids on the one hand, as carriers of information, and proteins on the other hand, as generators of the phenotype. In living systems nowadays nucleic acids and proteins mutually need each other. As division of labor the former store the heritable information and the latter read and express it. Once again: which came first? Arguments in favor of RNA coming first can be found in [6]. RNA seems to be able to bridge the gap between chemistry and biology. It can display also catalytic activity [21], and can be made to evolve also new catalytic activities through molecular Darwinian selection [22], but the question remains whether chemical evolution itself could have produced replicative RNA, since the spontaneous generation of RNA seems to be very problematic under prebiotic conditions [23]. Once it is there, however, the game can go on.

One of the alternative approaches is a lipid world to start with. Lipids are amphiphiles with a capacity to spontaneously self-organize into supramolecular structures such as droplets, micelles, bilayers, and vesicles. Lipids are very likely to occur. Later they can give rise to more complex biopolymers, and there seems to be evidence that catalysis is not restricted to proteins and RNA, but also lipids can evolve catalytic capacities and precursors of metabolism in the end. While in the RNA world the first occurrence of RNA appears as a discontinuous event, since it is quite unlikely to be produced under prebiotic conditions, here one seems to end up at a discontinuity between catalytic lipid aggregates and biopolymer-based cellular life that we observe nowadays [24]. An ansatz has been followed that information is coded in the form of specific molecular compositions [25], but the question remains how progress in evolution can be achieved without an alphabet-based coding system. Was the lipid world scenario just the precursor of the RNA world? The final answer is not yet known.

Two chapters, 11 and 12, deal with contemporary lab experiments that attempt to imitate the step from the prebiotic to the biotic world via the production of protocells. In two different approaches, these chapters illustrate the struggle for the right choice of initial ensemble or initial ingredients.

## 1.6 A Quick Look into the Nanoworld

Nanotechnology makes it possible to have a look into the nanoworld, it is the world on the scale of nanometers ($10^{-9}$ m), which is the typical diameter of constituents of cells. Eukaryotic cells have a typical diameter of 10–100 $\mu$m ($1\mu = 10^{-6}$ m), procaryotic cells a factor of ten less; they live in the "microworld", but looking at their constituents, we find the diameter of the DNA double helix or of ribosomes, the width of microtubules, or intermediate filaments to be on the order of some nanometers. A detailed study of ongoing processes within the cell reveals fascinating insights into cells as well structured, hierarchically organized, and well functioning nano-factories. These factories are robust against many kind of attacks from outside, stable in an ever-fluctuating environment, equipped with molecular machines with a remarkable degree of efficiency in transforming chemical to mechanical energy, and endowed with an intracellular traffic system. Compared to all these achievements, manmade copies are poor with respect to their degree of efficiency and robustness. In intracellular traffic, some molecular motors transport cargo to designated locations [26], others are involved in the transmission of genetic information [28], some motors create tracks, others destroy them, they also may change lanes along the microtubule network [26]. In theoretical modeling and single-molecule experiments it is meanwhile possible to predict their average velocity (a few base pairs per second, see, e.g., [27]), the stall force [28–30], and many other single-motor properties [31]. The experimentally measured stall force for motors such as kinesin and myosin is on the order of a few piconewtons ((pN), that is $10^{-12}$ N) [29]. Traffic jams or failure of the traffic regulation may lead to diseases. Alzheimer's disease is nowadays explained in terms of malfunction of the intracellular traffic [26].

So we find a high level of organization already in the nanoworld with many parallels to organization in societies. It is natural to compare production logistics in companies with biologistics in cellular networks. Therefore parallels in cells and societies are at hand (see Chap. 16). It is certainly a great challenge for the future to search for mechanisms of how the hierarchical organization of cells as factories could ever have evolved (merely on the basis of dynamical laws, that is, in a self-organized way).

## 1.7 Playing the Tape Again

Let us assume we could rewind the tape of evolution in its whole breadth, rewind it a million years, a billion years, 14 billion years back to the beginning of the universe and press the button to rerun evolution, to allow free play afterwards. Would galactic structures form again, providing conditions for a blue planet, would life unfold in the same way, ending up with *Homo sapiens* as the pride of creation in roughly the same amount of time? In its full breadth such an experiment is science fiction, but as we shall see, it is not fictitious to rerun small sequences of the tape today to achieve some partial answers.

Let us start with a "sequence" that in real evolution was played shortly after the big bang, when the universe had cooled down to a temperature of $\sim 10^{12}$ K and the hot plasma of free quarks and gluons transformed into the phase in which quarks and gluons were confined into mesons and hadrons once and for all, in particular into protons and neutrons, constituents of matter's nuclei today. Recently (end of March 2010) the relaunch of the Large Hadron Collider (LHC) experiments at the nuclear research center CERN in Geneva featured in the top daily news with the comment that researchers had succeeded in mimicking the big bang. What is so exciting about these experiments? When protons (more generally heavy ions) are smashed together in the large colliders, for a very short instant of time (on the order of some $10^{-23}$ s) and in a very tiny region of space (on the order of some $10^{-13}$ m) the energy densities created are so high that even the strongly bound nuclei melt into their basic constituents; the quarks and gluons are then no longer confined and the resulting state resembles the exotic state of matter present during an early phase of the universe. (In the experiments at CERN protons were smashed head on at an energy of 7 TeV (1 TeV $= 10^{12}$ eV).) Details of the transformation back into the confinement phase are certainly different under lab conditions, compared to the conditions in a slowly cooling early universe, but alone the production of this exotic state at extreme energy densities is exciting enough not only for experts in particle physics.

The next sequence on the tape we want to consider is dated about 4 billion years ago, at the transition from the prebiotic phase to the origin of life. For some of us an old dream, for others a nightmare: if man were able to create man from scratch (if scratch stands for nonliving ingredients) and it is up to man to design man. In more realistic terms we refer to artificial life and the current attempts to construct precursors of cells, or even more realistic, certain functioning compartments of protocells out of nonliving components (see Chaps. 11 and 12). Drawbacks and successes in these attempts certainly improve our understanding of what mechanisms are essential for making a cell alive, and as a byproduct, such studies can lead to applications in medical treatment of diseases, based on an understanding in terms of microbiology and biochemistry. Artificial life is probably the most ambitious attempt to rerun some part of evolution.

The third sequence on the tape that we select could have started a little later, about 3.5 billion years ago, with populations of *Escherichia coli* bacteria. Over two decades recently (1988–2008) a remarkable experiment [32] was run in which generations of generations (altogether some 10,000 generations) of *E. coli* bacteria were grown under well controlled initial conditions in vitro in a certain nutrition background. The experiment should shed some light on a basic puzzle that is at the core of evolution: the tension between contingency and necessity. The results point out the importance of historical contingency for the evolution of key innovations. In these experiments a key innovation amounts to the capacity of *E. coli* to exploit citrate for nutrition in a glucose-limited medium that also contains citrate, but "normally" would not be used by *E. coli*. The key innovation happened in one population after 31,500 generations, and, as an innovation, it led to an increase in population size and diversity. Explanations are available according to which this innovation was caused by one big mutation, or alternatively, by an accumulation of small effects along a certain historical path. How sensitive are innovations to the prior history of an evolving population? Representatives of opposite viewpoints are Stephen Jay Gould [33] and Simon Conway Morris [34]. According to Gould, each change on an evolutionary path has some causal relation to the circumstances in which it arose, but outcomes must eventually depend on the history, that is, on the long chains of antecedent states. Therefore rerunning the tape would lead to a world quite different from ours. According to Conway Morris, "the evolutionary routes are many, but the destinations are limited" [34], so that gross features after rerunning the tape would be same, while inherent contingency would be confined to minor details in which the re-created world would differ from ours.

Related to the puzzle on contingency and necessity is the tension and linkage between stochastic and deterministic processes, which is currently in the main focus of research on dynamical systems in biological applications. It is a major puzzle how the robust and reliable functioning of systems such as the networks of genes or cells is possible in spite of the inherent and unavoidably fluctuating background, with fluctuations of different origin and on different scales. It would be surprising if nature did not exploit the various forms of "noise", as it is omnipresent and allows a number of possible distinct effects. Noise as disturbing background to signals or a source of destabilization are our usual negative thoughts, but this is just one side. Counterintuitive effects have been demonstrated in which noise increases the order of the system, induces transitions to qualitatively new states, or leads to regular propagation of signals in excitable media such as neural networks. For the impact of noise on stationary states in populations such as those of *E. coli* we refer to Chap. 13, where noise stands for stochastic reactions and fluctuations in the finite population size. In the *E. coli* experiments of [32], "noise" is realized in the form of mutations, and the interest is in the linkage between the deterministic part of evolution and the stochastic element of mutations. There the emphasis is on the kind of mutation. Was it one big mutation or an accumulation of small effects, after a small fluctuation at the beginning of a chain of events? In this book, the fundamental role of noise for evolution in relation to the evolution speed and to robustness against mutations is addressed by Kunihiko Kaneko in Chap. 10. Here "noise" stands for genotypic and phenotypic fluctuations, where the phenotypic fluctuations may have two origins: at the genetic and the epigenetic level.

## 1.8 There is More than Intuition

It is very easy to generate extremely small numbers for the probabilities of creating complex behavior out of a random, well-stirred soup, whatever complexity refers to. The small probabilities correspond to astronomically long times to produce the complex behavior, and these times easily exceed the age of the universe. Attempts at explaining the origin of life often have to face the prejudice that life is too complex to be ever produced by reproducible mechanisms in a finite time. On the other hand, rules, dynamical laws, and mechanisms are known that create ordered or even complex structures out of a random start, see Chap. 5. A mere guess is certainly not sufficient to decide how realistic proposed explanations for the origin of life are.

In general it is impossible to predict by mere intuition how many small effects accumulate in the course of time. One may think about large sums of small terms that individually approach zero the later the term appears in the sum. What is the value of the sum in the limit of infinitely many terms? From first courses in calculus one knows that the sum diverges if the terms decay to zero only according to $1/n$ for $n \to \infty$, but converges to a finite value for faster decay such as $\propto 1/n^2$. Intuition does not tell us the solution.

Evolution is in particular a history of bursts and extinctions of populations of species of all kinds. A sudden rise or decrease in the number of species or their diversity amounts to a phase transition (see also Chap. 4). Phase transitions are ubiquitous in nature. Some of these transitions happen suddenly and are accompanied by dramatic changes; they come as a surprise to those who rely on naive continuous extrapolations of ongoing small changes. Other transitions proceed more smoothly, but show what are called critical phenomena, in which small perturbations spread over large scales due to long-range correlations in the transition region. Transitions such as extinctions of populations or other catastrophes are rare events, but how rare they are, so rare that they never occur or that they do occur if we wait long enough, is a matter of quantitative estimate. Similar considerations apply to rare mutations.

Another caveat are linear extrapolations in systems with intrinsically nonlinear dynamics. The rich dynamical behavior of evolved turbulence of a viscous fluid, including its dynamical hierarchies, would never be intuitively expected from the uniform interaction of fluid elements. In all these cases mathematical modeling is the only means to predict the final outcome of how a large number of individually small or locally linear effects sum up. The contribution of Schuster in Chap. 2 contains an appeal to extend the successful mathematical modeling of physics and chemistry to biology.

## 1.9 Reduction of Complexity

There is no doubt, life is complex and so are living systems, and even nonliving ones may exhibit a high degree of complexity, whatever definition of complexity is used: structural (as for patterns that are neither random nor regular), functional (as in biological systems), algebraic (as in mathematics), or algorithmic (as in computer science). Still, there is a chance to reduce complexity whenever processes decouple and are well separated on different scales in space and time. From the viewpoint of long time scales, processes running on time scales that are several orders of magnitude smaller average out like fluctuations. On the coarse-grained scale they may be dropped, although zooming into the short time scale may uncover a rich dynamics. Similarly, one may neglect the compositional substructure of objects from the perspective of a very coarse scale in space. It is then a matter of convenience to describe these objects as "elementary" with effective degrees of freedom on the larger scale. Even the partition into disciplines such as subnuclear, nuclear, atomic, and molecular physics is just a reflection of the possible decoupling of phenomena on different scales. Collective behavior (as observed in large populations of oscillatory units in synchrony, or chains of always co-acting enzymes) may also lead to a reduction of complexity; a collective variable is then sufficient to describe the large ensemble when it behaves as one unit. Therefore the reduction of complexity – whenever it is possible – should not be mixed up with the arrogance of simplicity, which is a common prejudice against mathematical modeling.

## 1.10 How This Book Is Organized

### 1.10.1 Background

The background to this book is a summer school on *Steps in Evolution: Perspectives from Physics, Biochemistry and Cell Biology 150 Years after Darwin* that we organized at the occasion of the Darwin year 2009 on the campus of Jacobs University in Bremen. Participants with a background in life sciences and an interest in mathematical modeling, as well as from mathematics, physics, and chemistry with an interest in biological applications, enjoyed a school with stimulating lectures which provided the basis for the contributions to this book. Worldwide the ongoing controversial discussion about Darwinism and the modern theory of evolution was revived in many contributions in the media with different degrees of objectiveness. Therefore we found a collection of articles based on these lectures on progress in the theory of evolution a valuable contribution to this revived interest. The articles are at the cutting edge of different disciplines. We have restricted our viewpoint to the natural sciences, since the progress in the understanding of evolution that has been achieved there is already so rich and versatile that an extension to social sciences is beyond the scope of a single book. Therefore achievements in the cultural evolution of man are only touched on, for example, linguistics (Chap. 3), game theory (Chap. 14), and amazing parallels of organizational structures between natural and social systems (Chaps. 15 and 16).

### 1.10.2 Rationale

Controversial discussions on Darwinism versus creationism are accompanied by strong emotions, and the scientific approach to the theory of evolution has to face prejudices of the kind that its reductionism demystifies nature, that it is the arrogance of simplicity of scientists to oversimplify the description of nature, ending up with a claim to explain "the world as a whole". Moreover, an attempt to create artificial life is interpreted as hubris, as lack of respect for nature and its creatures. Sometimes such criticism is based upon ignorance of the real statements, achievements, and goals of research, or upon naive extrapolations of seemingly obvious facts.

As a contribution to the ongoing discussion, the main messages of our book may be summarized as follows. All contributions follow some kind of scientific reductionism in order to go beyond the superficial level of a mere description. Deriving a theory is more than fitting data. They try to explain features on a given level in terms of more basic features underlying a set of common observations, but nowhere is the reductionistic viewpoint pushed to its extreme, extreme in the pretense to explain phenomena of life science in terms of elementary particle physics. Apart from the fact that such an endeavor is unrealistic, it is not even desirable whenever transparency gets lost. Instead, the field of evolution theory intrinsically requires interdisciplinary collaborations from the various fields of natural science, from mathematics and computer science to wet experiments in physics, chemistry, and biology, since the emerging features of a derivation from one discipline may require notions from the other discipline at the next level of description.

The contributions amount to an appeal for mathematical modeling, since more than intuition is needed for an appropriate understanding. Common folklore is often based on naive extrapolations and shortcuts when they refer to the final accumulation of a sequence of ongoing events. Here a reliable prediction should be based on a mathematical description. Shortcuts in a line of arguments may lead to statements about the extremely low probability of man coming into existence, so that the occurrence of life needs the action of an external creator, but why such a line of argument amounts to a shortcut is pointed out in Chaps. 5 and 6.

In contrast to the common belief held by skeptics of the scientific approach, a deeper understanding, a reduction of complexity, or a break-up of chicken-and-egg dilemmas are usually accompanied by fascination and amazement. Amazement results, for example, from a deeper insight into the sophisticated, well-functioning, highly flexible, and very robust performance of processes in things alive. Any information scientist who knows the struggle to program robots to perform a certain task will certainly appreciate the performance of the well-controlled movement of flies in spite of their tiny brains, the coordinated motion of a school of fish or flock of birds and their unerring ability to locate prey. Neuroscientists involved in optimizing the performance of artificial neural networks will not hold the progress in learning of a 3-year-old in lower esteem than someone who considers the growing up of a child an automatic process.

In contrast to other books on evolution, the selection of topics here is more diverse. Topics such as cosmology, population genetics, game theory, and artificial life, touched on in this book, easily each deserve a book of their own, staying within the frame of one discipline. A detailed understanding of all the results presented in this book and related to the theory of evolution would require an education ranging in the extreme case from mathematical physics and quantum field theory to experimental biochemistry and biophysics. Therefore the indicated technicalities will not be understandable in full detail to experts from other fields of research, no doubt about that. Still, it may be inspiring to the non-experts to see a similar kind of struggle occurring in a different discipline. In most parts of the book, when dealing with basic principles of evolution, population genetics, and game theory, parallels in questions and their analysis are evident. It is just one of the main aims to point out the common concepts such as self-organization, the many facets of chicken-and-egg problems, universal processes, and universal laws (but definitely not the universal theory of "everything" (including life)).

Along with universality aspects there is obviously a need for a common language to describe such different objects as molecules, bacteria, agents, or words with one and the same type of variable. The chapters on game theory and population dynamics may convince the skeptical reader that such a language is possible. When only the initial abundances of species matter and it is otherwise the rules of the game that determine the final fate of being a winner or loser, the realization of the species as a bacterium or human agent becomes irrelevant and can be dropped from the description. The existence of such a common theoretical framework gives a hint to why it is neither hopeless, nor predestined to a superficial level, to put such versatile aspects of evolution together in one book.

In relation to the theory of evolution it is quite easy to ask questions that are simple, evocative, and touching at the same time: What is the fate of the universe? What is life, how did it start on earth, what defines its very beginning and its very end on the individual level? Is Darwin right after all? Simple questions induce cascades of further questions. Since we do not know any simple answers, the reader will find answers to more specific questions, so specifically posed that they allow an answer from the perspective of the natural sciences.

This book should leave an overall impression that the theory of evolution itself is open to further evolution with ongoing challenges in various disciplines. No end is in sight for possibilities of replaying sequences of the tape in new rounds of this never boring game; we are certain of this.

### *1.10.3 About the Articles*

The articles are organized in four parts: *Principles of Evolution* deals with models on basic processes such as replication, mutation, and selection; the second part *From Random to Complex Structures: The Concept of Self-Organization for Galaxies, Asters, and Spindles* is about one of the main overarching approaches to explaining the occurrence of complex structures from a random start; the third part is devoted to the basic units of life, more precisely to *Protocells In Silico and In Vitro*, to then conclude with intimate parallels ranging *From Cells to Societies* in a fourth part. The following extended summaries of the articles mainly serve to embed their content in the wider context of the book.

Part I: Principles of Evolution

*P. Schuster* ("Physical Principles of Evolution") Throughout the book the notions of population, species, evolution, mutation, selection, and fitness are frequently used, in various contexts and with various meanings. So the reader may wonder whether precise definitions in mathematical terms are available, or whether the notions mainly serve as metaphors for generic processes to which one inevitably assigns a vague meaning. Therefore in the first section Peter Schuster introduces the basic notions of the theory of evolution for the simplest available systems, such as evolvable molecules in cell-free assays. He presents precise definitions in mathematical terms with an emphasis on the need for and use of such a mathematical approach to the theory of evolution. Schuster's first section reviews historical aspects, starting with Darwin's notions of multiplication, variation, and selection, and next describing the dawning of mathematical modeling in biology. Along with progress on the theoretical side, new experimental techniques arose in the second half of the last century, such as high-throughput measurements, which led to a large amount of data that needs to be structured and understood by means of theoretical models. Nowadays it is systems biology that ideally should bridge the gap between a reductionistic and a holistic view.

The second section deals with the selection equation and its precursors, in particular the Verhulst equation, which describes exponential growth combined with limited resources. For the selection equation it is proven that the term with the largest assigned fitness value (highest reproduction rate) dominates all other species at sufficiently long times, so that selection in Darwin's sense is realized. In general, evolution is addressed to mutations in the genotype, while selection acts on the phenotype. Accordingly, Schuster explains next what evolution in genotype space means. Evolutionary dynamics is described as change of the population vector, whose components are the numbers of individuals for the different species as a function of time. The time dependence is determined by deterministic differential equations or stochastic equations, to be considered later. The genotypes are DNA and RNA sequences and the genotype space is therefore the space of all polynucleotide sequences. As a space in the mathematical sense, it can be endowed with a metric based, for example, on the Hamming distance. Related notions such as the consensus sequences of populations and a formal space for recombinations are introduced as well. As mentioned in several chapters throughout this book, the final aim is to relate features of phenotype space to evolutionary dynamics in genotype space. In general the mapping between genotype and phenotype is rather complex, but Schuster considers a case in which genotypes unfold unambiguously into a unique phenotype. He considers a genotype space at constant sequence length $l$ over four and two letter alphabets. A concrete example is provided by in vitro evolution of RNA.

Schuster's contribution is on *physical* principles of evolution, and it is molecular *physics* that provides the tools for modeling the folding of molecules into structures. One tool is the concept of conformation space, in which a free energy is assigned to each conformation of a molecule by means of a potential energy function. Furthermore the conformation space concept enables us to talk about landscapes for the evolution of RNA molecules. Structure predictions then search for the most stable structures or, equivalently, for the global minimum of this landscape. Chemical kinetics of evolution is then first described by a deterministic equation for the evolution of a population; now, in addition to replication and selection, also mutation is included in the dynamics. The deterministic description applies whenever the population size $N$ is large, mutation rates high enough, and other stochastic effects negligible. The focus here is on the time evolution of the (quasi)species as a function of the mutation rate. There exists a critical mutation rate at which replication errors accumulate, ending up in random replication and a uniform distribution of species. Above this threshold, populations migrate, evolution is no longer possible, and fitness differences in the fitness become irrelevant.

Experiments on sequence–structure mappings indicate that many genotypes lead to the same phenotype and identical fitness values. What replaces natural selection in this case? The answer is random selection as in the neutral theory of evolution, but only for sufficiently distant (master) sequences. In principle, evolution is a stochastic process, and fluctuations in the reaction rates and in the population size are not always negligible (see also Chap. 13). The deterministic population variables are then replaced by their corresponding probabilities, and the probabilities obey a chemical master equation. In the proposal of this equation, it is not enough to add a term that just compensates for the population growth, a detailed model is required. One of these models based on a real experimental device is the so-called flow reactor, which Schuster describes for the evolution of RNA molecules. In order to evaluate the impact of stochastic evolution on the phenotype, the sequence–structure map has to be an integral part of the computational model in order to determine the landscape of phenotypes and their properties on the fly. Two procedures were carried out in the simulations reported by Schuster: the optimization of properties such as replication rates on a conformation landscape, and a search for a specific target structure in shape space. According to the results and common to both procedures, the progress in evolution proceeds stepwise rather than continuously, and short adaptive phases are interrupted by long quasi-stationary epochs. Moreover, and interestingly, different computer simulations with identical initial conditions lead to different structures with similar values for the optimized rate parameters, giving rise to contingency in evolution. Here Schuster refers to the recent experiments with *E. coli* of [32] that we discussed also in Sect. 1.7: these experiments can be understood in terms of random searches on a neutral network. Only one of the twelve bacteria colonies happened to come close to a position in sequence space from where a small mutation was enough to lead to a big innovation in the end. The reader may compare the in vitro experiments of [32] with the theoretical modeling reported by Schuster.

Naively one may expect that higher mutation rates accelerate evolution and the related optimization process, without limitation. Unreflected extrapolation is again misleading. Schuster refers to simulations that demonstrate the existence of a threshold: optimization of evolution becomes more efficient with increasing error rate in the replication only until a threshold value is reached. Above this threshold optimization breaks down. How sharp the threshold is depends on the fitness landscape.

Schuster finally addresses the origin of complexity in evolution: It is not evolutionary dynamics itself that is complex (this is in agreement with results of Chap. 6, where it is the "rules" that are simple and lead to complex structures). The reason for complexity here is the genotype-to-phenotype map and the influence of the environment.

*R. Blythe* ("The Interplay of Replication, Variation and Selection in the Dynamics of Evolving Populations") Richard Blythe focuses on two basic questions that are also of central importance for the theory of evolution in general. The first concerns the existence of a single framework for the description of evolution in applications as different as genetics, ecology, and linguistics. More precisely, without such a framework, the description would amount to a collection of superficial analogies between processes of molecules (genetics), species (ecology), and languages (linguistics). The second question refers to the concept of selection and what the object of selection really is. Blythe summarizes a formulation of evolutionary dynamics due to Hull (see his chapter) and its utility for its interpretation. As it turns out, there is a level of abstraction, expressed in terms of mathematical models, that allows a unified description of evolutionary processes and provides a deeper understanding of what the essential mechanisms are, which sometimes are hidden behind the superficial appearances. Blythe characterizes evolution as a theory of change via replication. What replicated DNA is in genetics, are reproduced species in ecology and intangible behavior in culture in linguistics. During replication variation occurs via mutation or recombination. Some of these variations may lead to some organisms being more successful than others (in the sense of having more offspring), this in turn affects the relative frequencies of different genotypes in a population; this process defines "selection". As indicated in our introduction, simple questions may induce cascades of further questions. Here the question of whether selection is all that determines evolution is such an example. The immediately induced question relates to what the units of selection are, for example, where it takes place, at the level of the genotype, or the level of the phenotype (as usually assumed), or even at the level of groups or species. Hull's abstract formulation was motivated by this debate. Blythe illustrates Hull's analysis with molecular evolution, community ecology, and language change.

For the mathematical analysis of selection he discusses the Price equation. The Price equation allows one to distinguish between the selective component of a change and the component due to the generation of variation in the replication process. More precisely it determines the variation of the mean of some quantitative character (called trait) after one generation of reproduction, assuming we know all the offspring numbers and the changes in the trait of each of the individual offspring. Its familiar applications are kin and group selection. Blythe demonstrates that "survival of the fittest" is only part of the whole prediction of the Price equation. In reality, offspring numbers and changes in the trait are random numbers, since reproduction and mutations (here in the general sense) are stochastic processes. Already these fluctuations may cause effects that are misinterpreted as selection. Since so many emotional debates are concerned with selection, it is quite important to disentangle from a single evolutionary process a systematic effect of selection from purely random effects. What is an appropriate standard for comparison? The answer is given in terms of genetic drift, which describes the outcome of neutral demographic fluctuations. These are fluctuations due to stochastic birth and death processes, for which all replicator types have the same number of offspring. Therefore it is essential to first analyze the statistics of neutral population dynamics. How to do this is explained in Blythe's sections on the fixation probability, the mean fixation time, the experimental observation of genetic drift, and the effect of immigration and mutation in neutral models. (Immigration and mutation are responsible for maintaining variability in a population of finite size with faithful replication.) When it comes to nonrecurrent immigration (where every immigration event introduces a new replicator type to the population), one formulation (different from a master-equation approach) turns out to be more convenient. This is a backwards-time dynamics that evolves from observations about the diversity of a present-day population backwards to an unknown initial condition. This so-called ancestral formulation of neutral evolution finally leads to testable hypotheses in favor or against neutral evolution. Neutral models therefore provide the null models in a quantitative analysis of evolution. A detailed analysis of the omnipresent fluctuations from various stochastic sources is essential. Moreover, once the target of selection is known (which is a nontrivial issue as we learn in this chapter), the results of evolutionary processes can be predicted.

We would like to emphasize that Blythe's contribution includes aspects of cultural evolution, in particular mathematical models for language change. Evolution of languages certainly amounts to a major transition in communicative abilities and in evolution as a whole. It is intimately related to another simple question: "what does it mean to be human?", as Blythe concludes his chapter.

*S. Thurner* ("A Simple General Model of Evolutionary Dynamics") Stefan Thurner's concept of evolution is not restricted to the genetic level, but is intended to apply to biological evolution, technical and industrial innovation, economics, finance, sociodynamics, opinion formation and ecological dynamics. Inspired by observed universality classes in statistical physics (see Sect. 1.3), the aim is to derive universal systemic properties of populations such as their proneness to collapse or their potential for diversification. As is shown, phases of relative stability in terms of diversity are followed by phases of pronounced restructuring, phases with high and low diversity are separated by sharp transitions, and statistical characteristics of respective time series turn out to be common to various systems. The variety of applications becomes comprehensible by noting that the $N$-dimensional time-dependent vector with binary entities 0 or 1 merely denotes the presence or absence of elements, which represent species, goods, "things", according to Thurner. Furthermore the recombination and production of new elements is specified in terms of a production table stating that a production of element $k$ from $i$ and $j$ is possible (1) or not (0). As typical processes for evolution, the dynamics should account for selection, competition, and destruction. All allowed destructive combinations of elements are fixed in another table with binary components. The dynamics is then formulated in terms of an update rule for the vector of elements: if there exist more production than destruction processes associated with a particular element, it gets produced, otherwise not, or it gets destroyed if it already exists. Whether a certain production is actually active depends on the availability of the corresponding elements. This leads to the notion of an active production network that captures the set of active production processes at a given time. To furthermore account for spontaneous appearance or disappearance of elements, existing elements are spontaneously annihilated and nonexisting ones created with a certain probability. In the associated tables it remains to fix the topology of the productions and destructions which are allowed in principle, as the table entries refer to (im)possible productions of $k$ from $i$ and $j$. Thurner chooses these tables with randomly distributed entries of 0 or 1. The tables are therefore fixed by the number of constructive and destructive rule densities. As in other examples of evolutionary dynamics in this book, the rules are quite simple, and the evolving elements here have just a single degree of freedom (being there or not being there). Numerical simulations of this model show then two phases, characterized by an almost constant set of existing elements, and a phase of massive restructuring. Along with that, plateaus of constant product diversity are separated by restructuring periods with large fluctuations. The diversity is measured as a normalized sum over all present elements at time $t$. Obviously, the duration of the plateau and restructuring period depends on the "innovation" rate, here meaning the newly created or deleted elements per time (in contrast to more specifically defined innovations by Jain and Krishna in Chap. 5). The degree of diversity depends on the densities of destructive and constructive rules. Thurner lists a number of model variations that have marginally no effect on the results. The model allows also an analytical approach. Its stochastic generalization can be solved within a mean-field approximation. Special cases of the diversity dynamics amount to projections on macroeconomic instruments, chemical reaction networks, and lifetime distribution of species. According to Thurner, model predictions and corresponding data from experiments share in all three cases gross features, such as power-law behavior in certain distributions. Thurner emphasizes that fitness in this diversity dynamics is an emergent feature rather than an a priori assigned property of evolving bitstrings.

*S. Jain and S. Krishna* ("Can We Recognize an Innovation? Perspective from an Evolving Network Mode") Jain and Krishna consider special events of evolution that often have a long-lasting impact on the subsequent history: innovations. Innovations are easily identified in retrospect. Usually they stand for progress in evolving societies and evolution of life. Is it possible to identify innovations as they emerge? The authors give the answer in structural changes of graphs that represent the interacting system of species. The framework is a dynamical network, consisting of one kind of nodes, one kind of links, and one kind of variables. The variables, assigned to the nodes, are relative population concentrations. The network is dynamic in two respects: one concerns the dynamics assigned to the node variables, the other the dynamics of rewiring the links so that the network topology becomes a dynamical quantity as well. The combined dynamics includes two essential ingredients of evolutionary dynamics: growth and selection, selection according to success, and lowest success means lowest population concentration; the corresponding node with the lowest concentration gets eliminated from time to time and replaced by a new node according to certain rules. Since it is only the relative concentration of species that enters the description, the nodes may represent a species in an ecosystem, a substrate in a metabolic network or an agent in a society, depending on how realistic the description of the assigned interactions with other nodes is. Jain and Krishna motivate their dynamics with a metaphor: A pond in the prebiotic world, full of chemical species, gets flooded from time to time by a river or the sea. The flooding drives some species out of the pond and new species in. Between the flooding the species grow. Remarkably, this combined dynamics, although again simple, is rich enough to produce a dense graphical structure of connected species out of random connectivity at the start. All that happens in a short time during a rapid growth phase. After the initial growth phase, the population fluctuates between a densely connected set of grown species and a state of being extinct, afterwards to grow again from scratch. These transitions in the evolution of populations can be traced back to the (first) occurrence of certain graphical structures (so-called autocatalytic sets). Moreover, innovations can be characterized "on the fly" in terms of irreducible subgraphs, generated by an incoming node, and changing core and periphery structures due to the newcomers; the authors classify and analyze the set of all relevant changes in detail. The changing graphical structure is responsible for the involved change of the dynamical performance and therefore provides a suitable criterion for recognizing events as innovations. One of the interesting lessons drawn by the authors from this modeling is the ambivalent impact of innovations: what leads first to the innovation's success (e.g., the burst of growth in populations of species), later leads to its destruction (e.g., the extinction of these species). Although originally inspired by the metaphor of a prebiotic pond, the mechanisms for innovations may be quite similar in social and economic systems.

Part II: From Random to Complex Structures: The Concept of Self-Organization for Galaxies, Asters and Spindles

*G.M. Schütz* ("How Stochastic Dynamics Far from Equilibrium Can Create Non-random Patterns") Do you want to know the probability that "a monkey who wildly hacks symbols into a computer accidentally types 64 characters out of a poem of Shakespeare"? Gunter Schütz calculates this number to be $2^{-384}$. The probability is the same as generating a DNA sequence of 192 letters, or a periodically alternating sequence $010101\ldots$ of binary numbers of length 384. What this small number for the probability actually means is illustrated with a Gedankenexperiment, designed to be most efficient and extremely fast, but truly random. To be successful at least once in $2^{373}$ attempts, performed over the age of the earth, that is over roughly 10 billion years, the probability is still not more than $10^{-4}$. As mentioned in Sect. 1.10.2, these small probabilities are often taken as an argument against any scientific explanation that life could have come into existence and in favor of the need of external interference. Why arguments based on these extremely small probabilities fail and intuition is misleading is illustrated by Schütz with predictions based on "paper, pencil, and PC", that is, on analytical calculations, combined with stochastic processes that make use of a random number generator. In these studies, complex structures such as Shakespeare's poem or the DNA sequence will be represented by the alternating binary sequence, and the wildly hacking monkey (which is equivalent to tossing an unbiased coin or throwing symmetric dice) will be replaced by a totally asymmetric simple exclusion process (TASEP). The TASEP is a prototype of models that describe out-of-equilibrium physics. Imagine a ring with a discrete number of sites, and a given number of particles, such that at most one particle can occupy a site. Each particle tries to jump to its right neighboring site after an exponentially distributed random time interval. It is only allowed to perform the jump if the neighboring site is empty, otherwise it does not jump. The binary numbers therefore correspond to possible occupation numbers of these sites, being either 0 or 1. This sounds like a simple rule, but the process allows a rich dynamical behavior, various applications, and rigorous mathematical predictions. Schütz considers the TASEP in four versions. Common to them is the so-called driven dynamics, which maintains an out-of-equilibrium steady state and short-range interactions, but two of them are stochastic and two are deterministic versions. Here we only mention two of the results, the most counterintuitive ones. Firstly, the random dynamics (specified in Schütz's contribution) generates an ordered state out of a disordered start within a time that scales only proportional to the size of the system, not exponentially to its size, and secondly, starting from an ordered state, the random dynamics preserves this order. Together with results for the other versions of the model, one is led to the following conclusions: In contrast to what one would like to do, one cannot conclude from the emergence of ordered or disordered patterns that the underlying dynamics is deterministic or stochastic. (In the light of these results it becomes less surprising that a basic and intricate task in population dynamics is to correctly interpret fluctuations in the data and to uncover hidden rules of selection if they were really at work during the generation of patterns, see Chap. 3.) Moreover, it is true that throwing symmetric dice or tossing an unbiased coin would never create complex structures ("never" means that it would take astronomically long times); tossing biased coins, however, say in the form of the TASEP, it becomes possible to create complex structures relatively fast owing to the hidden rules in this process. If it is not the TASEP as discussed in Schütz's contribution, but a combination of rules or laws from physics and chemistry that govern the (partially stochastic, partially deterministic) processes of nature, one may find it less incomprehensible that even complex structures such as life have emerged.

Referring to the ongoing debate of Darwinism versus Creationism, we would like to quote Schütz from his contribution: "Evolution is not the right place to find God".

*M. Bartelmann* ("Structure Formation in the Universe") Structure formation in the universe can be understood as a self-organized process in the framework of the cosmological standard model. In the first part of his contribution, Matthias Bartelmann summarizes the empirical and theoretical foundations of this model. The model is meanwhile well established and based on Einstein's field equations. Einstein's field equations are nonlinear and therefore not solvable in full generality. Thus, their solutions are usually constructed on the basis of simplifying symmetry assumptions. It is then the assumption of spatial isotropy and homogeneity that leads to the class of Friedmann solutions and the Friedmann model of the universe. It is argued in what sense the universe may be considered as being isotropic, although a quick look at the sky reveals an anisotropic distribution of stars. Once the symmetries are implemented in Einstein's equations it is only an equation of state that is missing.

The equation of state relates the pressure to the density of matter, and depends on the very type of matter. The combined set of equations constitutes the class of Friedmann models. Based on the observation that the universe is expanding, the Friedmann model predicts a finite age of our universe and an inevitable big bang prior to some finite time.

In view of experimental imprints of the universe's early evolution, Bartelmann explains the origin of the cosmic microwave background (CMB) radiation, nowadays one of the most important sources of data from the early phase, well after the entire universe acted as a big fusion reactor and hydrogen was converted via deuterium to helium-4. Thermal radiation should be left over from this time. When the universe was about 400,000 years old, it became transparent to this radiation, which from that time on could freely propagate through the universe. The radiation was predicted and experimentally confirmed to have a blackbody spectrum, corresponding nowadays to a temperature of 2.726 K. Resolving the CMB with differential microwave radiometers, the expected amplitude fluctuations of the temperature were not confirmed, leading Jim Peebles to the proposal of dark matter (in contrast to our familiar ordinary baryonic matter). After all, dark matter makes up the majority of matter and acts as the dominant source in an equation that describes the structure formation (see below).

As a summary of Bartelmann's first part, and as a great success of the reductionistic approach, the cosmological standard model is compatible with data that probe the physical state of the universe at several instances, ranging from a few minutes after the big bang until today, 14 billion years later. While the entering model assumptions may appear speculative to nonexperts in the field, the great success of experimental verification seems to justify them.

Bartelmann's second section on structure formation deals with self-organization. While Turing patterns are coded in differential equations for reaction–diffusion mechanisms, structure formation of the (dark) matter density (more precisely its contrast with respect to the background density) is included in an equation combined from the continuity equation for mass conservation, Euler's equation for momentum conservation and the Poisson equation for Newtonian gravity. Bartelmann argues why Newtonian hydrodynamics is a justified approximation (not even Einstein's equations are needed to explain structure formation). In the following sections linear and nonlinear structure evolution are discussed, where the linear equation in the density contrast holds only for small fluctuations about the average background. As it turns out, structures such as galaxies, galaxy clusters, and even larger structures can only have arisen if dark matter is the dominant contribution to matter. It is further argued in which sense dark matter should be cold and how the statistics of cold dark matter fluctuations can be predicted. If the primordial matter density fluctuations were Gaussian, the statistics of a Gaussian random field then predict filamentary structures that must form before they fragment into smaller objects. These filamentary structures are actually observed in large-scale galaxy surveys. An upper limit to the mass of an object can be derived for stars to form at all.

In a last section, Bartelmann presents arguments in favor of an inflationary phase in the early universe. One argument is the almost perfect isotropy of the CMB, the other concerns an explanation of the origin of structure in the CMB. (Here we see the recurring question about the appropriate initial conditions and their origin.) Luckily, the hypothesis of inflation leads to a testable prediction on the scaling of the power spectrum of the density fluctuations with the wave number, and the prediction is in agreement with measurements. The assumed vacuum fluctuations that are thought to have been amplified during inflation fall in the quantum epoch with quantum cosmology as an appropriate framework; only after the quantum epoch does classical cosmology become the appropriate theoretical framework, which is the topic of Bartelmann's chapter.

*C. Kiefer* ("The Need for Quantum Cosmology") In Sect. 1.1 on the chronology of evolution we argued already the need to unify quantum mechanics with gravitational theory to give quantum gravity, since the classical notions of space and time lose their meaning from a certain mass density on. Although to date no unique candidate for quantum gravity exists, and indeed no candidate that satisfies all the requirements of such a theory, it makes sense to discuss such candidates in order to see which questions they are able to answer and which not.

One candidate is quantum geometrodynamics, which is introduced by C. Kiefer in Chap. 8. As the Schrödinger equation is representative for nonrelativistic quantum mechanics, the Wheeler–DeWitt equation is the basic equation of quantum geometrodynamics. Although the Schrödinger equation is neither relativistic nor a field-theoretic equation, it makes sense to discuss quantum mechanics in this approximation. Similarly, as Kiefer argues, it is sensible to answer some questions of quantum cosmology from the Wheeler–DeWitt equation. This equation describes the evolution of a scalar field as the simplest representative of all kinds of matter, but what is most remarkable about this equation is the replacement for "time". Time, which plays the role of an external parameter in quantum mechanics and a dynamical parameter in general relativity, is replaced by a scale parameter called "intrinsic time". Our familiar notion of time fades away and becomes an induced, emerging concept. The implications are far reaching and concern an appropriate choice of initial and boundary conditions, which differ from familiar choices in quantum mechanics and general relativity. Kiefer discusses two choices for the boundary conditions, so-called "no-boundary" and "tunneling" proposals. Along with the formal obstacles comes intrinsic trouble with the very interpretation of the equation. It belongs to the basic mode of our understanding to describe phenomena in space and time, where space and time provide the background for the observer's description. This splitting into the observer's background and the observed phenomena is lacking without an external time parameter, and therefore the very interpretation of the wave function as a solution of the Wheeler–DeWitt equation remains open. The bridge between quantum and classical cosmology is provided by a semiclassical approximation of the Wheeler–DeWitt equation. According to Kiefer, classical geometry emerges from quantum gravity via decoherence. (After all, we are now living in a world that allows classical notions of space and time, so there is a need for explaining their emergence.)

Another provoking and fundamental question concerns the omnipresent arrow of time. It is not only inherent in aging of all living beings. Some classes of phenomena are also not invariant under time reversal, so that a direction is distinguished, and it is natural to trace back the master arrow of time to cosmology. Again, remarkably, the Wheeler–DeWitt equation is fundamentally asymmetric with respect to the intrinsic scale parameter, as Kiefer explains. Coming from physics, one would like to relate the arrow of time to an increase of entropy, as one is used to from classical physics. Although there is no general expression for the entropy of the gravitational field, Kiefer summarizes that it is possible to define an entanglement entropy, which increases with increasing scale parameter of the Wheeler–DeWitt equation. In the semiclassical limit, "our" extrinsic time can be constructed in terms of the scale parameter, leading to an increase of entanglement entropy also with our classical notion of time and defining its arrow. With a few lines in the end of Chap. 8, Kiefer indicates the connection of quantum cosmology to structure formation in the universe as described by Bartelmann, more precisely to the origin of the inhomogeneities in the CMB data, which are the seeds for the structures that appear later.

As shown in Chap. 8, there is definitely a need to construct quantum gravity. Unavoidably, however, when it comes to *the* very beginning, available theories become speculative to a certain degree, and along with the notion of time the interpretation of the equations becomes blurred.

*L. Dehmelt and P. Bastiaens* ("Self-Organization in Cells") Although reporting on cells in vivo, the authors use a notion of self-organization that is compatible with its use in physics and chemistry. Self-organization is a process in which a pattern at the global level emerges merely from many dynamic interactions of units on the local level, and the interaction rules are followed on the basis of local information within the cell. Within the cell, the authors distinguish self-organization from other organizational principles such as "master regulators" and "templates, blueprints, or recipes". Master regulators, often controlled by global feedback, steer other subordinated units. Examples of such regulators are growth factors, their receptors, and immediate signal mediators as they relay signals to many target regulators. However, as outlined later in Dehmelt and Bastiaens's chapter, many cellular regulators do not follow a hardcoded recipe for performing a certain process . They do not have a blueprint as an architect would have. Templates, on the other hand, are typically coded in DNA, where the information is stored in the base sequence and used to generate mRNA. Self-organization is also seen in contrast to self-assembly, in which building blocks are combined into a larger stable nondynamic structure.

Three forms of self-organization are distinguished by the authors: (a) dynamic activity gradients based on feedback systems, such as calcium waves (based on similar reaction–diffusion mechanisms as wave patterns in sand dunes); (b) directed transport and growth systems with feedback regulation (an example here is neurite outgrowth and the process is seen in analogy to the building process in termite colonies); (c) dynamic structures, which are formed by complex force feedback interactions such as the mitotic spindles (they are seen in analogy to convection patterns in Bénard cells). These three cases are illustrated with a number of further examples and discussed in detail in the following sections. Here we want to mention one particular distinction made by the authors, which refers to how self-organization actually proceeds. The more familiar way is the situation in which the local units directly interact, and this interaction leads to patterns on the global scale. The less familiar may be "stigmergy" which plays a role in directional morphogenetic growth or transport processes together with feedback regulation. Stigmergy is the organization of a building process based on the work in progress, as used in termite nest building or wall and chamber building by ants. The main point here is that the interaction of the local units is indirect via the work in progress, for example via positive feedback between fluctuating entities and a developing structure.

Moving next to structures of higher complexity within the cell, emerging structures are discussed that arise at least partially by means of self-organization of microtubules and associated motors. To these belong aster and spindle formation. Finally, Dehmelt and Bastiaens describe mechanisms of self-organization that are at work in actin-rich structures such as lamellipodia. The lamellipodia are an example of a dense branching network that is constantly rebuilt during treadmilling. The model for this process shows the basic ingredients of evolution: inheritance, mutation, and selection, as the authors outline. Their outlook summarizes challenges for future research. Let us mention one typical challenge and consider the Golgi apparatus as example of a cellular compartment. A typical open question is whether Golgi fragments are essential templates, required for the formation of the Golgi body, or whether the Golgi apparatus can be formed de novo, that is, in the absence of Golgi-derived vesicles. Stated differently: does this sophisticated inner-cellular module emerge solely via self-organizing mechanisms, or is at least some structural information hardcoded in preexisting templates?

Part III: Protocells In Silico and In Vitro

*K. Kaneko* ("Approach of Complex-Systems Biology to Reproduction and Evolution") Let us recall that one candidate in the discussion of "which came first" in the prebiotic world was "metabolism-first models" (Sect. 1.1), where metabolism is described in terms of catalytic reaction networks. In the first part of his contribution, Kaneko discusses three types of catalytic reaction networks as an attempt to bridge the gap between chemistry and biology, where biology is represented by a reproducing cell. The approach is again in the spirit of reductionism. All properties that are assumed to be irrelevant for an observed qualitative behavior are stripped off so that the models for the protocells do not even intend to be realistic, or to imitate cellular functions. The cell state is just characterized by the number of chemical species. These numbers change through mutual reactions, and the reaction dynamics is described as a catalytic reaction network. In its simplest version, reversible two-body reactions between catalysts are assumed, and the dynamics satisfies detailed balance. Within this model, Kaneko studies the question of how an out-of-equilibrium state can be maintained over a long period by the network itself. A cell in chemical and thermodynamical equilibrium would be dead; the out-of-equilibrium condition is considered as a necessary ingredient for life. According to Kaneko, it is possible to understand the basic mechanisms for a slow or even prevented relaxation to equilibrium as due to a negative correlation between an excess chemical and its catalyst. In a second version of the catalytic reaction networks, catalysis may proceed through several steps, leading to higher order catalysis. Resource chemicals can be transformed into others, and a cell can grow. In this framework, Kaneko analyzes the consistency between cell reproduction and molecule replication. Under certain conditions the cell maintains the composition of chemicals during reproduction when the speed of growth is optimized. Cell-to-cell fluctuations in the chemical composition show a log-normal distribution. The third type of catalytic reaction network, in addition to the former properties, consists of replicating units. Positive feedback is implemented as an autocatalytic process to synthesize each molecule species and to consider replication reactions. The resulting network is a replication reaction network, which is then used to study a possible origin of genetic information as a result of so-called minority control.

In the second part, Kaneko focuses on the relation between fluctuations and robustness during evolution. Fluctuations occur on the genetic and epigenetic level, accordingly one should distinguish the corresponding robustness with respect to fluctuations. In particular the effect of phenotypic fluctuations on evolution is considered, where the phenotypic fluctuations are of isogenic origin. So they cannot be due to corresponding fluctuations on the genetic level, but only due to epigenetic effects (change in the environment or developmental processes). Although these fluctuations may be naively assumed not to be inheritable, the result is that they do influence the speed of evolution. The claim is a positive correlation between the evolution speed and isogenic phenotypic fluctuations. The relation is called the fluctuation response relationship, inspired by fluctuation–dissipation theorems of physics. On the other hand, another theorem has been proposed which states that the evolution speed is proportional to the variance of phenotypic fluctuations due to genetic variations. Both relationships are therefore only consistent if the two variances are proportional to each other. This is analyzed in a model with developmental dynamics of a phenotype. A positive effect of sufficiently high phenotypic noise may be higher robustness with respect to mutations. In the last section, Kaneko presents a phenomenological model (i.e., not derived from first principles), to explain the observed relations between isogenic phenotypic and genetic variances.

Needless to say, the mapping between genotype and phenotype is rather complex in general. In summary of the second part, Kaneko's focus is on supposed relations between the associated fluctuations at the two levels.

*H. Fellermann* ("Wet Artificial Life: The Construction of Artificial Living Systems") The reader may wonder whether first forms of artificial life have been created or not, after all. According to Fellermann, the answer may be "almost". The question is about cell-like chemical systems that are able to self-replicate and evolve. To recall from our introduction, a cell needs three basic ingredients: inheritable information, a container, and metabolism. To date, there are no cell-free "naked" replicators, but the whole container-metabolism-information system replicates itself. In the first section of Chap. 11, these three ingredients are explained in more detail. For example, what are the basic processes that are required for replication of biopolymers without the assistance of enzymes? Candidates for protocells are lipids of which one part is solvable in water and another in oil. Lipids are able to self-assemble into supramolecular structures and provide the name for the lipid world that we mentioned in Sect. 1.5 in connection with the puzzle "which came first in the prebiotic world?". As Fellermann outlines, here the currently discussed candidates for containers are much simpler in their composition than in real cells, where the cellular membrane contains both lipids and proteins. An example of such a simpler candidate is the vesicle used by the group of Schwille, as discussed in detail in Chap. 12.

Protocell metabolism refers to a network of chemical reactions that allows a protocell to produce its own building blocks (container, information coding unit, and the maintenance of a(n) (auto)catalytic network) from the provided nutrients and energy. Here it turned out to be a big challenge to design protocellular metabolism from scratch that leads to autocatalytic closure. In a following section, Fellermann gives a comprehensive review of bottom-up approaches to artificial cells, from historical to current approaches, before he focuses on the "minimal protocell of Rasmussen and coworkers". The functional molecules for information and metabolism are placed at the exterior interface of a lipid container and not inside the vesicle. The information-containing component affects metabolism via electrochemical properties and not via enzymes. The whole life cycle of the envisioned protocell is explained in detail. The experiments are accompanied by theoretical investigations and in silico simulations, to test for alternatives in the design and interpret experimental results against the background of knowledge of soft-matter physics and minimal replicator systems. The simulations are done in the framework of dissipative particle dynamics (DPD), based on Navier–Stokes equations under inclusion of thermal fluctuations. This way one can capture thermodynamic and hydrodynamic features of the system, which from the theoretical point of view is described as a complex fluid such as an oil–water mixture. DPD describes the system on a mesoscopic level (in contrast to the microscopic level of molecular dynamics simulations that we mentioned in Sect. 1.2). Accordingly, the evolution of the system can be followed over a longer time period than is typical in molecular dynamics simulations. In the DPD framework it is of the order of microseconds. According to Fellermann, in silico it is possible to show that each individual step of the protocell life cycle is feasible, but also some obstacles have to be overcome when the steps are combined.

Obviously, a complex chemical reaction network is needed to couple the three basic ingredients of the minimal protocell. How this is achieved is explained in some detail in the following section. (Here it is certainly not sufficient to stay on a single level with one kind of variable and use a description with binary numbers for Boolean functions.) Also part of the answer to why evolution prefers cellular organisms over naked replicators will be found. The reader may notice that what is easily claimed as basic ingredients are by far not unique, as a comparison between different designs of protocells reveals. Certainly, the mapping between chemical information and metabolic regulation is rather complex, so it should not come as a surprise if with increasing complexity also the evolutionary potential of the various designs differs. As reviewed by Fellermann, it ranges from mere adaptation (towards the most effective catalyst) to what John von Neumann termed "universal construction", or from limited to rich variability in the phenotype. Division of labor, as in storing information (in the genome) physically well separated from its action (in the proteome), appears to be an essential step towards a rich evolutionary potential.

(Here we would like to point out an analogy from variational calculus in mathematics. Say we want to search for the optimal solution of a given variational problem, posed as a combination of different tasks at the same time. We can search for the best solution in more or less restricted solution spaces. Imagine we search for the best solution of the combined problem in only a subspace of functions in which the optimal solution for a single task lies. The restriction to a certain subspace constrains the quality of the approximative solution that is supposed to approximate the true solution in full space; this is obvious. Division of labor inside cells seems to have led to a considerably increased "solution space" that evolution could explore to find a solution that optimizes several different tasks at the same time.)

Fellermann's hints on the varying evolutionary potential of different artificially produced protocells are certainly at the center of interest in the theory of evolution. To date, it does not appear that artificial protocells will ever seriously rival our cells (in vivo) in evolutionary potential.

*P. Schwille* ("Towards a Minimal System for Cell Division") Petra Schwille distinguishes two concepts of synthetic biology. The first is microbial engineering, which combines biological units of "hardware" and "software" in some analogy to electrical engineering and is triggered by the progress in nanotechnology. One typical goal is the functional integration of large protein machines to achieve tasks related to environmental or medical applications. The second concept aims at a better understanding of cellular systems from the biophysical point of view, following the bottom-up approach as in Chap. 11. Striving for the artificial cell is here specified as "engineering a specific functionality by employing a set of biological devices, for example, proteins". The aim is not to reproduce complex objects such as the cells of today, but to consider simpler systems in order to concentrate on the basic underlying mechanisms. Here the focus is on minimal systems that allow the study of cell division, one of the most important transformations in biological systems. Molecular modules are listed that will be needed for the construction of a minimal divisome machinery that allows a controlled division of biomolecular compartments. Schwille and her coworkers concentrate on a special vesicle, the giant unilamellar vesicle (GUV). After generating the vesicles, the next step is to include factors in this system that induce division of the cell-like compartments in a controlled way and to combine the compartments with information units that should be reproduced during division. After adding mechanical stability via creation of an artificial cytoskeleton, the next goal is to define a division site, that is, to tell the vesicle not only to divide, but to divide into equal halves. Here the idea is to use two division mechanisms known from *E. coli* bacteria and to reconstitute bacterial divisome machineries in vesicles. One of these mechanisms is provided by the Min system, which we mention here explicitly since it is a classical energy-consuming self-organized system that leads to dynamic pattern formation. Schwille's group has observed traveling wave patterns on two-dimensional open planes. So the next step is to study the behavior of Min waves when the membrane surface is closed. (Here we remark that the influence of the geometry on the propagation of waves in excitable media is an important topic in theoretical modeling as well.) The construction of artificial cellular modules with defined properties, such as division at a given site, may also improve the understanding of possible precursors of our cells of today, that is, of possible forms of life in an early stage of evolution. In addition, synthetic biology of minimal systems can ultimately lead to experimental tests of quantitative predictions from the theoretical side, such as nonlinear or statistical theoretical physics.

Part IV: From Cells to Societies

*E. Frey and T. Reichenbach* ("Bacterial Games") Frey and Reichenbach's contribution is about evolutionary game theory. The players of the games are microbial systems, in particular bacteria. Bacteria assemble into large communities such as biofilms, they compete for nutrients, they cooperate by providing public goods needed for the maintenance of their group, and they communicate via secretion and deletion of certain substances. They even show coordinated behavior, like social groups. All these features make it natural to describe their action in terms of game theory. Although interacting bacteria may show nice forms of self-organized pattern formation, this is not the main aspect in this chapter. Instead, one would like to determine the games' losers and winners and the time it takes until a certain species goes extinct. Moreover, at the core of evolution and ecology are two questions: the first is about the origin of cooperation – the phenomenon of cooperation provides a puzzle that may be summarized in the form of the prisoner's dilemma (see also Chap. 14) – and the second one concerns the origin and maintenance of biodiversity. Naively one may expect that after some transient time always the strongest species will win and dominate all others in the end, so that the only task would be to determine the strongest and calculate the time when it will win the struggle for survival. Instead, a look at nature immediately shows a plethora of coexisting organisms. A more focused look at a Petri dish with three strains of *E. coli* bacteria also reveals their possible coexistence, as the authors report, while only one strain survives if the bacteria are put into a flask with additional stirring. Therefore, spatial organization seems to be essential for the maintenance of biodiversity. Another aspect, often neglected in a first approach, is the inherent and ubiquitous stochasticity of processes in various ways. The authors distinguish between phenotypic noise (which itself is of different origin (see also Chap. 10)), interaction noise (leading to different numbers of interacting species), and external noise (from fluctuations in the environment). It is the role of the interaction noise that is further considered in this contribution.

Implementing this noise in a master equation approach, one can calculate only the probability of finding a certain system size at a certain time, rather than the size itself (as in a deterministic approach). Within the same approach it is also possible to predict the extinction time of a species as a function of the system size. As the authors show, this allows coexistence stability to be classified, and in particular to distinguish neutral from selection-dominated evolution in this context. (In general, it is a nontrivial task, as we mentioned in connection with Chap. 3.) Finally, the mobility of bacteria should be taken into account. Mobility acts as a mechanism that competes with localized spatial interactions; it effectively acts as a stirring mechanism. The authors give an example of a sharp mobility threshold, such that diversity is maintained below this threshold but destroyed above it.

As already summarized in Sect. 1.3, Frey and Reichenbach point out the increasing complexity of the set of equations that is needed when more and more possible processes are taken into account: beyond the direct interaction processes, also those that are due to finite and fluctuating population size, spatially arranged interactions, diffusing or migrating species. Correspondingly, the equations range from ordinary differential equations (with linear or nonlinear interactions) to master equations for stochastic processes to stochastic partial differential equations. Therefore, from the methodological point of view, tools from nonlinear dynamics and nonequilibrium statistical physics are required.

Finally, one may wonder whether bacteria play similar games to "us" (humans). According to the examples of Frey and Reichenbach, they play a public-goods game, a snowdrift game, and cyclic three-strategy games. The last can be realized within the cyclic Lotka–Volterra model as the "rock–paper–scissors" game. This game is played on many scales, from genes to bacteria to a hand game between two or more human players. So it appears to be a universal motif for players' cyclic competition.

*K. Sigmund and C. Hilbe* ("Darwin and the Evolution of Human Cooperation") Sigmund and Hilbe give a comprehensive overview of game theory for games played by humans. In contrast to their biological counterparts such as bacteria, humans are assumed to follow rational strategies, based on rational decisions. The framework of game theory was originally developed for analyzing rational behavior in the context of economics. The games can be played between two or more players who follow a certain strategy that is thought to optimize their payoff. Payoff now replaces the concept of fitness in evolutionary game theory. The payoff is characterized in terms of a payoff matrix. For two players, the columns label the strategies of the first player, and the rows label the strategies of the second player, the matrix entries stand for the payoff. Strategies may be based on the anticipated gain or loss, on imitation, or background information. Players can also change their strategies during a game and retain some memory of experiences in the past.

In relation to evolution, the occurrence of animal and human societies is considered as one of the major transitions (see Sect. 1.1). In particular the evolution of cooperation as a frequent form of organization in societies is regarded as a central puzzle. From the viewpoint of an individual player, it is always better to defect. The better outcome for the society as a whole is, however, mutual cooperation. The conflict between the common interest and the selfish interest is usually expressed as the prisoner's dilemma. Sigmund and Hilbe explain two approaches to analyzing the phenomenon of cooperation. The first corresponds to the "selfish gene" view, realized in the theory of kin selection (for which the degree of relatedness can be made more precise), but kin selection cannot be the only reason for cooperation.

The second reason is of economic nature. It is formulated in the theory of reciprocal altruism, which may be sketched by the phrase "to give and to receive"; if benefit is larger than cost, one cooperates. The simplest strategy to reciprocate good with good and bad with bad in repeated games goes under the name "tit for tat". The direct reciprocity may be extended to an indirect one. In all these cases the goal is to determine the "winner", if there is a single one (winner in the sense of a subset of the total population that follows the most successful strategy); otherwise, if there is no single winner, the goal is to determine the conditions under which players with different strategies may coexist in a stable way. The authors also discuss models for competition of moral systems. A simple moral system is called SCORING, where it is always considered as bad to refuse to help, independently of whether the help is refused to a good or a bad player. According to Sigmund and Hilbe, the evolution of moral faculties, human language, and social intelligence was driven by indirect reciprocity. Conceptually more demanding are games with more than two players. An example here is the public goods game, another expression of a social dilemma with competition between individual and group benefit. In the last part the authors refer to the long controversy between individual selection and group selection, quoting also Darwin and his (often opposed and criticized) emphasis of the importance of group survival.

The contribution contains a number of citations from the original literature, in particular from Darwin's *On the Origin of Species*. This is certainly appreciated in a book that deals so much with implications of Darwin's insights and ideas and their later influence on the science of evolution.

*B. Kahng, D. Lee, and P. Kim* ("Similarities Between Biological and Social Networks in Their Structural Organization") Kahng and coworkers' contribution is from current research on dynamical networks. Networks stand for graphs of nodes, connected via links with an assigned meaning. Here they represent data sets from three different databanks, two from biology, one from social science. In the first example, the phylogenetic tree, the nodes represent organisms of the GenBank database, which are connected via a link if the organisms are connected in a taxonomic tree. In the second example, nodes stand for proteins; they are connected via a link if the proteins physically interact with each other. Finally, in the coauthorship network two nodes are connected if the corresponding authors have a paper in common. These rules for connecting nodes generate graphical representations of the datasets. The authors now identify a specific graphical structure in common to these three graphs of very different origin, that is the structure of a critical scale-free branching tree. The definition of this characteristic is explained in Chap. 15. In general, the possible underlying mechanisms that may have generated a given graphical structure of a dataset are far from unique. (There is more than the mechanism of preferential attachment that leads to a graph with a scale-free degree distribution.) If, however, the graphical structure is very specific, such as that of a scale-free critical branching tree, one may be tempted to conclude that also similar mechanisms have been at work when the datasets evolved (although the time scales for their generation are quite different in the three considered cases). The similar mechanisms here are called the "self-organized burst manner" by the authors.

*A.S. Mikhailov* ("From Swarms to Societies") In several chapters (7, 9, 13) preceding this contribution by A.S. Mikhailov, the concept of self-organization is illustrated with examples from physics and biology; often it is realized in reaction–diffusion systems with local interactions and diffusive flow. Such processes occur in fluids, but also in single cells and cell populations. Mikhailov poses the question whether similar mechanisms of self-organization are sufficient for describing social systems. Distinct from passively diffusing particles in a reaction–diffusion system, groups of actively moving units are called "swarms", examples being schools of fish, flocks of birds, and the like. Swarms show some kind of collective behavior that looks intelligent, for example, in their reaction to a predator attack, and the behavior of human drivers in highway traffic appears on the same level of complexity as bacteria in biofluids. What makes, however, a clear difference between simple biological organisms and members of a society is the internal complexity. Agents can organize themselves in space, but on top of their coordinates in space and time they exhibit a rich complexity in "internal" space. Internal space may represent the inside of a cell with all its molecular components and functions, or the social activities of a human agent. While interactions between particles are expressed in terms of (locally acting) physical forces, interactions in internal space are mediated via "communication", and, as Mikhailov points out, this communication is in general nonlocal in internal space. That makes a big difference in view of the implications and for the description. (Most sophisticated are obviously the communication tools of humans: their language and telecommunication. To a large extent these tools make them independent of their actual location in space and time.)

An important form of cooperation in all kind of societies is synchronization, synchronization of their individual oscillation, active motion, or any other kind of response. More generally one observes the development of coherence. In particular coherence can refer to internal space. If these forms of cooperation refer not to the whole population, but just to some part of it, it is called "dynamical clustering". In dynamical clustering, structures grow around emerging seeds of coherently operating groups, and their evolution depends on the intensity of interactions. According to Mikhailov, these phenomena may be considered as a paradigm of primary social self-organization.

In a section on hierarchies, Mikhailov stresses their important role in complex systems. If complex systems are hierarchically structured, their complexity can be reduced. Usually, a certain level of a hierarchical organization can be characterized by the time scale of direct interactions between the units at this level, and so far, self-organization mostly refers to these interactions at a fixed level, see Sect. 1.4. In contrast, interactions between units from different hierarchy levels are only indirectly possible. The overall system may then be decomposed into different levels, on each of which simple laws are sufficient to describe the dynamics. Therefore there is not only in physical and biological systems a way to reduce complexity, but also in social systems, in spite of all their facets and multitude of internal degrees of freedom.

Dynamical systems with various types of spatial interactions (all-to-all, only with nearest neighbors, with a random subset of all other units) leads to the concept of dynamical networks, on which coherent network activity up to network turbulence may be studied. To preserve a certain degree of coherent behavior, it needs feedback and control mechanisms, which may be more or less "democratic".

Mikhailov concludes with a short section on social evolution. Social evolution proceeds via the cultural transfer of innovations from generation to generation (here generations of populations in a society rather than generations of genes as in biology). "Balancing at the edge of chaos" is often thought to ensure optimal performance of dynamical systems, which allows different organizational forms ranging from totally ordered to completely chaotic states. Mikhailov suggests this mode as an optimal control strategy for societies, providing a prospect of open problems at the end.

# References


1. P. Schuster, Complexity **15/6**, 7 (2010)
2. A. Linde, *Particle Physics and Inflationary Cosmology* (Harwood Academic, London 1990)
3. M. Roos, *Introduction to Cosmology* (Wiley, Chichester, 1994)
4. E.W. Kolb, M.S. Turner, *The Early Universe* (Westview, Boulder, CO, 1994)
5. F. Press, R. Siever, *Understanding Earth*, 3rd edn. (W.H. Freeman, New York, NY, 2001)
6. J. Maynard Smith, E. Szathmáry, *The Major Transitions in Evolution* (Oxford University Press, Oxford, 1995)
7. H. Weyl *Space, Time, Matter*, translated from the 4th German edition (London, Methuen 1922) [*Raum, Zeit, Materie* 8. Aufl. (Springer, Berlin, Heidelberg, 1993)]
8. L. O'Raifeartaigh, *The Dawning of Gauge Theory* (Princeton University Press, Princeton, NJ, 1997)
9. K.G. Wilson, Rev. Mod. Phys. **47**, 773 (1975)
10. L.P. Kadanoff, Rev. Mod. Phys. **49**, 267 (1977)
11. A. Samal, S. Jain, BMC Syst. Biol. **2**, 21 (2008). A. Samal, S. Singh, V. Giri, S. Krishna, N. Raghuram, S. Jain, BMC Bioinformatics **7**, 118 (2006)
12. P.L. Freddolino, A.S. Arkhipov, S.B. Larson, A. McPherson, K. Schulten, Structure **14**, 437 (2006)
13. M.B. Elowitz, S. Leibler, Nature **403**, 335 (2000)
14. T. Reichenbach, M. Mobilia, E. Frey, Nature **448**, 1046 (2007)
15. N.F. Britton, *Essential Mathematical Biology*, 1st edn. (Springer, Berlin, Heidelberg, 2003)
16. A. Turing, Philos. Trans. R. Soc. Lond. B **237**, 37 (1952)
17. J.D. Murray, *Mathematical Biology*, 2nd edn. (Springer, Berlin, Heidelberg, 1993)
18. H. Meinhardt, *Wie Schnecken sich in Schale werfen* (Springer, Berlin, Heidelberg, 1997)
19. Y. Xu, C.M. Vest, J.D. Murray, Appl. Opt. **22**, 3479 (1983)
20. T. Surrey, F. Nédélec, S. Leibler, E. Karsenti, Science **292**, 1167 (2001)
21. T.R. Cech, Gene **135** (1–2), 33 (1993)
22. M. Eigen, P. Schuster, J. Mol. Evol. **19**(1), 47 (1982)
23. R. Shapiro, Orig. Life Evol. Biosph. **14**, 565 (1984)
24. D. Segré, D. Ben-Eli, D.W. Deamer, D. Lancet, Orig. Life Evol. Biosph. **31**(1–2), 119 (2001)

25. D. Segré, D. Lancet, Chemtracts Biochem. Mol. Biol. **12**, 382 (1999)
26. M. Schliwa (ed.), *Molecular Motors* (Wiley-VCH, Weinheim, 2003). K. Nishinari, Y. Okada, A. Schadschneider, D. Chowdhury, Phys. Rev. Lett.**95**, 118101 (2005). P. Greulich, A. Garai, K. Nishinari, A. Schadschneider, D. Chowdhury, Phys. Rev. E **75**, 041905 (2007)
27. A. Garai, D. Chowdhury, D. Chowdhury, T.V. Ramakrishnan, Phys. Rev. E **80**, 011908 (2009). A. Garai, Ph.D. Thesis, IIT Kanpur, India (2010)
28. A. Garai, D. Chowdhury, M.D. Betterton, Phys. Rev. E **77**, 061910 (2008)
29. K. Svoboda, C.F. Schmidt, B.J. Schnapp, S.M. Block, Nature **365**, 721 (1993). J.T. Finer, R.M. Simmons, J.A. Spudich, Nature **368**, 113 (1994)
30. M.D. Wang, M.J. Schnitzer, H. Yin, R. Landick, J. Gelles, S.M. Block, Science **282**, 902 (1998). D. Sinha, U. Bhalla, G.V. Shivashankar, Appl. Phys. Lett. **85**, 4789 (2004)
31. J.D. Wen, L. Lancaster, C. Hodges, A.C. Zeri, S.H. Yoshimura, H.F. Noller, C. Bustamante, I. Tinoco, Jr., Nature **452**, 598 (2008). A. Garai, D. Chowdhury, arXiv:1004.4327v2 [phys. bio-ph]
32. Z.D. Blount, C.Z. Borland, R.E. Lenski, Proc. Natl. Acad. Sci. USA **105**, 7899 (2008)
33. S.J. Gould, *The Structure of Evolutionary Theory* (Belknap, Cambridge, MA, 2002)
34. S. Conway Morris, *Life's Solution* (Cambridge University Press, Cambridge, UK, 2003)
35. A.S. Mikhailov, V. Calenbuhr, *From Cells to Societies* (Springer, Berlin, Heidelberg, 2002)

# Part I
# Principles of Evolution

# Chapter 2
# Physical Principles of Evolution

**Peter Schuster**

**Abstract** Theoretical biology is incomplete without a comprehensive theory of evolution, since evolution is at the core of biological thought. Evolution is visualized as a migration process in genotype or sequence space that is either an adaptive walk driven by some fitness gradient or a random walk in the absence of (sufficiently large) fitness differences. The Darwinian concept of natural selection consisting in the interplay of variation and selection is based on a dichotomy: All variations occur on genotypes whereas selection operates on phenotypes, and relations between genotypes and phenotypes, as encapsulated in a mapping from genotype space into phenotype space, are central to an understanding of evolution. Fitness is conceived as a function of the phenotype, represented by a second mapping from phenotype space into nonnegative real numbers. In the biology of organisms, genotype–phenotype maps are enormously complex and relevant information on them is exceedingly scarce. The situation is better in the case of viruses but so far only one example of a genotype–phenotype map, the mapping of RNA sequences into RNA secondary structures, has been investigated in sufficient detail. It provides direct information on RNA selection in vitro and test-tube evolution, and it is a basis for testing *in silico* evolution on a realistic fitness landscape. Most of the modeling efforts in theoretical and mathematical biology today are done by means of differential equations but stochastic effects are of undeniably great importance for evolution. Population sizes are much smaller than the numbers of genotypes constituting sequence space. Every mutant, after all, has to begin with a single copy. Evolution can be modeled by a chemical master equation, which (in principle) can be approximated by a stochastic differential equation. In addition, simulation tools are available that compute trajectories for master equations. The accessible population sizes in the range of $10^7 \leq N \leq 10^8$ molecules are commonly too small for problems in chemistry but sufficient for biology.

P. Schuster (✉)
Institut für Theoretische Chemie der Universität Wien, Währingerstraße 17, A-1090 Wien, Austria
e-mail: pks@tbi.univie.ac.at

H. Meyer-Ortmanns, S. Thurner (eds.), *Principles of Evolution*,
The Frontiers Collection, DOI 10.1007/978-3-642-18137-5_2,
© Springer-Verlag Berlin Heidelberg 2011

## 2.1 Mathematics and Biology

The beginning of modern science in the sixteenth century was initiated by the extremely fruitful marriage between physics and mathematics. Nobody has expressed the close relation between mathematics and physics more clearly than Galileo Galilei in his famous statement [1]: *Philosophy (science) is written in this grand book, the universe, . . . . It is written in the language of mathematics, and its characters are triangles, circles and other geometric features . . . .* Indeed, physics and mathematics have cross-fertilized each other from the beginnings of modern science until the present day. Theoretical physics and mathematical physics are highly respected disciplines and no physics journal will accept empirical observations without an attempt to bring it into a context that allows for quantification and interpretation by theory. General concepts and successful abstractions have a high reputation in physics and the reductionists' program[1] is the accepted scientific approach towards complex systems. This view is common in almost all subdisciplines of contemporary physics and, in essence, is shared with chemistry and molecular biology.

Conventional biology, in this context, is very different: Great works of biology, such as Charles Darwin's *Origin of Species* [2] or, in recent years, Ernst Mayr's *Growth of Biological Thought* [3], do not contain a single mathematical expression; theoretical and mathematical biology had and still have a bad reputation among macroscopic biologists; special cases are preferred over generalizations, which are looked upon with scepticism; and holistic views are commonly more appreciated than reductionists' explanations, whether or not they are in a position to provide insight into problems. A famous and unique exception among others is Charles Darwin's theory of *natural selection* by reproduction and variation in finite populations. Although not cast in mathematical equations, the theory is based on a general concept whose plausibility is erected upon a wealth of collected and carefully interpreted empirical observations. Darwin's strategy has something in common with the conventional mathematical approach based on observation, abstraction, conjecture, and proof: On different islands of the Galapagos archipelago Darwin observed similar-looking species in different habitats and concluded correctly that these different species are closely related and owe their existence to histories of adaptation to different environments on the individual islands. The occurrence of adaptations has been attributed to natural selection as a common mechanism through abstraction from specific cases. Darwin's conjecture combines three facts known in his time:

---

[1] The reductionist program, also called methodological reductionism, aims at an exploration of complex objects through breaking them up into modular, preferentially molecular parts and studying the parts in isolation before reassembling the object. Emergent properties are assumed to be describable in terms of the phenomena from and the processes by which they emerge. The reductionist program is different from ontological reductionism, which denies the idea of ontological emergence by the claim that emergence is merely a result of the system's description and does not exist on a fundamental level.

1. *Multiplication*: All organisms multiply by cell division, (parthenogenesis or sexual reproduction), multiplication is accompanied by inheritance – "progeny resembles parents", and under the condition of unlimited resources multiplication results in exponential growth of population size.
2. *Variation*: All natural populations show variance in phenotypic properties, either continuously varying features, such as body size, or discontinuously varying features, such as the number of limbs, the number of digits, color of flowers, skin patterns, or seed shapes, and it is straightforward to relate variation to inheritance.[2]
3. *Selection*: Exponential growth results in overpopulation of habitats,[3] only a small fraction of offspring can survive and have progeny of their own, and this stringent competition prevents less efficient variants from reproduction.

Taking together these three items and introducing the notion of fitness for the number of offspring that reach the age of fertility, the conjecture could be formulated in the following way:

**Natural selection**: In nonhomogeneous populations the frequencies of variants with fitness values below the population average will decrease, while those with fitness values above average will increase and consequently the population average itself will increase until it reaches the maximum value corresponding to a homogeneous population of the best adapted or fittest variant.

Darwin's *Origin of Species* is an overwhelming collection of observations from nature, from animal breeders, and from nursery gardens that provide strong evidence for the correctness of Darwin's conjecture. This enormous collection in a way is the empirical substitute for a mathematical proof.

Although Gregor Mendel analyzed his experiments on inheritance in peas by mathematical statistics and found thereby the explanatory regularities, mathematics did not become popular in biology. On the contrary, Mendel's work was largely ignored by the biological community for more than 30 years. Then Mendel was rediscovered and genetics became an important discipline of biology. Population genetics was founded by the three scholars Ronald Fisher [4], J.B.S. Haldane, [5] and Sewall Wright [6]. In the 1930s they succeeded in uniting Mendelian genetics and Darwin's natural selection, and to cast evolution in a rigorous mathematical frame, but conventional geneticists and evolutionary biologists continued to fight until the completion of the synthetic theory almost 20 years later [3].

Modeling in biology became an important tool for understanding complex dynamical phenomena. Representative for many other approaches we mention here

---

[2] Gregor Mendel was the first to investigate such relations experimentally [7–9] and discovered the transmittance of properties in discrete packages from the parents to offspring. His research objects were the pea (*Pisum*) from where he derived his rules of inheritance and the hawkweed (*Hieracium*), which was rather confusing for him, because it is apomictic, i.e., it reproduces asexually. Charles Darwin, on the other hand, had a mechanism of inheritance in mind that was entirely wrong. It was based on the idea of blending of the parents' properties.

[3] According to his own records Charles Darwin was influenced strongly by Robert Malthus and his demographic theory [10].

only three: (i) Modeling of coevolution in a predator–prey system was introduced by Alfred Lotka [11] and Vito Volterra [12] by means of differential equations that were borrowed from chemical kinetics. In a way, they were the pioneers of theoretical ecology, which was developed by the brothers Howard and Eugene Odum [13] and became a respectable field of applied mathematics later [14]. (ii) A model for pattern formation based on the reaction–diffusion (partial differential) equation with a special chemical mechanism was suggested and analyzed by Alan Turing [15]. Twenty years later the Turing model was applied to biological morphogenesis [16, 17] and provided explanations for patterns formed during development [18, 19]. (iii) Based on experimental studies of nerve pulse propagation in the squid giant axon, Alan Hodgkin and Andrew Huxley formulated a mathematical model for nerve excitation and pulse propagation [20] that became the standard model for single nerve dynamics in neurobiology. They were both awarded the Nobel Prize in Medicine in 1963. A second breakthrough in understanding neutral systems came from modeling networks of neurons. John Hopfield conceived an exceedingly simple model of neurons in networks [21] that initiated a whole new area of scientific computing: computation with *neutral networks*, in particular modeling and optimization of complex systems. Despite these undeniable and apparent successes, the skepticism of biologists with respect to theory and mathematics nevertheless continued for almost the entire remainder of the twentieth century.

The advent of molecular biology in the 1950s brought biology closer to chemistry and physics, and changed the general understanding of nature in a dramatic way [22]. Inheritance received a profound basis in molecular genetics and reconstruction of phylogenies became possible through comparison of biopolymer sequences from present-day organisms. Structures of biomolecules at atomic resolution were determined by refined techniques from physical chemistry and they provided deep insights into biomolecular functions. Spectroscopic techniques, in particular nuclear magnetic resonance, require a solid background in mathematics and physics for conceiving and analyzing conclusive experiments. A novel era of biology was initiated in the 1970s when the highly efficient new methods for DNA sequencing developed by Walter Gilbert and Frederick Sanger became available [23, 24]. Sequencing whole genomes became technically within reach and financially affordable. The first two complete bacterial genomes were published in 1995 [25] and the following years saw a true explosion of sequencing data. High-throughput techniques using chip technology for genome-wide analysis of translation and transcription products known as proteomics and transcriptomics followed, and an amount of data was created that had never been seen before. In this context it is worth citing the Nobel laureate Sydney Brenner, [26] who made the following statement in 2002 to characterize the situation in molecular biology:

> I was taught in the pre-genomic era to be a hunter. I learnt how to identify the wild beasts and how to go out, hunt them down and kill them. We are now, however, being urged to be gatherers. To collect everything lying about and put it into storehouses. Someday, it is assumed someone will come and sort through the storehouses, discard the junk and keep the rare finds. The only difficulty is how to recognize them.

Who else but a theorist should this "someone" be? The current development seems to indicate that "someday" is not too far away. The flood of data and the urgent need for a comprehensive theory have driven back the biologists' aversion to computer science and mathematics. Modern genetics and genome analysis without bioinformatics are unthinkable, and understanding network dynamics without mathematics and computer modeling is impossible.

The new discipline of systems biology has the ambitious goal to find holistic descriptions for cells and organisms without giving up the roots in chemistry and physics. Although still in its infancy and falling into one trap after another, modeling in systems biology is progressing slowly towards larger and more detailed models for regulatory modules in cell biology. New techniques are being developed and applied. Examples are flux-balance analysis [27] and application of inverse methods [28], whereby the primary challenge is up-scaling to larger systems such as whole organisms. Recent advances in experimental evolution allow for an extension of detailed models to questions of evolution, which is of central importance in biology, as Theodosius Dobzhansky encapsulated in his famous sentence: "Nothing in biology makes sense except in the light of evolution" [29]. From a conceptional point of view, theoretical biology is in a better position than theoretical physics, where attempts at unification of two fundamental theories, quantum mechanics and relativity theory, have not been successful so far. Biology has one comprehensive theory, the theory of evolution, and present-day molecular biology is building the bridge to chemistry and physics. Lacking are a proper language and efficient techniques to handle the enormous complexity and to build proper models.

## 2.2 Darwin's Theory in Mathematical Language

If Charles Darwin had been a mathematician, how might he have formulated his theory of natural selection? Application of mathematics to problems in biology has a long history. The first example that is relevant to evolution dates back to medieval times. In the famous *Liber Abaci* written in the year 1202 by Leonardo Pisano, also known as Fibonacci (*filius Bonacci*), we find a counting example of the numbers of pairs of rabbits in subsequent time spans. Every adult pair is assumed to give birth to another pair, newborn rabbits have to wait one time interval before they become fertile adults. Starting from a single couple yields the following series:

$$(0) \quad 1 \ \ 1 \ \ 2 \ \ 3 \ \ 5 \ \ 8 \ \ 13 \ \ 21 \ \ 34 \ \ 55 \ \ 89 \ \ \ldots .$$

Every number is the sum of its two precursors and the Fibonacci series is defined by the recursion

$$F_{i+1} = F_i + F_{i-1} \quad \text{with} \quad F_0 = 0 \text{ and } F_1 = 1 . \tag{2.1}$$

It is straightforward to show that the Fibonacci series can be approximated well by exponential functions as upper and lower limits (Fig. 2.1). The exponential

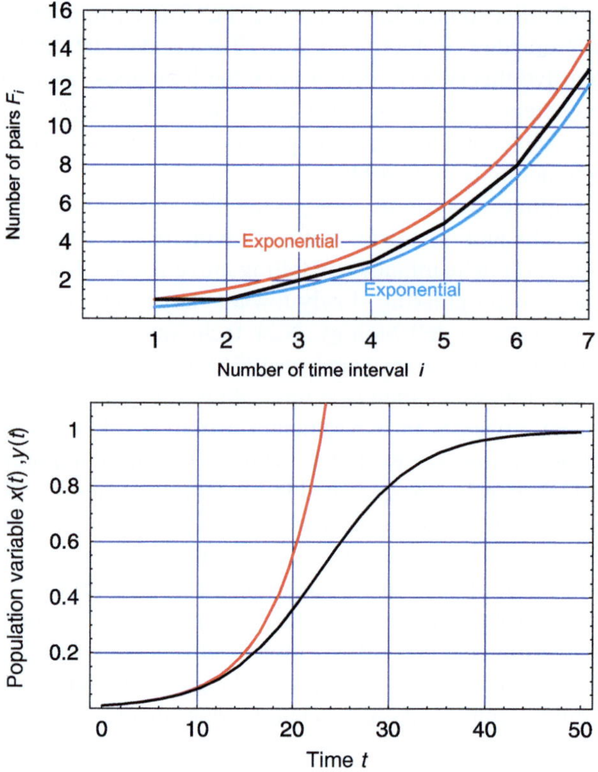


**Fig. 2.1** Fibonacci series, exponential functions, and limited resources. The Fibonacci series (*black; upper plot*) is embedded between two exponential functions in the range $0 < i \leq 10$: $n_{\text{upper}}(t) = \exp\big(0.4453(t - 1)\big)$ (*red*) and $n_{\text{lower}}(t) = \exp\big(0.5009(t - 2)\big)$ (*blue*), where the time $t$ is the continuous equivalent to the discrete (generation) index $i$. The *lower plot* compares the exponential function, $y(t) = y_0 \exp(r\,t)$ for unlimited growth (*red*; $y_0 = 0.02$, $r = 0.1$) with the normalized solution of the Verhulst equation ($x(t)$, *black*; $x_0 = 0.02$, $r = 0.1$, and $C = 1$ by definition)

function, however, was not known before the middle of the eighteenth century; it was introduced in the fundamental work of the Swiss mathematician Leonhard Euler [30]. Robert Malthus – although he lived 50 years later – still used a geometric progression, $2, 4, 8, 16, \ldots$ , for the unlimited growth of populations [10]. The consequences of unlimited growth for demography are disastrous and, as said, Malthus's work was influential on Darwin's thoughts.

A contemporary of Charles Darwin, the mathematician Pierre-François Verhulst [31], formulated a model based on differential equations combining exponential growth and limited resources (Fig. 2.1):

$$\frac{dN}{dt} = \dot{N} = r\,N \left(1 - \frac{N}{C}\right) \tag{2.2}$$

with $N(t)$ describing the number of individuals at time $t$, $r$ being the Malthusian parameter, and $C$ the carrying capacity of the ecosystem. Equation (2.2) consists of two terms: (i) the exponential growth term, $rN$, and (ii) the constraint to finite population size expressed by the term $-rN^2/C$. In other words, the ecosystem can only support $N = C$ individuals and $\lim_{t\to\infty} N(t) = C$. The solution of the differential equation (2.2) is of the form

$$N(t) = \frac{N_0\, C}{N_0 + (C - N_0)\exp(-rt)}. \tag{2.3}$$

Here $N_0 = N(0)$ is the initial number of individuals. It is straightforward to normalize the variable to the carrying capacity, $x(t) = N(t)/C$, yielding

$$x(t) = \frac{x_0}{x_0 + (1 - x_0)\exp(-rt)} \tag{2.3'}$$

with $x_0 = N_0/C$. It will turn out to be useful to write the term representing the constraint in the form $N\,\phi(t)/C = x\,\phi(t)$. Then we obtain for the Verhulst equation

$$\frac{dx}{dt} = \dot{x} = x\left(r - \phi(t)\right) \quad \text{with} \quad \phi(t) = x\,r \tag{2.2'}$$

being the (mean) reproduction rate of the population.

Finally, we generalize to the evolution of $n$ species or variants[4] in the population $\Xi = \{X_1, X_2, \ldots, X_n\}$. The numbers of individuals are now denoted by $[X_i] = N_i$ with $\sum_{i=1}^{n} N_i = N$ and the normalized variables $x_i = N_i/N$ with $\sum_{i=1}^{n} x_i = 1$. Each variant has its individual Malthus parameter or fitness value $f_i$, and for the selection constraint leading to constant population size we find now $\phi(t) = \sum_{i=1}^{n} x_i\, f_i$, which is the mean reproduction rate of the entire population. The selection constraint $\phi(t)$ can be used for modeling much more general situations than constant population size by means of the mean reproduction rate. As we shall see in Sect. 2.5, the proof for the occurrence of selection can be extended to very general selection constraints $\phi(t)$ as long as the population size does not become zero, $N > 0$.

The kinetic differential equation in the multispecies case, denoted as the selection equation,

$$\dot{x}_j = x_j\left(f_j - x_j \sum_{i=1}^{n} x_i\, f_i\right) = x_j\left(f_j - x_j\,\phi(t)\right), \quad j = 1, 2, \ldots, n\,, \tag{2.4}$$

can be solved exactly by the integrating factors transform ([32], pp. 322ff.)

---

[4] In this chapter we shall not consider sexual reproduction or other forms of recombination. In asexual reproduction a strict distinction between variants and species is neither required nor possible. We shall briefly come back to the problem of bacterial or viral species in Sect. 2.7.

$$z_j(t) = x_j(t) \cdot \exp\left(\int_0^t \phi(\tau)d\tau\right) . \tag{2.5}$$

Insertion into (2.4) yields

$$\dot{z}_j = f_j z_j \text{ and } z_j(t) = z_j(0) \cdot \exp(f_j t) ,$$

$$x_j(t) = x_j(0) \cdot \exp(f_j t) \cdot \exp\left(-\int_0^t \phi(\tau)d\tau\right) \text{ with}$$

$$\exp\left(\int_0^t \phi(\tau)d\tau\right) = \sum_{i=1}^n x_i(0) \cdot \exp(f_i t) ,$$

where we have used $z_j(0) = x_j(0)$ and the condition $\sum_{i=1}^n x_i = 1$. The solution finally is of the form

$$x_j(t) = \frac{x_j(0) \cdot \exp(f_j t)}{\sum_{i=1}^n x_i(0) \cdot \exp(f_i t)} ; \quad j = 1, 2, \ldots, n . \tag{2.6}$$

The interpretation is straightforward. The term with the largest fitness value, $f_m = \max\{f_1, f_2, \ldots, f_n\}$, dominates the sum in the denominator after sufficiently long time[5]:

$$\sum_{i=1}^n x_i(0) \cdot \exp(f_i t) \rightarrow x_m(0) \cdot \exp(f_m t) \text{ for large } t \text{ and } x_m(t) \rightarrow 1 .$$

Optimization in the sense of Charles Darwin's principle of selection of the fittest variant, $X_m$, takes place.

The occurrence of selection in (2.4) can be verified also without knowing the solution (2.6). For this goal we consider the time dependence of the constraint $\phi$, which is given by

$$\begin{aligned}
\frac{d\phi}{dt} &= \sum_{i=1}^n f_i \dot{x}_i = \sum_{i=1}^n f_i \left(f_i x_i - x_i \sum_{j=1}^n f_j x_j\right) = \\
&= \sum_{i=1}^n f_i^2 x_i - \sum_{i=1}^n f_i x_i \sum_{j=1}^n f_j x_j = \\
&= \overline{f^2} - \left(\overline{f}\right)^2 = \mathrm{var}\{f\} \geq 0 .
\end{aligned} \tag{2.7}$$

Since a variance is always nonnegative, (2.7) implies that $\phi(t)$ is a nondecreasing function of time. The value $\mathrm{var}\{f\} = 0$ implies a (local) maximum of $\phi$ and

---

[5] We assume here that the largest fitness value $f_m$ is non-degenerate, i.e., there is no second species having the same (largest) fitness value. In Sect. 2.5 we shall drop this restriction.

hence, $\phi$ is optimized during selection. Zero variance is tantamount to a homogeneous population containing only one variant. Since $\phi$ is at a maximum, this is the fittest variant $X_m$.

## 2.3 Evolution in Genotype Space

Evolution can be visualized as a process in an abstract genotype or sequence space, $\mathcal{Q}$. At constant chain lengths $\ell$ of polynucleotides the sequence space is specified as $\mathcal{Q}_\ell^{\mathcal{A}}$, where $\mathcal{A}$ is the alphabet, for example $\mathcal{A} = \{\mathbf{0}, \mathbf{1}\}$ or $\mathcal{A} = \{\mathbf{G}, \mathbf{C}\}$ is the binary alphabet and $\mathcal{A} = \{\mathbf{A}, \mathbf{U}, \mathbf{G}, \mathbf{C}\}$ the natural nucleotide alphabet. The gains of such a comprehensive view of genotypes are generality and the framework for reduction to the essential features; the shortcomings, obviously, are lack of detail. Building a model for evolution upon a space that fulfills all requirements required for the molecular view of biology and which may, eventually, bridge microscopic and macroscopic views, is precisely what we are aiming for here. The genotypes are DNA or RNA sequences and the proper genotype space is sequence space. The concept of a static sequence space [33, 34] was invented in the early 1970s in order to bring some ordering criteria into the enormous diversity of possible biopolymer sequences. Sequence space $\mathcal{Q}_\ell^{\mathcal{A}}$, as long we are only dealing with reproduction and mutation, is a metric space with the Hamming distance[6] serving as the most useful metric for all practical purposes. Every possible sequence is a point in the discrete sequence space and in order to illustrate the space by a graph, sequences are represented by nodes and all pairs of sequences with Hamming distance one by edges (Fig. 2.2 shows a space of binary sequences as an example. Binary sequence spaces are hypercubes of dimension $\ell$, $\ell$ being the length of the sequences).

Two properties of sequence spaces are important: (i) All nodes in a sequence space are equivalent in the sense that every sequence has the same number of nearest neighbors with Hamming distance $d_H = 1$, next nearest neighbors with Hamming distance $d_H = 2$, and so on, which are grouped properly in mutant classes. (ii) All nodes of a sequence space are at the boundary of the space or, in other words, there is no interior. Both features are visualized easily by means of hypercubes[7]: All points are positioned at equal distances from the origin of the (Cartesian) coordinate system. What makes sequence spaces difficult to handle are neither internal structures

---

[6] The Hamming distance $d_H(X_i, X_j)$ [35] counts the number of positions at which two aligned sequences $X_i$ and $X_j$ differ. It fulfills the four criteria for a metric in sequence space: (i) $d_H(X_i, X_j) \geq 0$ (nonnegativity), (ii) $d_H(X_i, X_j i) = 0$ if and only if $X_i = X_j$ (identity of indiscernibles), (iii) $d_H(X_i, X_j) = d_H(X_i, X_j)$ (symmetry), and (iv) $d_H(X_i, X_j) \leq d_H(X_i, X_k) + d_H(X_i, X_k)$ (triangle inequality). For sequences of equal chain length $\ell$, end-to-end alignment is the most straightforward alignment, although it may miss close relatedness that is a consequence of deletions and insertions, which are mutations that alter sequence length.

[7] An $\ell$-dimensional hypercube in the Cartesian space of dimension $\ell$ is the analogue of a (three-dimensional) cube. The $\ell$-dimensional hypercube is constructed by drawing $2\ell$ (hyper)planes of dimension $(\ell - 1)$ perpendicular to the coordinate axes at the positions $\pm a$. The corners of the hypercubes are the $2\ell$ points where $\ell$ planes cross.

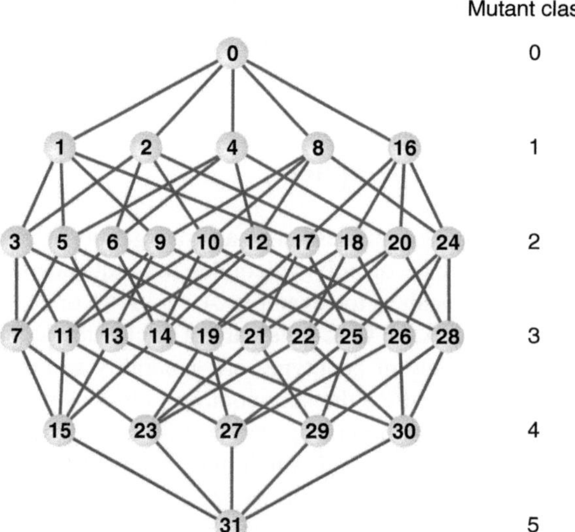



**Fig. 2.2** Sequence space of binary sequences of chain length $\ell = 5$. The sequence space $\mathcal{Q}_5^{\{0,1\}}$ comprises 32 sequences. Each sequence is represented by a *point*. The *numbers* in the *yellow balls* are the decimal equivalents of the binary sequences and can be interpreted as sequences of two nucleotides, "**0**" $\equiv$ "**C**" and "**1**" $\equiv$ "**G**". Examples are $0 \equiv$ **00000** $\equiv$ **CCCCC**, $14 \equiv$ **01110** $\equiv$ **CGGGC** or $29 \equiv$ **11101** $\equiv$ **GGGCG**. All positions of a (binary) sequence space are equivalent in the sense that each sequence has $\ell$ nearest neighbors, $\ell(\ell - 1)/2$ next nearest neighbors, etc. Accordingly, sequences are properly grouped in mutant classes around the reference sequence, here 0

nor construction principles but the hyper-astronomically large numbers of points: $|\mathcal{Q}_\ell^{\mathcal{A}}| = \kappa^\ell$ for sequences of length $\ell$ over an alphabet of size $\kappa$ with $\kappa = |\mathcal{A}|$.

The population $\varXi = \{X_1, X_2, \ldots, X_n\}$ is represented by a vector with the numbers of species as elements $\mathbf{N} = (N_1, N_2, \ldots, N_n)$, the population size is the $L_1$-norm:

$$N = \|\mathbf{N}\|_1 = \sum_{i=1}^{n} |N_i| = \sum_{i=1}^{n} N_i \, ,$$

where absolute values are dispensable since particle numbers are real and non-negative by definition. Normalization of the variables yields $\mathbf{x} = \mathbf{N}/\|\mathbf{N}\|$ or $x_i = N_i/N$ and $\sum_{i=1}^{n} x_i = 1$, respectively. A population is thus represented by an $L_1$-normalized vector $\mathbf{x}$ and the population size $N$. An important property of a population is its *consensus sequence*, $\bar{X}$, consisting of a nucleotide distribution at each position of the sequence. This consensus sequence can be visualized as the center of the population in sequence space.

A sequence is conventionally understood as a string of $\ell$ symbols chosen from some predefined alphabet with $\kappa$ letters, which can be written as

$$X_j \;=\; \left(b_1^{(j)}, b_2^{(j)}, \ldots, b_\ell^{(j)}\right) \quad \text{with} \quad b_i^{(j)} \in \mathcal{A} = \{\alpha_1, \ldots, \alpha_\kappa\}\,.$$

The natural nucleotide alphabet contains four letters: $\mathcal{A} = \{\mathbf{A}, \mathbf{U}, \mathbf{G}, \mathbf{C}\}$, but RNA molecules with catalytic functions have been derived also from three- and two-letter alphabets [36, 37]. For the forthcoming considerations it is straightforward to adopt slightly different definitions: A sequence $X_j$ results from the multiplication of the alphabet vector $\boldsymbol{\alpha} = (\alpha_1, \ldots, \alpha_\kappa)$ with a $\kappa \times \ell$ matrix $\mathcal{X}_j$ having only 0 and 1 as entries:

$$X_j \;=\; \boldsymbol{\alpha} \cdot \mathcal{X}_j \;=\; \boldsymbol{\alpha} \cdot \left(\boldsymbol{\beta}_1^{(j)}, \boldsymbol{\beta}_2^{(j)}, \ldots, \boldsymbol{\beta}_\ell^{(j)}\right) \quad \text{with}$$

$$\boldsymbol{\beta}_i^{(j)} \in \left\{ \begin{pmatrix} 1 \\ 0 \\ \vdots \\ 0 \end{pmatrix}, \begin{pmatrix} 0 \\ 1 \\ \vdots \\ 0 \end{pmatrix}, \ldots, \begin{pmatrix} 0 \\ 0 \\ \vdots \\ 1 \end{pmatrix} \right\}\,. \tag{2.8}$$

In other words, the individual nucleotides in the sequence $X_j$ are replaced by products of two vectors, $b_i^{(j)} = \boldsymbol{\alpha} \cdot \boldsymbol{\beta}_i^{(j)}$.

With the definition (2.8) it is straightforward to compute the consensus sequence of a population $\boldsymbol{\Xi}_k$:

$$\boldsymbol{\Xi}_k \;=\; \boldsymbol{\alpha} \cdot \sum_{j=1}^{n} x_j^{(k)}\, \mathcal{X}_j\,, \tag{2.9}$$

and the distribution of nucleotides at position "$i$" is given by

$$\boldsymbol{b}_i^{(k)} \;=\; \boldsymbol{\alpha} \cdot \sum_{j=1}^{n} x_j^{(k)}\, \boldsymbol{\beta}_i^{(j)}\,. \tag{2.9$'$}$$

It is important to note the difference between $b_i^{(j)}$ and $\boldsymbol{b}_i^{(k)}$: The former refers to the nucleotide at position "$i$" in a given sequence whereas the latter describes the nucleotide distribution at position "$i$" in the population. If one nucleotide is dominant at all positions, the distribution can be collapsed to a single sequence, the consensus sequence.

The internal structure of every sequence space $\mathcal{Q}_\ell^{\mathcal{A}}$ is induced by point mutation and this is essential for inheritance because it creates a hierarchy in the accessability of genotypes. Suppose we have a probability $p$ of making one error in the reproduction of a sequence then, provided mutation at different positions is assumed to be independent, the probability of making two errors is $p^2$, of making three errors is $p^3$, etc. Inheritance requires sufficient accuracy of reproduction – otherwise children would not resemble their parents – and this implies $p$ has to be sufficiently small. Then, $p^2$ is smaller and the power series $p^{d_\mathrm{H}}$ decreases further with increasing distance from the reference sequence. This ordering of sequences according to a probability criterion that is intimately related to the Hamming metric (Sect. 2.5). As a matter of fact, mutation is indeed a fairly rare event in evolution and populations are commonly dominated by a well-defined single consensus sequence since single nucleotide exchanges that occur at many different positions do not contribute significantly to the average.

Evolutionary dynamics is understood as change of the population vectors in time: $\mathbf{N}(t)$ or $\mathbf{x}(t)$. This change can be modeled by means of differential equations (Sect. 2.5) or stochastic processes (Sect. 2.6). A practical problem concerns the representation of genotype space. Complete sequence space, $\mathcal{Q}_\ell^A$ has the advantage of covering all possible genotypes but its extension is huge and, since the numbers of possible genotypes exceed even the largest populations by far, we are confronted with the problem that most degrees of freedom are empty and very likely will never be populated during the evolutionary process described. Alternatively the description could be restricted to those genotypes that are actually present in the population and that constitute the *population support* $\boldsymbol{\Phi}(t)$, which is defined by

$$\boldsymbol{\Phi}(t) \doteq \{X_j | N_j(t) \geq 1\}. \tag{2.10}$$

The obvious advantage is a drastic reduction in the degrees of freedom to a tractable size but one has to pay a price for this simplification: The population support is time dependent and changes whenever a new genotype is produced by mutation or an existing one goes extinct [38]. Depending on population size, population dynamics on the support can either be described by differential equations or modeled as a stochastic process. Support dynamics, on the other hand, are intrinsically stochastic since every mutant starts from a single copy.

Finally, it is important to mention that recombination without mutation can be modeled successfully as a process in an abstract recombination space [39–41] and plays a major role in the theory of genetic algorithms [42, 43]. A great challenge for theorists is the development of a genotype space for both mutation and recombination. Similarly, convenient sequence spaces for genotypes with variable chain lengths are not at hand.

## 2.4 Modeling Genotype–Phenotype Mappings

The unfolding of genotypes to yield phenotypes is studied in developmental biology and provides the key to understanding evolution and, in particular, the origin of species. For a long time it has been common knowledge that the same genotype can develop into different phenotypes, depending on differences in the environmental conditions and epigenetic effects.[8] Current molecular biology provides explanations for several epigenetic observations and reveals mechanisms for the inheritance of

---

[8] *Epigenetics* was originally used as a term subsuming phenomena that could not be explained by conventional genetics.

properties that are not encoded by the DNA of the individual. Genetics is still shaping the phenotypes – otherwise progeny would not resemble parents – but epigenetics and environmental influences provide additional effects that are indispensable for understanding and modeling the relations between genotypes and phenotypes. Here we shall adopt the conventional strategy of physicists and consider simple cases in which the genotypes unfolds unambiguously into a unique phenotype. This condition is fulfilled, for example, in evolution in vitro when biopolymer sequences form (the uniquely defined) minimum free energy structures as phenotypes. Bacteria in constant environments provide other cases of simple genotype–phenotype mappings (the long-term experiments of Richard Lenski [44–46] may serve as examples; see Sect. 2.6). Under this simplifying assumption genotype–phenotype relations can be modeled as mappings from an abstract genotype space into a space of phenotypes or *shapes*. A counter example in a simple system is provided by biopolymers with metastable suboptimal conformations, which can serve as models where a single genotype – a sequence – can give rise to several phenotypes – molecular structures [47].

Since only point mutations will be considered here, the choice of an appropriate genotype space is straightforward. It is the sequence space $\mathcal{Q}_\ell^{\mathcal{A}}$ with the Hamming distance $d_\mathrm{H}$ as metric. The phenotype space or shape space $\mathcal{S}_\ell$ is the space of all phenotypes formed by all genotypes of chain length $\ell$. Although the definition of a physically or biologically meaningful distance between phenotypes is not at all straightforward, some kind of metric can always be found. Accordingly the genotype–phenotype mapping $\psi$ can be characterized by

$$\psi : \left\{ \mathcal{Q}_\ell^{(\mathcal{A})} ;\ d_\mathrm{H}(X_i, X_j) \right\} \xRightarrow{\text{mfe}} \{ \mathcal{S}_\ell;\ d_S(S_i, S_j) \} \quad \text{or} \quad S_k = \psi(X_k) . \quad (2.11)$$

where mfe indicates minimum free energy. The map $\psi$ need not be invertible. In other words, several genotypes can be mapped onto the same phenotype when we are dealing with a case of neutrality.

An example of a genotype–phenotype mapping that can be handled straightforwardly by analytical tools is provided by in vitro evolution of RNA molecules [48–50]. RNA molecules are transferred to a solution containing activated monomers as well as a virus-specific RNA replicase. The material consumed by the replication reaction is replenished by serial transfer of a small sample into fresh solution. The replicating ensemble of RNA molecules optimizes the mean RNA replication rate of the population in the sense of Darwinian evolution [see (2.6)]. The interpretation of RNA evolution in vitro identifies the RNA sequence with the genotype. The RNA structure, the phenotype, is responsible for binding to the enzyme and for the progress of reproduction, since the structure of the template molecules has to open in order to allow replication [51–53]. In the case of RNA aptamer selection[9] the binding affinity is a function of molecular structure, and

---

[9] An aptamer is a molecule that binds to a predefined target molecule. Aptamers are commonly produced by an evolutionary selection process [57].

sequence–structure mapping is an excellent model for the relation between genotype and phenotype.

RNA sequences fold spontaneously into secondary structures consisting of double-helical stacks and single-stranded stretches. Within a stack, nucleotides form base pairs that are elements of a pairing logic $\mathcal{B}$, which consists of six allowed base pairs in the case of RNA structures: $\mathcal{B} = \{\mathbf{AU}, \mathbf{UA}, \mathbf{GC}, \mathbf{CG}, \mathbf{GU}, \mathbf{UG}\}$. Further structure formation, very often initiated by the addition of two-valent cations, mostly $Mg^{2+}$, folds secondary structure into three-dimensional structures by means of sequence specific *tertiary* interactions of nucleotide bases called motifs [54, 55]. Secondary structures have the advantage of computational and conceptional simplicity, allowing the application of combinatorics to global analysis of sequence–structure mappings [47, 56]. A conventional RNA secondary structure consists exclusively of base pairs and unpaired nucleotides and can be represented in a formal three-letter alphabet with the symbols '·', '(',')' for unpaired nucleotides, downstream-bound, and upstream-bound nucleotides, respectively (Fig. 2.3). A

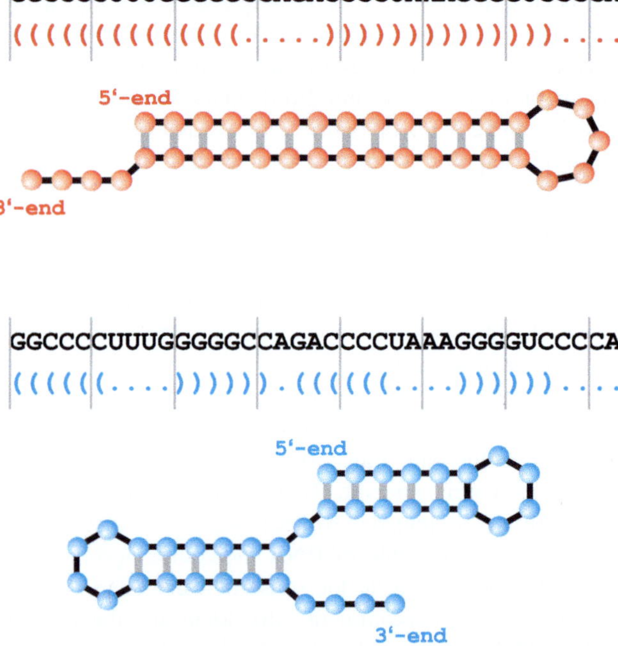



**Fig. 2.3** Symbolic notation of RNA secondary structures. RNA molecules have two chemically different ends, the $5'$- and the $3'$-end. A general convention determines that all strings corresponding to RNA molecules (sequences, symbolic notation, etc.) start from the $5'$-end and have the $3'$-end at the right-hand side (rhs). The symbolic notation is equivalent to graphical representation of secondary structures. Base pairs are denoted by parentheses, where the opening parenthesis corresponds to the nucleotide closer to the $5'$-end and the closing parenthesis to the nucleotide closer to the $3'$-end of the sequence. In the figure we compare the symbolic notation with the conventional graphical representations for two structures formed by the same sequence

straightforward way to annotate pairs in structures is given by the *base pair count* $S_i = [\gamma_1^{(i)}, \ldots, \gamma_\ell^{(i)}]$, which we illustrate here by means of the lower (blue) structure in the figure as an example[10]:

$$S_i = [1,2,3,4,5,6,0,0,0,0,6,5,4,3,2,1,0,7,8,9,10,11,12,0,0,0,0,12,11,10,9,8,7,0,0,0,0,0]$$

Consecutive numbers are assigned to first nucleotides of base pairs corresponding to an opening parenthesis in the sequence, in which they appear in the structure, and the same number is assigned to the corresponding closing parenthesis lying downstream. Unpaired nucleotides are denoted by '0'. In total the structure contains $n_p$ base pairs and $n_s$ single nucleotides with $2n_p + n_s = \ell$.

Molecular physics provides an excellent tool for modeling folding of molecules into structures, the concept of *conformation space*: A free energy is assigned to or calculated for each conformation of the molecule. Commonly, the variables of conformation space are continuous, bond lengths, valence angles or torsion angles may serve as examples. The free energy (hyper)surface or free energy landscape of a molecule presents the free energy as a function of the conformational variables. The mfe structure corresponds to the global minimum of the landscape, metastable states to local minima. In the case of RNA secondary structures conformation space and shape space are identical, and they are discrete spaces, since a nucleotide is either paired or unpaired. Whether a given conformation, a given base pairing pattern, is a local minimum or not depends also on the set of allowed moves in shape space $S$. The move set defines the distance between structures, the metric $d_S(S_i, S_j)$ in (2.11). An appropriate move set for RNA secondary structures comprises three moves: (i) base pair closure, (ii) base pair opening, and (iii) base pair shift [47, 62]. The first two moves need no further explanation; the shift move combines base pair opening and base pair formation with neighboring unpaired nucleotides. This set of three moves corresponds to a metric $d_S(S_i, S_j)$, which is the Hamming distance between the symbolic notations of the two structures $S_i$ and $S_j$.

Conventional structure prediction deals with single structures derived from single sequence inputs. Structure formation depends on external conditions such as temperature, pH value, ionic strength, and the nature of the counter-ions; in order to obtain a unique solution these conditions have to be specified. Commonly the search is for the most stable structure, the mfe structure, which corresponds to the global minimum of the conformational free energy landscape of the RNA molecule. In Fig. 2.4 the mfe structure $S_0 = \psi(X)$ is a single long hairpin shown (in red) at the lhs of the picture. A sequence that forms a stable mfe structure $S_0$ (free

---

[10] The base pair count is another equivalent representation of RNA secondary structures. In the case of conventional secondary structures, the symbolic notation is converted into the base pair count by an exceedingly simple algorithm: Starting with zero at the $5'$-end and proceeding from left to right a positive integer counting the number of open parenthesis is assigned to every position along the sequence. The base pair count is not only more convenient for base pair assignments but also more general. It is, for example, applicable to RNA structures with pseudoknots.

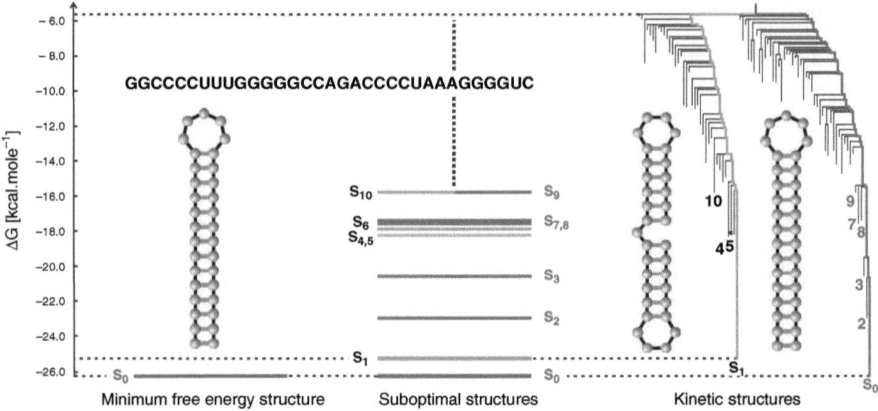


**Fig. 2.4** Secondary structures of ribonucleic acid molecules (RNAs). Conventional RNA folding algorithms compute the mfe structure for a given sequence [58, 59]. Hairpin formation is shown as an example on the lhs of the figure. In addition, the sequence can fold also into a large number of suboptimal conformations (diagram in the *middle* of the figure), which are readily computed by efficient computer programs [60, 61]. If a suboptimal structure is separated from the mfe structure by a sufficiently high activation barrier, the structure is metastable. The metastable structure in the example shown here is a double hairpin (rhs of the figure). The activation energy of more than 20 kcal/mol does not allow interconversion of the two structures at room temperature. (For the calculation of kinetic structures see, for example, [62, 63])

energy of folding[11]: $\Delta G_{\text{fold}}(S_0) < 0$) commonly forms almost always a set of suboptimal conformations $\{S_1, S_2, \ldots, S_m\}$ with higher free energies of formation, $\Delta G_{\text{fold}}(S_i) > \Delta G_{\text{fold}}(S_0)$ for $i \neq 0$. In Fig. 2.4 (middle) the ten lowest suboptimal structures are listed; together with $S_0$ they represent the 11 lowest states of the spectrum of structures associated with the sequence $X$. Low-lying suboptimal conformations may influence the molecular properties, in particular when conformational changes are involved. The Boltzmann-weighted contributions of all suboptimal structures at temperature $T$ are readily calculated by means of the partition function of RNA secondary structures [59, 64]. Instead of base pairs the analysis of the partition function yields base pairing probabilities that tell how likely it is to find two specific nucleotides forming a base pair in the ensemble of structures at thermal equilibrium.

Although folding RNA sequences into secondary structures is, presumably, the simplest conceivable case of a genotype–phenotype map, it is at the same time an example of the origin of complexity at the molecular level. The base pairing interaction is essentially nonlocal since a nucleotide can pair with another nucleotide from almost any position of the sequence.[12] The strongest stabilizing contributions

---

[11] The free energy of folding is the difference in free energy between the structure $S_i$ and the unfolded (open) chain $\mathcal{O}$: $\Delta G_{\text{fold}}(S_i) = G(S_i) - G(\mathcal{O})$.

[12] Pairing with nearest neighbors is excluded for geometrical reasons. In other words, base pairs of two adjacent nucleotides have such a high positive free energy of formation that they are never observed.

to the free energy of structure formation come from neighboring base pairs and are therefore local. The combination of local and nonlocal effects is one of the most common sources of complex relations in mappings.

The relation of an RNA sequence and its suboptimal structures is sketched in Fig. 2.5 (lower part). A single sequence $X$ gives rise to a whole set of structures spread all over shape space. In principle, all structures that are *compatible* with the sequence appear in the spectrum of suboptimals but only a subset is stable in the sense that the structure $S_i$ $(i = 1, \ldots)$ corresponds to a local minimum of the conformational energy surface and the free energy of folding is negative $(\Delta G_{\text{fold}}(S_i) < 0)$. Using the base pair count, the set of all structures that are compatible with the sequence $X_h$ can be defined straightforwardly:

$$S_i \in \mathcal{C}(X_h) \text{ iff } \{\gamma_j^{(i)} = \gamma_k^{(i)} \implies b_j^{(h)} b_k^{(h)} \in \mathcal{B} \, \forall \, \gamma_j \neq 0, \, j = 1, \ldots, \ell\} \quad (2.12)$$

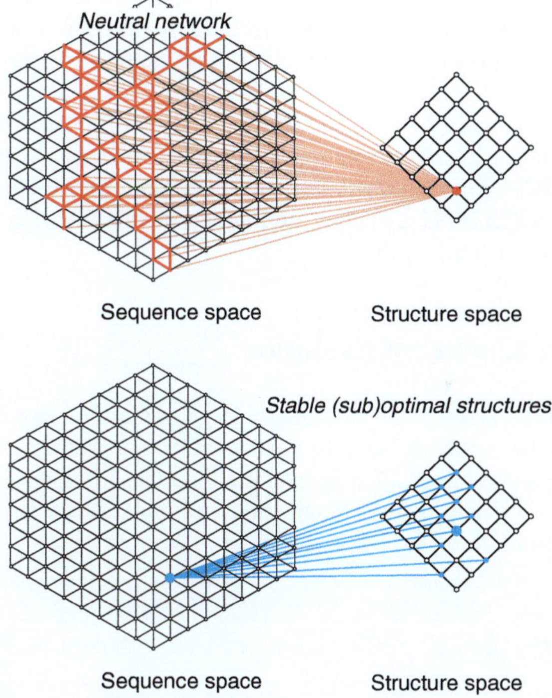


**Fig. 2.5** Mappings from sequence space onto shape space and back. The *upper part* of the figure sketchows schematically a mapping from sequence space onto structure or shape space. (Both sequence space and shape space are high-dimensional. The two-dimensional representation is used for the purpose of illustration only.) One structure is uniquely assigned to each sequence. The *drawing* shows the case of a mapping that is many-to-one and noninvertible: Many sequences fold into the same secondary structure and build a *neutral network*. The *lower part* of the figure illustrates the set of stable (sub)optimal structures that are formed by a single sequence. The mfe structure is indicated by a *larger circle*

In other words, a structure $S_i$ is compatible with a sequence $X_h$ if, and only if, two nucleotides that can form a base pair appear in the sequence at all pairs of positions that are joined by a base pair in the structure. For an arbitrary sequence the number of compatible structures is extremely large but the majority of them have either positive free energies of folding ($\Delta G_{\text{fold}}(S_i) > 0$) and/or represent saddle points rather than local minima of the conformational energy surface. Figure 2.5 indicates the relation between an RNA sequence, its mfe structure, and its stable suboptimal conformations.

Studies of mfe structures or suboptimal structures refer to a certain set of conditions – for example, temperature $T$, pH, ionic strength – but time is missing because free energy differences ($\Delta G$) or partition functions are equilibrium properties. The structures that are determined and investigated experimentally, however, refer always to some time window – we are not dealing with equilibrium ensembles but with metastable states. The finite time structures of RNA are obtained by kinetic folding (see, e.g., [62, 63]). The RNA example shown in Fig. 2.4 represents the case of a bistable molecule: The most stable suboptimal structure $S_1$, a double hairpin conformation (blue), is the most stable representative of a whole family of double hairpin structures forming a broad basin of the free energy landscape of the molecule. This basin is separated from the basin of the single hairpin structure $S_0$ by a high energy barrier of about 20 kcal/mol and this implies that practically no interconversion of the two structures will take place at room temperature. We are dealing with an RNA molecule with one stable and one metastable conformation, a so-called RNA switch. RNA switches are frequent regulatory elements in prokaryotic regulation of translation [65].

## 2.5 Chemical Kinetics of Evolution

Provided population sizes $N$ are sufficiently large, mutation rates are high enough, and stochastic effects are reduced by statistical compensation, evolution can be described properly by means of differential equations. In essence, we proceed as described in Sect. 2.2 and find for replication and mutation as an extension of the selection equation (2.4)

$$\frac{dx_j}{dt} = \sum_{i=1}^{n} Q_{ji} f_i x_i - \phi(t) x_j, \quad j = 1, \ldots, n \text{ with } \phi(t) = \sum_{i=1}^{n} f_i x_i$$

$$\text{or} \quad \frac{d\boldsymbol{x}}{dt} = \left( \mathrm{Q} \cdot \mathrm{F} - \phi(t) \right) \boldsymbol{x} = \left( \mathrm{W} - \phi(t) \right) \boldsymbol{x},$$

(2.13)

where $\boldsymbol{x}$ is an $n$-dimensional column vector and Q and F are $n \times n$ matrices. The matrix Q contains the mutation probabilities $Q_{ji}$, referring to the production of $X_j$ as an error copy of template $X_i$, and F is a diagonal matrix whose elements are the replication rate parameters or fitness values $f_i$.

Solutions of the mutation-selection equation (2.13) can be obtained in two steps: (i) integrating factor transformation allows the nonlinear term $\phi(t)$ to be eliminated and (ii) the remaining linear equation is solved in terms of an eigenvalue problem [66–69]:

$$x_j(t) \;=\; \frac{\sum_{k=1}^{n} b_{jk} \sum_{i=1}^{n} h_{ki}\, x_i(0)\, \exp(\lambda_k t)}{\sum_{l=1}^{n} \sum_{k=1}^{n} b_{lk} \sum_{i=1}^{n} h_{ki}\, x_i(0)\, \exp(\lambda_k t)}, \quad j = 1, \ldots, n. \qquad (2.14)$$

The new quantities in this equation, $b_{jk}$ and $h_{kj}$, are the elements of two transformation matrices:

$$B = \{b_{jk}; j = 1, \ldots, n; k = 1, \ldots, n\} \quad \text{and}$$

$$B^{-1} = \{h_{kj}; k = 1, \ldots, n; j = 1, \ldots, n\}.$$

The columns of B and the rows of $B^{-1}$ represent the right-hand and left-hand eigenvectors of the matrix $W = Q \cdot F$ with $B^{-1} \cdot WB = \Lambda$ being a diagonal matrix containing the eigenvalues of W. The elements of the matrix W are nonnegative by definition since they are the product of a fitness value or replication rate parameter $f_i$ and a mutation probability $Q_{ji}$, which are both nonnegative. If, in addition, W is a nonnegative primitive matrix[13] – implying that every sequence can be reached from every sequence by a finite chain of consecutive mutations – the conditions for the validity of the Perron–Frobenius theorem [70] are fulfilled. Two (out of six) properties of the eigenvalues and eigenvectors of W are important for replication-mutation dynamics:

(i).  The largest eigenvalue $\lambda_1$ is nondegenerate, $\lambda_1 > \lambda_2 \geq \lambda_3 \geq \ldots \geq \lambda_n$, and
(ii). the unique eigenvector belonging to $\lambda_1$ denoted by $\boldsymbol{\xi}_1$ has only positive elements, $\xi_j^{(1)} > 0 \,\forall\, j = 1, \ldots, n$.

After sufficiently long time the population converges to the largest eigenvector $\boldsymbol{\xi}_1$, which is therefore the stationary state of (2.13). Since $\boldsymbol{\xi}_1$ represents the genetic reservoir of an asexually replicating species it is called the *quasispecies* [68]. A quasispecies commonly consists of a fittest genotype, the *master sequence*, and a mutant distribution surrounding the master sequence in sequence space. Although the solution of the mutation-selection equation is straightforward, the experimental proof of the existence of a stationary mutant distribution and its analysis are quite involved [71]. The work has been conducted with relatively short RNA molecules (chain length: $\ell = 87$). Genotypic heterogeneity in virus populations was first detected in the 1970s [72]. Later, the existence of quasispecies in nature was demonstrated for virus populations (For an overview and a collection of reviews see [73, 74]). Since it

---

[13] A square nonnegative matrix $W = \{w_{ij}; i, j = 1, \ldots, n; w_{ij} \geq 0\}$ is called *primitive* if there exists a positive integer $m$ such that $W^m$ is strictly positive: $W^m > 0$, which implies $W^m = \{w_{ij}^{(m)}; i, j = 1, \ldots, n; w_{ij}^{(m)} > 0\}$.

is very hard, if not impossible, to prove that a natural population is in a steady state, the notion *virus quasispecies* was coined for virus populations observed in vitro and in vivo.

In order to explore quasispecies as a function of the mutation rate $p$, a crude or zeroth-order approximation consisting of neglect of backward mutations has been adopted [33]. The differential equation for the master sequence is then of the form

$$\frac{dx_m^{(0)}}{dt} = Q_{mm} f_m x_m^{(0)} - x_m^{(0)} \phi(t) = x_m^{(0)} \left( Q_{mm} f_m - \bar{f}_{-m} - x_m^{(0)} (f_m - \bar{f}_{-m}) \right),$$

with $\bar{f}_{-m} = \left( \sum_{j=1, j \neq m}^{n} f_j x_j \right)/(1 - x_m)$. We apply the uniform error approximation and assume that the mutation rate per nucleotide and replication event, $p$, is independent of the nature of the nucleotide (**A**, **U**, **G** or **C**) and the position along the sequence. We find for the elements of the mutation matrix Q

$$Q_{jj} = (1 - p)^\ell \text{ and } Q_{ji} = (1 - p)^\ell \left( \frac{p}{1 - p} \right)^{d_{\mathrm{H}}(X_i, X_j)}, \tag{2.15}$$

and obtain for the stationary concentration of the master sequence

$$\bar{x}_m^{(0)} = \frac{Q_{mm} - \sigma_m^{-1}}{1 - \sigma_m^{-1}} = \frac{1}{\sigma_m - 1} \left( \sigma_m (1 - p)^\ell - 1 \right),$$

where $\sigma_m = f_m / \bar{f}_{-m} > 1$ is the *superiority* of the master sequence and $\bar{f}_{-m}$ is defined by

$$\bar{f}_{-m} = \frac{1}{1 - x_m} \sum_{i=1, i \neq m}^{n} x_i f_i .$$

In this zeroth-order approximation the stationary concentration $\bar{x}_m^{(0)}(p)$ vanishes at the critical value (Fig. 2.6)

$$p_{\mathrm{cr}} \approx 1 - (\sigma_m)^{-1/\ell} . \tag{2.16}$$

Needless to say, zero concentration of the master sequence is an artifact of the approximation, because the exact concentration of the master sequence cannot vanish by the Perron–Frobenius theorem as long as the population size is nonzero. In order to find out what really happens at the critical mutation rate $p_{\mathrm{cr}}$ computer solutions of the complete equation (2.13) were calculated for the single peak fitness landscape.[14] These calculations [75] show a sharp transition from the ordered quasispecies to the uniform distribution, $\bar{x}_j = \kappa^{-\ell} \forall j = 1, \ldots, \kappa^\ell$. At the critical

---

[14] The single peak fitness landscape is a kind of mean field approximation: A fitness value $f_m$ is assigned to the master sequence, whereas all other variants have the same fitness $f_0$. For this

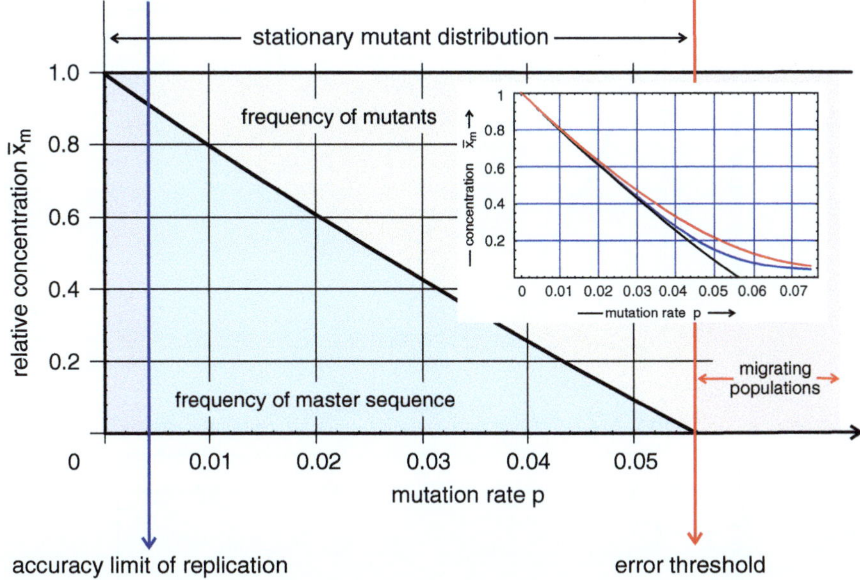



**Fig. 2.6** The error threshold in RNA replication. The stationary frequency of the master sequence $X_m$ is shown as a function of the mutation rate $p$. In the zeroth-order approximation neglecting mutational backflow the function $\bar{x}_m^{(0)}(p)$ is almost linear in the particular example shown here. In the *inset* the zeroth-order approximation (*black*) is shown together with the exact function (*red*) and an approximation applying the uniform distribution to the mutational cloud ($\bar{x}_j = (1-\bar{x}_m)/(n-1)$ $\forall\, j \neq m$; *blue*), which is exact at the mutation rate $p = 0.5$ for binary sequences. The error rate $p$ has two natural limitations: (i) the physical accuracy limit of the replication process provides a lower bound for the mutation rate and (ii) the error threshold defines a minimum accuracy of replication that is required to sustain inheritance and sets an upper bound for the mutation rate. Parameters used in the calculations: binary sequences, $\ell = 6$, $\sigma = 1.4131$

mutation rate $p_{\mathrm{cr}}$, replication errors accumulate and (independently of initial conditions) all sequences are present at the same frequency in the long-time limit, as is reflected by the uniform distribution. The uniform distribution is the exact solution of the eigenvalue problem at equal probabilities for all nucleotide incorporations ($\mathbf{A}\rightarrow\mathbf{A}$, $\mathbf{A}\rightarrow\mathbf{U}$, $\mathbf{A}\rightarrow\mathbf{G}$, and $\mathbf{A}\rightarrow\mathbf{C}$) occurring at $\tilde{p} = \kappa^{-1}$. The interesting aspect of the *error threshold* phenomenon consists in the fact that the quasispecies approaches the uniform distribution at a critical mutation rate $p_{\mathrm{cr}}$ that is far below the random mutation value $\tilde{p}$. As a matter of fact, the appearance of an error threshold and its shape depend on details of the fitness landscape [76, pp. 51–60]. Some landscapes show no error threshold at all but a smooth transition to the uniform distribution [77]. More realistic fitness landscapes with a distribution of fitness values reveal a much more complex situation: For constant superiority the value of $p_{\mathrm{cr}}$ becomes

---

particular landscape the position $\bar{x}_m^{(0)} = 0$ calculated within the zeroth-order approximation almost coincides with the position of the critical change in the population structure (Fig. 2.7).

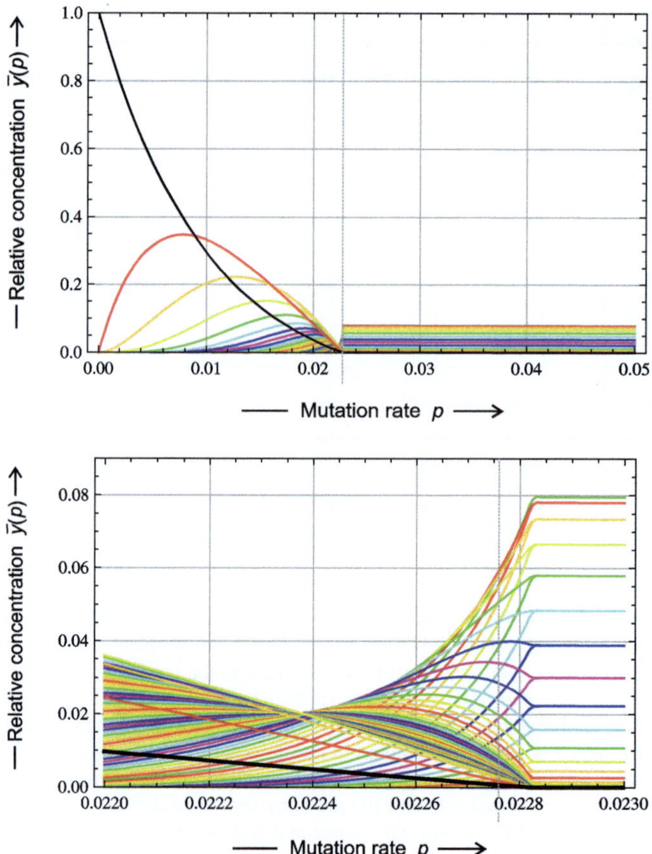

**Fig. 2.7** The error threshold on single peak fitness landscapes. The *upper part* of the figure shows the quasispecies as a function of the mutation rate $p$. The variables $\bar{y}_k(p)$ ($k = 0, 1, \ldots, \ell$) represent the total concentrations of all sequences with Hamming distance $d_H = k$: $\bar{y}_0 = \bar{x}_m$ (*black*) is the concentration of the master sequence, $\bar{y}_1 = \sum_{i=1,d_H(X_i,X_m)=1}^{n} \bar{x}_i$ (*red*) is the concentration of the one-error class, $\bar{y}_2 = \sum_{i=1,d_H(X_i,X_m)=2}^{n} \bar{x}_i$ (*yellow*) that of the two-error class and, accordingly, we have $\bar{y}_k = \sum_{i=1,d_H(X_i,X_m)=k}^{n} \bar{x}_i$ for the $k$-error class. The *lower part* shows an enlargement. The position of the error threshold computed from the zeroth-order approximation (2.16) is shown as by a *dotted line* (*gray*). Choice of parameters: $\kappa = 2$, $\ell = 100$, $f_m = 10$, $f_0 = 1$ and hence $\sigma_m = 10$ and $p_{cr} = 0.02276$

smaller with increasing variance of fitness values. The error threshold phenomenon can be split into three different observations that coincide on the single peak landscape: (i) vanishing of the master sequence $x_m$, (ii) phase-transition-like behavior, and (iii) transition to the uniform distribution. On suitable model landscapes the three observations do not coincide and thus can be separated [78, 79].

How do populations behave at mutation rates above the error threshold? In reality a uniform distribution of variants as required for the stationary state cannot be realized. In RNA selection experiments population sizes hardly exceed $10^{15}$ molecules, the smallest aptamers have chain lengths of $\ell = 27$ nucleotides [80] and this implies $4^{27} \approx 18 \times 10^{15}$ different sequences. Even in this most favorable case we are dealing with more sequences than molecules in the population: a uniform distribution cannot exist. Although the origin of the lack of selective power is completely different – high mutation rates wiping out the differences in fitness values versus fitness differences being zero or too small for selection – the scenarios most likely to occur are migrating populations similar to evolution on a flat landscape [81]. Bernard Derrida and Luca Peliti find that the populations break up into clones, which migrate into different directions in sequence space. Migrating populations are unable to conserve a genotype over generations, and unless a large degree of neutrality allows a phenotype to be maintained despite changing genotypes, evolution becomes impossible because inheritance breaks down.

Because of high selection pressure resulting from the hosts' defense systems, virus populations operate at mutation rates as high as possible in order to allow fast evolution, and this is just below the error threshold [82]. Increasing the mutation rate should drive the virus population beyond threshold, where sufficiently accurate replication is no longer possible. Therefore virus populations are doomed to die out at mutation rates above threshold, and this suggested a novel antiviral strategy that has led to the development of new drugs [83]. A more recent discussion of the error threshold phenomenon tries to separate the error accumulation phenomenon from mutation-caused fitness effects leading to virus extinction, known as *lethal mutagenesis* [84, 85]. In fact lethal mutagenesis describes the error threshold phenomenon for variable population size $N$ as required for $\lim N \to 0$, but an analysis of population dynamics without and with stochastic effects at the onset of migration of populations is still lacking. In addition, more detailed kinetic studies on replication in vitro near the error threshold are required before the mechanism of virus extinction at high mutation rates can be understood.

Sequence–structure mappings of nucleic acid molecules (Sect. 2.4) and proteins provide ample evidence for neutrality in the sense that many genotypes give rise to the same phenotype and identical or almost identical fitness values that cannot be discriminated by natural selection. The possible occurrence of neutral variants was even discussed by Charles Darwin [2, chapter iv]. Based on the results of the first sequence data from molecular biology, Motoo Kimura formulated his neutral theory of evolution [86, 87]. In the absence of fitness differences between variants, random selection occurs because of stochastic enhancement through autocatalytic processes: more frequent variants are more likely to be replicated than less frequent ones. Ultimately a single genotype becomes *fixated* in the population. The average time of replacement for a dominant genotype is the reciprocal mutation rate, $\nu^{-1} = (\ell p)^{-1}$, which, interestingly, is independent of the population size. Are Kimura's results valid also for large population sizes and high mutation rates, as they occur, for example, with viruses? Mathematical analysis [88] together with recent computer studies [78] yields the answer: Random selection in the sense of Kimura occurs only for sufficiently distant (master) sequences. In full agreement with the exact result in the limit $p \to 0$ we find that two fittest sequences of Hamming distance

$d_{\mathrm{H}} = 1$, two nearest neighbors in sequence space, are selected as a strongly coupled pair with equal frequency of the two members. Numerical results demonstrate that this strong coupling occurs not only for small mutation rates but extends over the whole range of $p$ values from $p = 0$ to the error threshold $p = p_{\mathrm{cr}}$. For clusters of more than two sequences with $d_{\mathrm{H}} = 1$, the frequencies of the individual members of the cluster are given by the components of the largest eigenvector of the adjacency matrix. Pairs of fittest sequences with Hamming distance $d_{\mathrm{H}} = 2$, i.e., two next-nearest neighbors with two sequences in between, are also selected together but the ratio of the two frequencies is different from one. Again coupling extends from zero mutation rates to the error threshold. Strong coupling of fittest sequences manifests itself in virology as systematic deviations from consensus sequences of populations, as indeed observed in nature. For two fittest sequences with $d_{\mathrm{H}} \geq 3$ random selection chooses arbitrarily one of the two and eliminates the other one, as predicted by the neutral theory.

The function $\phi(t)$ was introduced as the mean fitness of a population in order to allow straightforward normalization of the population variables. A more general interpretation considers $\phi(t)$ as a flux out of the system. Then the equation describing evolution of the column vector of particle numbers $N = (N_1, \ldots, N_n)$ is of the form [89]

$$\frac{dN_j}{dt} = F_j(N) - \frac{N_j}{C(t)} \phi(t), \ i = 1, \ldots, n,$$

where $F_j(N)$ is the function of unconstrained reproduction. An example is provided by (2.13): $F_j(N) = \sum_{i=1}^{n} Q_{ji} f_i N_i$. Explicit insertion of the total concentration $C(t) = \sum_{i=1}^{n} N_i(t)$ yields

$$\phi(t) = \sum_{i=1}^{n} F_i(N) - \frac{dC}{dt} \ \text{or} \ C(t) = C_0 + \int_0^t \left( \sum_{i=1}^{n} F_i(N) - \phi(\tau) \right) d\tau.$$

Either $C(t)$ or $\phi(t)$ can be chosen freely; the second function is then determined by the equation given above. For normalized variables we find

$$\frac{dx_j}{dt} = \frac{1}{C(t)} \left( F_j(N) - x_j \sum_{i=1}^{n} F_j(N) \right).$$

For a large number of examples and for most cases important in evolution, the functions $F_j(N)$ are homogeneous functions in $N$. For homogeneity of degree $\gamma$ we have $F_j(N) = F_j(C \cdot N) = C^\gamma F_j(x)$ and find

$$\frac{dx_j}{dt} = C^{\gamma-1} \left( F_j(x) - x_j \sum_{i=1}^{n} F_j(x) \right), \ j = 1, \ldots, n. \tag{2.17}$$

Two conclusions can be drawn from this equation: (i) For $\gamma = 1$, e.g., the selection equation (2.4) or the replication-mutation equation (2.13), the dependence on the total concentration $C$ vanishes and the solution curves in normalized variables $x_j(t)$ are the same in stationary ($C = $ const) and nonstationary systems as long as $C(t)$ remains finite and does not vanish, and (ii) if $\gamma \neq 1$ the long-term behavior determined by $\dot{\mathbf{x}} = 0$ is identical for stationary and nonstationary systems unless the population dies out $C(t) \to 0$ or explodes $C(t) \to \infty$.

## 2.6 Evolution as a Stochastic Process

Stochastic phenomena are essential for evolution – each mutant after all starts out from a single copy – and a large number of studies have been conducted on stochastic effects in population genetics [90]. Not so much work, however, has been devoted so far to the development of a general stochastic theory of molecular evolution. We mention two examples representative for others [91, 92]. In the latter case the reaction network for replication and mutation was analyzed as a multi-type branching process and it was proven that the stochastic process converges to the deterministic equation (2.13) in the limit of large populations. What is still lacking is a comprehensive treatment, for example by means of chemical master equations [93]. Then the deterministic population variables $x_j(t)$ are replaced by stochastic variables $\mathcal{X}_j(t)$ and the corresponding probabilities

$$P_k^{(j)}(t) = \mathrm{Prob}\{\mathcal{X}_j = k\}, \ k = 0, 1, \dots, N; \ j = 1, \dots, n. \tag{2.18}$$

The chemical master equation translates a mechanism into a set of differential equations for the probabilities. The pendant of (2.13), for example, is the master equation

$$\begin{aligned} \frac{dP_k^{(j)}}{dt} &= \left( \sum_{i=1}^{n} Q_{ji} f_i \sum_{s=1}^{n} s\, P_s^{(i)} \right) P_{k-1}^{(j)} - \phi(t)\, P_k^{(j)} \\ &\quad - \left( \sum_{i=1}^{n} Q_{ji} f_i \sum_{s=1}^{n} s\, P_s^{(i)} \right) P_k^{(j)} + \phi(t)\, P_{k+1}^{(j)}. \end{aligned} \tag{2.19}$$

The only quantity that has to be specified further in this equation is the flux term $\phi(t)$. For the stochastic description it is not sufficient to have a term that just compensates the increase in population size due to replication, a detailed model of the process is required. Examples are (i) the Moran process [94–96] with strictly constant population size and (ii) the flow reactor (continuous stirred tank reactor, CSTR) with a population size fluctuating within the limits of a $\sqrt{N}$ law [97, 98].[15] The

---

[15] All thermodynamically admissible processes obey a so-called $\sqrt{N}$ law: For a mean population size of $N$ the actual population size fluctuates with a standard deviation proportional to $\sqrt{N}$.

Moran process assumes that for every newborn molecule one molecule is instantaneously eliminated. Strong coupling of otherwise completely independent processes has the advantage of mathematical simplicity but it lacks a physical background. The flow reactor, on the other hand, is harder to treat in the mathematical analysis but it is based on solid physical grounds and can be easily implemented experimentally. In computer simulation both models require comparable efforts and for molecular systems preference is given therefore to the flow reactor.

For evolution of RNA molecules through replication and mutation in the flow reactor, the following reaction mechanism has been implemented:

$$
\begin{aligned}
* \quad &\xrightarrow{\;a_0\,r\;} \quad A\,, \\
A + X_i \quad &\xrightarrow{\;Q_{ji}\,f_i\;} \quad X_i + X_j\,;\ i,j = 1,\ldots,n\,, \\
A \quad &\xrightarrow{\;r\;} \quad \varnothing\,,\quad \text{and} \\
X_j \quad &\xrightarrow{\;r\;} \quad \varnothing\,;\ j = 1,\ldots,n\,.
\end{aligned}
\tag{2.20}
$$

Stock solution flows into the reactor with a flow rate $r$ and it feeds the reactor with the material required for polynucleotide synthesis – schematically denoted by $A$ and consisting, for example, of activated nucleotides, **ATP**, **UTP**, **GTP**, and **CTP**, as well as a replicating enzyme – into the system. The concentration of $A$ in the stock solution is denoted by $a_0$. The molecules $X_j$ are produced by the second reaction either by correct copying or by mutation. The third and fourth reactions describe the outflux of material and compensate the increase in volume caused by the influx of stock solution. The reactor is assumed to be perfectly mixed at every instant (CSTR). For a targeted search the stochastic process in the reactor is constructed to have two absorbing states (Fig. 2.8): (i) extinction – all RNA molecules are diluted out of the reaction vessel – and (ii) survival – the predefined target structure has been produced in the reactor. The population size determines the outcome of the computer experiment: Below population sizes of $N = 13$ the reaction in the CSTR almost certainly goes extinct, but it reaches the target with a probability close to one for $N > 20$. The probability of extinction is very small for sufficiently large populations, and for population sizes $N \geq 1,000$, as reported here, extinction has been never observed.

In order to simulate the interplay between mutation acting on the RNA sequence and selection operating on RNA structures, the sequence–structure map has to be turned into an integral part of the model [97–99]. The simulation tool starts from a population of RNA molecules and simulates chemical reactions corresponding to replication and mutation in a CSTR according to (2.20) by using Gillespie's algorithm [100–102]. Molecules replicate in the reactor and produce both correct copies and mutants, the materials to be consumed are supplied by the continuous influx of stock solution into the reactor, and excess volume is removed by means of the outflux of reactor solution. Two kinds of computer experiments were performed: Optimizations of properties on a landscape derived from the sequence–structure

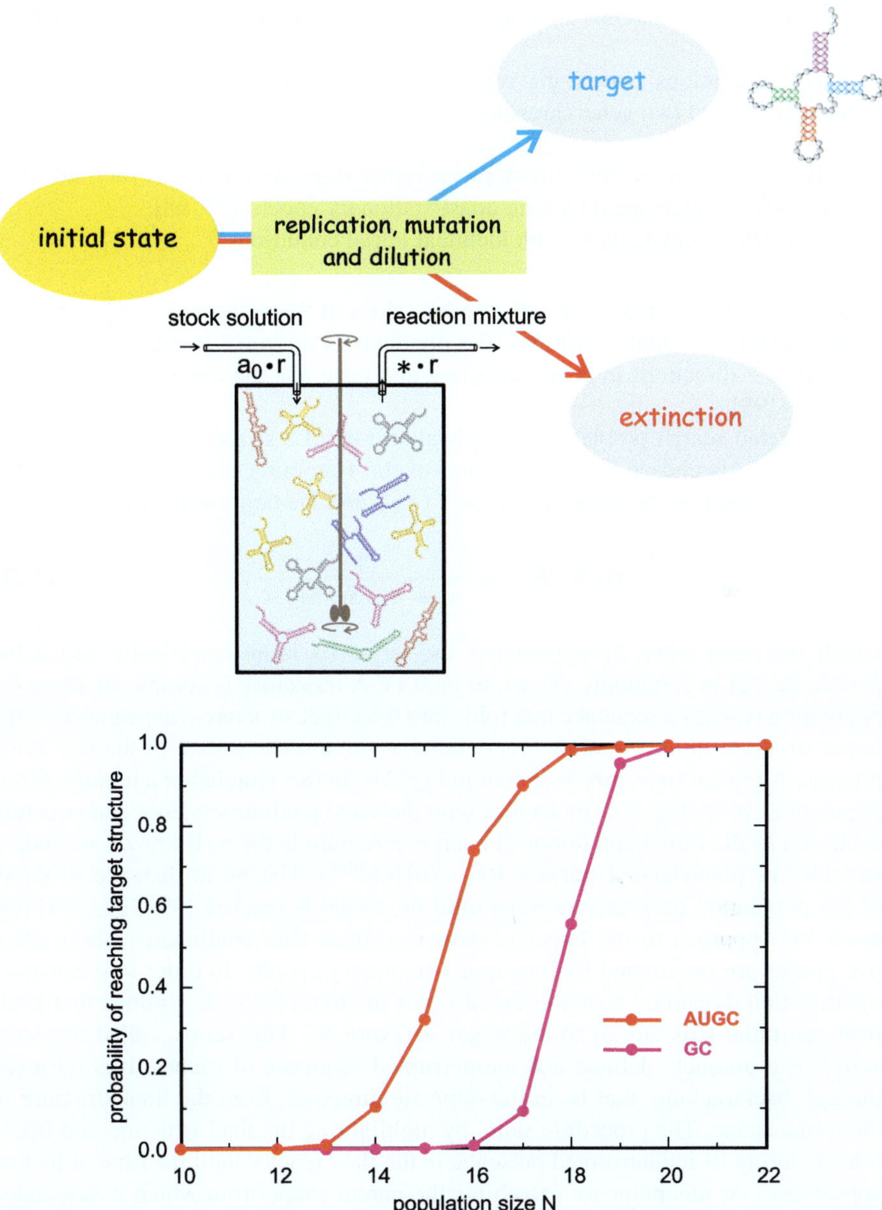


**Fig. 2.8** Survival in the flow reactor. Replication and mutation in the flow reactor are implemented according to the mechanism (2.20). The stochastic process has two absorbing states: (i) extinction, $\mathcal{X}_j = 0 \, \forall \, j = 1, \ldots, n$, and (ii) a predefined target state – here the structure of tRNA$^{\text{phe}}$. A rather sharp transition in the long-time behavior of the population is shown in the *lower plot*: populations of natural sequences (**AUGC**) switch from almost certain extinction to almost certain survival in the range $13 \leq N \leq 18$ and for binary sequences (**GC**) the transition is even sharper but requires slightly larger population sizes

map and targeted searches in shape space where the target is some predefined structure.

Early simulations optimizing replication rates in populations of binary **GC**-sequences yielded two general results:

(i) The progress in evolution is stepwise rather than continuous, as short adaptive phases are interrupted by long quasi-stationary epochs [97, 98].

(ii) Different computer runs with identical initial conditions[16]

resulted in different structures with similar values of the optimized rate parameters. Despite identical initial conditions, the populations migrated in different – almost orthogonal – directions in sequence space and gave rise thereby to contingency in evolution [98].

In targeted search problems the replication rate of a sequence $X_k$, representing its fitness $f_k$, is chosen to be a function of the Hamming distance[17] between the structure formed by the sequence, $S_k = f(X_k)$, and the target structure, $S_T$,

$$f_k(S_k, S_T) = \frac{1}{\alpha + d_H(S_k, S_T)/\ell}, \tag{2.21}$$

which increases when $S_k$ approaches the target ($\alpha$ is an empirically adjustable parameter that is commonly chosen to be 0.1). A trajectory is completed when the population reaches a sequence that folds into the target structure – appearance of the target structure in the population is defined as an absorbing state of the stochastic process. A typical trajectory is shown in Fig. 2.9. In this simulation a homogeneous population consisting of $N$ molecules with the same random sequence and structure is chosen as the initial condition. The target structure is the well-known secondary structure of phenylalanyl-transfer RNA (tRNA$^{phe}$). The mean distance to target of the population decreases in steps until the target is reached [99, 103, 104] and again the approach to the target is stepwise rather than continuous: Short adaptive phases are interrupted by long quasi-stationary epochs. In order to reconstruct optimization dynamics, a time-ordered series of structures is determined that leads from an initial structure $S_I$ to the target structure $S_T$. This series, called the *relay series*, is a uniquely defined and uninterrupted sequence of shapes. It is retrieved through backtracking, that is, in the opposite direction, from the final structure to the initial shape. The procedure starts by highlighting the final structure and traces it back during its uninterrupted presence in the flow reactor until the time of its first appearance. At this point we search for the parent shape from which it descended by mutation. Now we record the time and structure, highlight the parent shape, and repeat the procedure. Recording further backwards yields a series of shapes

---

[16] *Identical* means here that everything in the computer runs was the same except the seeds for the random number generators and this implies different series of random events.

[17] The distance between two structures is defined here as the Hamming distance between the two symbolic notations of the structures.

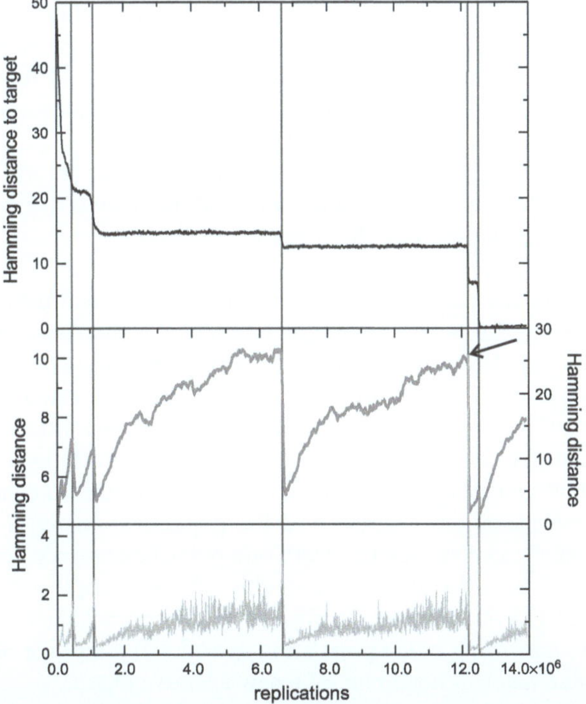

**Fig. 2.9** A trajectory of evolutionary optimization. The *topmost plot* presents the mean distance to the target structure of a population of 1,000 molecules, the plot in the *middle* shows the width of the population in Hamming distance between sequences, and the plot at the *bottom* is a measure of the velocity with which the center of the population migrates through sequence space. Diffusion on neutral networks causes spreading on the population in the sense of neutral evolution [105]). A remarkable synchronization is observed: At the end of each quasi-stationary plateau a new adaptive phase in the approach towards the target is initiated, which is accompanied by a drastic reduction in the population width and a jump in the population center. (The top of the peak at the end of the second long plateau is marked by an *arrow*.) A mutation rate of $p = 0.001$ was chosen, the replication rate parameter is defined in (2.21), and initial and target structures are shown in Table 2.1

and times of first appearance that ultimately ends in the initial population.[18] Use of the relay series and its theoretical background allows classification of transitions [99, 103, 106]. Inspection of the relay series together with the sequence record on the quasi-stationary plateaus provides strong hints for the distinction of two scenarios:

---

[18] It is important to stress two facts about relay series: (i) The same shape may appear two or more times in a given relay series series. Then, it was extinct between two consecutive appearances. (ii) A relay series is not a genealogy, which is the full recording of parent–offspring relations in a time-ordered series of genotypes.

(i) The structure is constant and we observe neutral evolution in the sense of Kimura's theory of neutral evolution [87]. In particular, the numbers of neutral mutations accumulated are proportional to the number of replications in the population, and the evolution of the population can be understood as a diffusion process on the corresponding neutral network [105].

(ii) The process during the quasi-stationary epoch involves several closely related structures with identical replication rates and the relay series reveals a kind of random walk in the space of these neutral structures.

The diffusion of the population on the neutral network is illustrated by the plot in the middle of Fig. 2.9, which shows the width of the population as a function of time [104]. The population width increases during the quasi-stationary epoch and sharpens almost instantaneously after a sequence has been created by mutation that allows the start of a new adaptive phase in the optimization process. The scenario at the end of the plateau corresponds to a *bottleneck* of evolution. The lower part of the figure shows a plot of the migration rate or drift of the population center and confirms this interpretation: Migration of the population center is almost always very slow unless the center "jumps" from one point in sequence space to a possibly distant point where the molecule initiating the new adaptive phase is located. A closer look at the three curves in Fig. 2.9 reveals coincidence of three events: (i) collapse-like narrowing of the population spread, (ii) jump-like migration of the population center, and (iii) beginning of a new adaptive phase.

It is worth mentioning that the optimization behavior observed in a long-term evolution experiment with *Escherichia coli* [46] can be readily interpreted in terms of random searches on a neutral network. Starting with twelve colonies in 1988, Lenski and his coworkers observed, after 31,500 generation or 20 years, a great adaptive innovation in one colony [45]: This colony developed a kind of membrane channel that allows uptake of citrate, which is used as a buffer in the medium. The colony thus conquered a new resource that led to a substantial increase in colony growth. The mutation providing citrate import into the cell is reproducible when earlier isolates of this particular colony are used for a restart of the evolutionary process. Apparently this particular colony has traveled through sequence space to a position from where the adaptive mutation allowing citrate uptake is within reach. None of the other eleven colonies gave rise to mutations with a similar function. The experiment is a nice demonstration of contingency in evolution: The conquest of the citrate resource does not happen through a single highly improbable mutation but by means of a mutation with standard probability from a particular region of sequence space where the population had traveled in one case out of twelve – history matters, or to repeat Theodosius Dobzhansky's famous quote: "Nothing makes sense in biology except in the light of evolution" [29].

Table 2.1 collects some numerical data sampled from evolutionary trajectories of simulations repeated under identical conditions. Individual trajectories show enormous scatter in the time or the number of replications required to reach the target. The mean values and the standard deviations were obtained from statistics of

**Table 2.1** Statistics of the optimization trajectories. The table shows the results of sampled evolutionary trajectories leading from a random initial structure $S_I$ to the structure of tRNA$^{phe}$, $S_T$ as the target.[a] Simulations were performed with an algorithm introduced by Gillespie [100–102]. The time unit is here undefined. A mutation rate of $p = 0.001$ per site and replication was used. The mean and standard deviation were calculated under the assumption of a log-normal distribution that fits well the data of the simulations

| | Population size | Number of runs | Real time from start to target | | Number of replications $[10^7]$ | |
|---|---|---|---|---|---|---|
| Alphabet | $N$ | $n_R$ | Mean value | $\sigma$ | Mean value | $\sigma$ |
| **AUGC** | 1,000 | 120 | 900 | +1,380–542 | 1.2 | +3.1–0.9 |
| | 2,000 | 120 | 530 | +880–330 | 1.4 | +3.6–1.0 |
| | 3,000 | 1,199 | 400 | +670–250 | 1.6 | +4.4–1.2 |
| | 10,000 | 120 | 190 | +230–100 | 2.3 | +5.3–1.6 |
| | 30,000 | 63 | 110 | +97–52 | 3.6 | +6.7–2.3 |
| | 100,000 | 18 | 62 | +50–28 | – | – |
| **GC** | 1,000 | 46 | 5,160 | +15,700–3,890 | – | – |
| | 3,000 | 278 | 1,910 | +5,180–1,460 | 7.4 | +35.8–6.1 |
| | 10,000 | 40 | 560 | +1,620–420 | – | – |

[a]The structures SI and ST used in the optimization were:

*SI*: ((.(((((((((((((............(((....)))......)))))).)))))))).))...(((......)))

*ST*: (((((...((((.........)))).(((((.......))))).....(((((.......))))).)))))....

trajectories under the assumption of log-normal distributions. Despite the scatter three features are unambiguously detectable:

(i) The search in **GC** sequence space takes about five time as long as the corresponding process in **AUGC** sequence space, in agreement with the difference in neutral network structure.

(ii) The time to target decreases with increasing population size.

(iii) The number of replications required to reach target increases with population size.

Combination of the results (ii) and (iii) allows a clear conclusion concerning time and material requirements of the optimization process: Fast optimization requires large populations whereas economic use of material suggests working with small population sizes just sufficient to avoid extinction.

A study of parameter dependence of RNA evolution was reported in a recent simulation [107]. Increase in mutation rate leads to an error threshold phenomenon that is closely related to one observed with quasispecies on a single-peak landscape as described above [69, 75]. Evolutionary optimization becomes more efficient[19] with increasing error rate until the error threshold is reached. Further increase in

---

[19] Efficiency of evolutionary optimization is measured by average and best fitness values obtained in populations after a predefined number of generations.

error rates leads to a breakdown of the optimization process. As expected the distribution of replication rates or fitness values $f_k$ in sequence space is highly relevant too: Steep decrease of fitness with the distance to the master structure represented by the target, which has the highest fitness value, leads to sharp threshold behavior, as observed on single-peak landscapes, whereas flat landscapes show a broad maximum of optimization efficiency without an indication of threshold-like behavior.

## 2.7 Concluding Remarks

Biology developed differently from physics because it refrained from using mathematics as a tool to analyze and unfold theoretical concepts. Application of mathematics enforces clear definitions and reduction of observations to problems that can be managed. Over the years physics became the science of abstractions and generalizations, biology the science of encyclopedias of special cases with all their beauties and peculiarities. Among others there is one great exception to the rule: Charles Darwin presented a grand generalization derived from a wealth of personal and reported observations together with knowledge from economics concerning population dynamics. In the second half of the twentieth century the appearance of molecular biology on the stage changed the situation entirely. A bridge was built from physics and chemistry to biology, and mathematical models from biochemical kinetics or population genetics became presentable in biology. Nevertheless, the vast majority of biologists still smiled at the works of theorists. By the end of the twentieth century molecular genetics had created such a wealth of data that almost everybody feels nowadays that progress cannot be made without a comprehensive theoretical foundation and a rich box of suitable computational tools. Nothing like this is at hand but indications for attempts in the right direction are already visible. Biology is going to enter the grand union of science that started with physics and chemistry and is progressing fast. Molecular biology started out with biomolecules in isolation and deals now with cells, organs, and organisms. Hopefully, this spectacular success will end the so-far fruitless reductionism versus holism debate.

Insight into the mechanisms of evolution reduced to the simplest conceivable systems was provided here. These systems deal with evolvable molecules in cell-free assays and are accessible by rigorous mathematical analysis and physical experimentation. An extension to asexual species, in particular viruses and bacteria, is within reach. The molecular approach provides a simple explanation of why we have species for these organisms despite the fact that there is neither restricted recombination nor reproductive isolation. The sequence spaces are so large that populations, colonies, or clones can migrate for the age of the universe without coming close to another asexual species. We can give an answer to the question of the origin of complexity: Complexity in evolution results primarily from genotype–phenotype relations and from the influences of the environment. Evolutionary dynamics may be complicated in some cases but it is not complex at all. This has been reflected already by the sequence–structure map of our toy example. Conformation spaces depending on the internal folding kinetics as well as on environmental conditions and compatible sets are metaphors for more complex features in evolution proper.

Stochasticity is still an unsolved problem in molecular evolution. The mathematics of stochastic processes encounters difficulties in handling the equations of evolution in detail. A comprehensive stochastic theory is still not available and the simulations lack more systematic approaches since computer simulations of chemical kinetics of evolution are at an early stage too. Another fundamental problem concerns the spatial dimensions: Almost all treatments assume spatial homogeneity but we have evidence of the solid-particle-like structure of the chemical factories of the cell. In the future, any comprehensive theory of the cell will have to deal with these structurally rich supramolecular structures too.

# References


1. G. Galilei, *Il Saggiatore*, vol. 6 (Edition Nationale, Florence, Italy, 1896), p. 232. English translation from Italian original
2. C. Darwin, *On the Origin of Species by Means of Natural Selection or the Preservation of Favoured Races in the Struggle for Life* (John Murray, London, 1859)
3. E. Mayr, *The Growth of Biological Thought. Diversity, Evolution, and Inhertiance* (The Belknap Press of Harvard University Press, Cambridge, MA, 1982)
4. R.A. Fisher, *The Genetical Theory of Natural Selection* (Oxford University Press, Oxford, 1930)
5. J.B.S. Haldane, *The Causes of Evolution* (Longmans Green, London, 1932). Reprinted 1990 in the Princeton Science Library by Princeton University Press, Princeton, NJ
6. S. Wright, *Evolution and the Genetics of Populations*, vol. 1–4 (University of Chicago Press, Chicago, IL, 1968, 1969, 1977, 1978)
7. G. Mendel, Verhandlungen des naturforschenden Vereins in Brünn **4**, 3 (1866)
8. G. Mendel, Verhandlungen des naturforschenden Vereins in Brünn **8**, 26 (1870)
9. G.A. Nogler, Genetics **172**, 1 (2006)
10. T.R. Malthus, *An Essay on the Principle of Population as it Affects the Future Improvement of Society* (J. Johnson, London, 1798)
11. A.J. Lotka, *Elements of Physical Biology* (Williams & Wilkins, Baltimore, MD, 1925)
12. V. Volterra, Mem. R. Accad. Naz. Lincei **Ser. VI/2**, 31 (1926)
13. E.P. Odum, *Fundamentals of Ecology* (W. B. Saunders, Philadelphia, PA, 1953)
14. R.M. May (ed.), *Theoretical Ecology. Principles and Applications* (Blackwell Scientific, Oxford, 1976)
15. A.M. Turing, Philos. Trans. R. Soc. Lond. Ser. B **237**, 37 (1952)
16. A. Gierer, H. Meinhardt, Kybernetik **12**, 30 (1972)
17. H. Meinhardt, *Models of Biological Pattern Formation* (Academic, London, 1982)
18. J.D. Murray, Sci. Am. **258**(3), 62 (1988)
19. J.D. Murray, *Mathematical Biology II: Spatial Models and Biomedical Applications*, 3rd edn. (Springer, New York, NY, 2003)
20. A.L. Hodgkin, A.F. Huxley, J. Physiology **117**, 500 (1952)
21. J.J. Hopfield, Proc. Natl. Acad. Sci. USA **79**, 2554 (1982)
22. H. Judson, *The Eighth Day of Creation. The Makers of the Revolution in Biology* (Jonathan Cape, London, 1979)
23. A. Maxam, W. Gilbert, Proc. Natl. Acad. Sci. USA **74**, 560 (1977)
24. F. Sanger, S. Nicklen, A. Coulson, Proc. Natl. Acad. Sci. USA **74**, 5463 (1977)
25. J. Dale, M. von Schantz, *From Genes to Genomes: Concepts and Applications of DNA Technology*, 2nd edn. (Wiley, Chichester, 2007)
26. S. Brenner, Scientist **16**(4), 14 (2002)

27. J.S. Edwards, M. Covert, B.O. Palsson, Environ. Microbiol. **4**, 133 (2002)
28. H.W. Engl, C. Flamm, P. Knugler, J. Lu, S. Müller, P. Schuster, Inverse Probl. **25**, 123014 (2009)
29. T. Dobzhansky, F.J. Ayala, G.L. Stebbins, J.W. Valentine, *Evolution* (W.H. Freeman, San Francisco, CA, 1977)
30. L. Euler, *Introductio in Analysin Infinitorum, 1748*. English Translation: John Blanton, *Introduction to Analysis of the Infinite*, vol. I, II (Springer, Berlin, Heidelberg, 1988)
31. P. Verhulst, Corresp. Math. Phys. **10**, 113 (1838)
32. D. Zwillinger, *Handbook of Differential Equations*, 3rd edn. (Academic, San Diego, CA, 1998)
33. M. Eigen, Naturwissenschaften **58**, 465 (1971)
34. J. Maynard-Smith, Nature **225**, 563 (1970)
35. R.W. Hamming, *Coding and Information Theory*, 2nd edn. (Prentice-Hall, Englewood Cliffs, NJ, 1986)
36. J. Rogers, G. Joyce, Nature **402**, 323 (1999)
37. J.S. Reader, G.F. Joyce, Nature **420**, 841 (2002)
38. P. Schuster, Physica D **107**, 351 (1997)
39. P. Gitchoff, G.P. Wagner, Complexity **2**(1), 37 (1998)
40. P.F. Stadler, R. Seitz, G.P. Wagner, Bull. Math. Biol. **62**, 399 (2000)
41. B.R.M. Stadler, P.F. Stadler, M. Shpak, G.P. Wagner, Z. Phys. Chem. **216**, 217 (2002)
42. D.E. Goldberg, *Genetic Algorithms in Search, Optimization, and Machine Learning* (Addison-Wesley, Reading, MA, 1989)
43. L.M. Schmitt, Theor. Comput. Sci. **259**, 1 (2001)
44. J.E. Barrick, D.S. Yu, H. Jeong, T.K. Oh, D. Schneider, R.E. Lenski, J.F. Kim, Nature **441**, 1243 (2009)
45. Z.D. Blount, Z. Christina, R.E. Lenski, Proc. Natl. Acad. Sci. USA **105**, 7898 (2008)
46. R.E. Lenski, M.R. Rose, S.C. Simpson, S.C. Tadler, Am. Nat. **38**, 1315 (1991)
47. P. Schuster, Rep. Prog. Phys. **69**, 1419 (2006)
48. D.R. Mills, R.L. Peterson, S. Spiegelman, Proc. Natl. Acad. Sci. USA **58**, 217 (1967)
49. S. Spiegelman, Q. Rev. Biophys. **4**, 213 (1971)
50. G.F. Joyce, Angew. Chem. Int. Ed. **46**, 6420 (2007)
51. C.K. Biebricher, M. Eigen, W.C. Gardiner, Jr., Biochemistry **22**, 2544 (1983)
52. C.K. Biebricher, M. Eigen, W.C. Gardiner, Jr., Biochemistry **23**, 3186 (1984)
53. C.K. Biebricher, M. Eigen, W.C. Gardiner, Jr., Biochemistry **24**, 6550 (1985)
54. A. Lescoute, N.B. Leontis, C. Massire, E. Westhof, Nucl. Acids Res. **33**, 2395 (2005)
55. N.B. Leontis, A. Lescoute, E. Westhof, Curr. Opin. Struct. Biol. **16**, 279 (2006)
56. I.L. Hofacker, P. Schuster, P.F. Stadler, Discr. Appl. Math. **89**, 177 (1998)
57. S. Klussmann (ed.), *The Aptamer Handbook. Functional Oligonucleotides and Their Applications* (Wiley-VCH, Weinheim, Germany, 2006)
58. M. Zuker, P. Stiegler, Nucl. Acids Res. **9**, 133 (1981)
59. I.L. Hofacker, W. Fontana, P.F. Stadler, L.S. Bonhoeffer, M. Tacker, P. Schuster, Monatsh. Chem. **125**, 167 (1994)
60. M. Zuker, Science **244**, 48 (1989)
61. S. Wuchty, W. Fontana, I.L. Hofacker, P. Schuster, Biopolymers **49**, 145 (1999)
62. C. Flamm, W. Fontana, I.L. Hofacker, P. Schuster, RNA **6**, 325 (1999)
63. M.T. Wolfinger, W.A. Svrcek-Seiler, C. Flamm, I.L. Hofacker, P.F. Stadler, J. Phys. A Math. Gen. **37**, 4731 (2004)
64. J.S. McCaskill, Biopolymers **29**, 1105 (1990)
65. M. Mandal, B. Boese, J.E. Barrick, W.C. Winkler, R.R. Breaker, Cell **113**, 577 (2003)
66. C.J. Thompson, J.L. McBride, Math. Biosci. **21**, 127 (1974)
67. B.L. Jones, R.H. Enns, S.S. Rangnekar, Bull. Math. Biol. **38**, 15 (1976)
68. M. Eigen, P. Schuster, Naturwissenschaften **64**, 541 (1977)
69. M. Eigen, J. McCaskill, P. Schuster, Adv. Chem. Phys. **75**, 149 (1989)

70. E. Seneta, *Non-negative Matrices and Markov Chains*, 2nd edn. (Springer, New York, NY, 1981)
71. N. Rohde, H. Daum, C.K. Biebricher, J. Mol. Biol. **249**, 754 (1995)
72. E. Domingo, D. Szabo, T. Taniguchi, C. Weissmann, Cell **13**, 735 (1978)
73. E. Domingo, J. Holland, in *RNA Genetics. Vol. III: Variability of Virus Genomes*, ed. by E. Domingo, J. Holland, P. Ahlquist (CRC Press, Boca Raton, FL, 1988), pp. 3–36
74. E. Domingo (ed.), *Quasispecies: Concepts and Implications for Virology* (Springer, Berlin, Heidelberg, 2006)
75. J. Swetina, P. Schuster, Biophys. Chem. **16**, 329 (1982)
76. P.E. Phillipson, P. Schuster, *Modeling by Nonlinear Differential Equations. Dissipative and Conservative Processes*, World Scientific Series on Nonlinear Science A, vol. 69 (World Scientific, Singapore, 2009)
77. T. Wiehe, Genet. Res. Camb. **69**, 127 (1997)
78. P. Schuster, Theory Biosci. **130**, 17 (2011)
79. P. Schuster. Quasispecies and error thresholds on realistic fitness landscapes (2010). Preprint
80. L. Jiang, A.K. Suri, R. Fiala, D.J. Patel, Chem. Biol. **4**, 35 (1997)
81. B. Derrida, L. Peliti, Bull. Math. Biol. **53**, 355 (1991)
82. J.W. Drake, Proc. Natl. Acad. Sci. USA **90**, 4171 (1993)
83. E. Domingo (ed.), Virus Res. **107**(2), 115 (2005)
84. J.J. Bull, L. Ancel Myers, M. Lachmann, PLoS Comput. Biol. **1**, 450 (2005)
85. J.J. Bull, R. Sanjuán, C.O. Wilke, J. Virol. **81**, 2930 (2007)
86. M. Kimura, Nature **217**, 624 (1968)
87. M. Kimura, *The Neutral Theory of Molecular Evolution* (Cambridge University Press, Cambridge, 1983)
88. P. Schuster, J. Swetina, Bull. Math. Biol. **50**, 635 (1988)
89. M. Eigen, P. Schuster, Naturwissenschaften **65**, 7 (1978)
90. R.A. Blythe, A. McKane, J. Stat. Mech. Theor. Exp. P07018 (2007). doi 10.1088/1742-5468/2007/07/P07018
91. B.L. Jones, H.K. Leung, Bull. Math. Biol. **43**, 665 (1981)
92. L. Demetrius, P. Schuster, K. Sigmund, Bull. Math. Biol. **47**, 239 (1985)
93. C.W. Gardiner, *Stochastic Methods. A Handbook for the Natural and Social Sciences*, 4th edn. Springer Series in Synergetics (Springer, Berlin, Heidelberg, 2009)
94. P. Moran, Proc. Camb. Philos. Soc. **54**, 60 (1958)
95. P. Moran, *The Statistical Processes of Evolutionary Theory* (Clarendon Press, Oxford, 1962)
96. M.A. Nowak, *Evolutionary Dynamics: Exploring the Equations of Life* (The Belknap Press of Harvard University Press, Cambridge, MA, 2006)
97. W. Fontana, P. Schuster, Biophys. Chem. **26**, 123 (1987)
98. W. Fontana, W. Schnabl, P. Schuster, Phys. Rev. A **40**, 3301 (1989)
99. W. Fontana, P. Schuster, Science **280**, 1451 (1998)
100. D.T. Gillespie, J. Comput. Phys. **22**, 403 (1976)
101. D.T. Gillespie, J. Phys. Chem. **81**, 2340 (1977)
102. D.T. Gillespie, Annu. Rev. Phys. Chem. **58**, 35 (2007)
103. W. Fontana, P. Schuster, J. Theor. Biol. **194**, 491 (1998)
104. P. Schuster, in *Evolutionary Dynamics – Exploring the Interplay of Accident, Selection, Neutrality, and Function*, ed. by J.P. Crutchfield, P. Schuster (Oxford University Press, New York, NY, 2003), pp. 163–215
105. M.A. Huynen, P.F. Stadler, W. Fontana, Proc. Natl. Acad. Sci. USA **93**, 397 (1996)
106. B.R.M. Stadler, P.F. Stadler, G.P. Wagner, W. Fontana, J. Theor. Biol. **213**, 241 (2001)
107. A. Kupczok, P. Dittrich, J. Theor. Biol. **238**, 726 (2006)

# Chapter 3
# The Interplay of Replication, Variation and Selection in the Dynamics of Evolving Populations

Richard A. Blythe

**Abstract** Evolution is a process by which change occurs through replication. Variation can be introduced into a population during the replication process. Some of the resulting variants may be replicated more rapidly than others, and so the characteristics of the population – and individuals within it – change over time. These processes can be recognised most obviously in genetics and ecology; but they also arise in the context of cultural change. We discuss two key questions that are crucial to the development of evolutionary theory. First, we consider how different application domains may be usefully placed within a single framework; and second, we ask how one can distinguish directed, deterministic change from changes that occur purely because of the stochastic nature of the underlying replication process.

Evolution is a theory of change by replication. When an organism reproduces, molecules of DNA are replicated and inherited by the offspring. The replicated DNA may be identical to that carried by the parent, or it may differ, for example, through mutation or recombination. Such differences at the molecular level (genotype) may in turn lead to variation in the macroscopic properties (phenotype) of the offspring (although such variation may also be due to other sources). Some of this variation may lead to some organisms being more successful (having more offspring) than others, which in turn leads to changes in the relative frequencies of different genotypes in a population, a process known as selection.

Molecular evolution is not the only process of change that results from a combination of replication, variation and selection. At a higher level, one can think of the population dynamics within an ecosystem of competing species as an evolutionary process, where selection might be more fruitfully thought of – at least in terms of modelling the dynamical process – as acting at the level of species rather than individual genes. Systems of learned human behaviour, such as language, the use of technologies or beliefs – collectively known as *culture* – also change over time by

R.A. Blythe (✉)
SUPA, School of Physics, University of Edinburgh
Mayfield Road, Edinburgh EH9 3JZ, UK
e-mail: r.a.blythe@ed.ac.uk

H. Meyer-Ortmanns, S. Thurner (eds.), *Principles of Evolution*,
The Frontiers Collection, DOI 10.1007/978-3-642-18137-5_3,
© Springer-Verlag Berlin Heidelberg 2011

means of a replication process. Here, it is intangible behaviour, rather than tangible molecules, that is replicated; nevertheless, differences in the behaviour may occur when the behaviour is replicated, and some of these differences may propagate more successfully than others, and so one has variation and selection in cultural evolution too.

That processes of biological and cultural evolution may be somewhat similar is not a new idea. Indeed, in his famous books *On the Origin of Species* [1] and *The Descent of Man* [2], Darwin used human language as an analogy to support his case that species should not be viewed as fixed categories, but as a dynamic classification where one species is defined in terms of its genealogical relationships with other species. For example, he wrote

> It may be worth while to illustrate this view of classification, by taking the case of languages. If we possessed a perfect pedigree of mankind, a genealogical arrangement of the races of man would afford the best classification of the various languages now spoken throughout the world; and if all extinct languages, and all intermediate and slowly changing dialects, had to be included, such an arrangement would, I think, be the only possible one. [1, p. 422]

Indeed, he identified further similarities between evolution in a biological and cultural (linguistic) setting:

> The formation of different languages and of different species, and the proofs that both have been developed through a gradual process, are curiously parallel. But we can trace the formation of many words further back than that of species, for we can perceive how they actually arose from the imitation of various sounds, as in alliterative poetry. We find in distinct languages striking homologies due to the community of descent, and analogies due to a similar process of formation. The manner in which certain letters or sounds change when others change is very like correlated growth. We have in both cases the reduplication of parts, the effects of long-continued use, and so forth. ... Dominant languages and dialects spread widely, and lead to the gradual extinction of other tongues. A language, like a species, when once extinct, never, as Sir C. Lyell remarks, reappears. [2, pp. 59–60].

Of course, in the 150 years since Darwin proposed his theory of evolution, much has been learned about the mechanics of the molecular processes that underpin biological evolution. At this level, biological and cultural evolution are rather dissimilar: the specifics of molecular processes such as meiosis have no analogue in the cultural evolutionary dynamics of language change, for example. However, there are recurrent questions that arise in different evolutionary contexts that are perhaps more easily recognised once one has found a common language to describe them all.

One such question relates to what the units of selection are: a topic that was at the centre of a long debate in biology (e.g., whether selection is best construed as taking place at the level of the genotype, phenotype or even some higher level such as groups or species [3]). In Sect. 3.1 below, we summarise a formulation of evolutionary dynamics due to Hull [4] that was in part motivated by this debate. Its utility for us is that it lays bare the essential components of an evolutionary dynamics, and allows one to recognise where superficially dissimilar processes are identical, and where they differ in a fundamental way.

Another recurrent question is of a more empirical nature, namely, accounting for a given evolutionary change in terms of selection, variation, stochasticity in birth and death, or some combination thereof. As we will discuss in Sect. 3.2, the Price equation [5] provides a rather simple, but general, means to distinguish changes due to selection from changes due to variation in replication. However, this leaves open the question of how much change is due to random effects alone. Theories of evolutionary change in the absence of selection are collectively referred to as *neutral theories*, have a long history in genetics [6], and have recently risen in prominence as null models for change in ecology [7, 8] and linguistics [9, 10]. It is rather important to establish the basic properties of neutral evolution, so that one can test for departure from a purely neutral theory. We therefore devote a large part of this chapter to discussing the predictions of neutral theory, and examining a few cases in which it has been shown to account for (or fail to account for) empirical data. As we will see, very similar mathematical models arise in different evolutionary contexts, but can lead to subtly different interpretations of the role that selection plays in these different contexts and what it means for evolution to be neutral.

Our aim in this chapter is to introduce some of the basic mathematical models of evolutionary dynamics and to point out some of their most important properties from the point of view of applications in genetics, ecology and cultural evolution. Readers who are interested in more comprehensive discussions of these mathematical models in their various formulations are encouraged to consult the excellent textbooks by Crow and Kimura [11], Ewens [12], Barton et al. [13] and Wakeley [14]. Readers with a background in physics may also find the reviews by Peliti [15], Baake and Gabriel [16], Drossel [17] and myself and McKane [18] useful.

## 3.1 Hull's General Analysis of Selection

The debate about the level at which selection may be recognised as operating is exemplified by an observation that troubled Darwin, who noted that neuter insects, whose traits, by definition, could not be inherited by future generations, could nonetheless display apparently selective adaptations [1, pp. 236–242]. Hull's general analysis of selection [4] was motivated in part to articulate more clearly the nature of this debate. His insight was to identify the key actors and processes through the role they play in the selection process itself. It is these general definitions that allow a precise formulation of nongenetic evolutionary processes such as the development of scientific theory [4] and language change [19].

Hull defines four key concepts, which we quote verbatim here [4, p. 408]:

- *replicator* – an entity that passes on its structure largely intact in successive replications.
- *interactor* – an entity that interacts as a cohesive whole with its environment in such a way that this interaction *causes* replication to be differential.
- *selection* – a process in which the differential extinction and proliferation of interactors *cause* the differential perpetuation of the relevant replicators.
- *lineage* – an entity that persists indefinitely through time either in the same or in an altered state as a result of replication.

This precise, formal definition requires some unpacking. The first main point is that the replicator and interactor need not be the same entity – that is, the different roles played in the selection process can be played by different actors. The second, and this is strongly emphasised by Hull [4, p. 404], is that selection is causal in a very specific way. It is the interaction of the interactor with its environment that causes different replicators to be replicated at different rates. Implicit in the term *relevant* replicators is that survivability in a given environment is heritable. Only when this causal link is in place can a change in the frequencies of different replicators be regarded as adaptive. Finally, we note two more minor details: the phrase *largely intact* indicates the place at which variation can be introduced into the process through the creation of new replicator types; and whilst replicators and interactors are transient objects, lineages persist through time and are the means by which replicators can be related to one another. This notion of lineage formalises Darwin's view of species classification.

### 3.1.1 Instances of Hull's General Analysis of Selection

In order to elucidate this generalised analysis of selection we shall discuss three concrete instances of it.

#### 3.1.1.1 Molecular Evolution

The most straightforward application of the general analysis of selection is to the asexually reproducing organisms (such as certain fungi, yeasts and plants). If we ignore the complications that arise from horizontal gene transfer [20], then the structure that is replicated largely intact is the organism's DNA. Changes in this structure may occur due to mutation. The most likely candidate for the interactor is the organism – differences in the phenotype and interaction with the environment will affect the reproductive success of different individuals. Those differences which are due to genetic variation will lead to an adaptation. A lineage can be drawn by considering the repeated replication of the DNA. The organism is not the only possible choice for the interactor: it could extend to include groups of organisms.

#### 3.1.1.2 Community Ecology

In contrast to genetics, ecology focuses more strongly on interactions between different species, and the particular qualities of a species that cause it to survive, or fail, in an ecosystem comprising many different species and possibly also abiotic factors. Ecologists often talk in terms of *niches*, the range of viable conditions under which the species can survive (see, e.g., [21]). Community ecologists are typically interested in the structure of and diversity within an ecosystem, for example, the distribution of species abundances or the number of species one expects to find in a given area. Although the underlying evolutionary mechanism is genetic, the primary consideration is whether two individuals are of the same species or not. Therefore, one way to model the evolution of an ecosystem is to identify both replicators and interactors as individuals of a species.

There is, however, a subtle distinction between replicator and interactor in terms of identity. Two replicators are identical if they are individuals of the same species. However, interactors may be distinguishable: they may exhibit individual differences (due to different life histories, for example) that can influence the survival of the species as a whole. However, these differences are (at least in this formulation) assumed not to be heritable, and so selection in this model takes place at the species level through the interactions of its individuals with those of other species.

In the generalised formulation, variation enters through the creation of new replicator types. In an ecological setting, this may occur through speciation, or through immigration from a pool external to the local community (this pool is sometimes called the *metacommunity*). Either way, the result is the same: an individual of a new species enters the community. In this model, lineages can be drawn connecting parent individuals to their offspring.

This is not the only way to define replicator and interactor in an ecological setting. For example, one can extend the interactor to include the species as a whole, or indeed multiple species fulfilling a common ecological function. However, the replicator is by definition restricted to those components of the interactions that are heritable, and are therefore most likely to be individuals.

### 3.1.1.3 Language Change

Croft [19] explicitly defines an evolutionary linguistic theory within Hull's scheme. Here, the replicators are tokens of linguistic structure, for example, vowel sounds, words or constructions like "the $X$er the $Y$er", where $X$ and $Y$ may be filled by many components to realise phrases like "the shorter the better". It is this type of structure that is replicated when humans speak. The analogue of the gene, called the *lingueme* by Croft [19], may exist in multiple variants when there is more than one way of saying the same thing (e.g., two different ways of pronouncing the same vowel in a set of words).

Hull's original analysis of selection, above, mandates that proliferation and extinction of the interactors is a central mechanism causing differential replication of the replicators. In order to place language change directly in Hull's framework, a rather subtle definition of the interactor is needed: Croft [19] identifies this as the speaker *combined* with her knowledge of the language (the *grammar*), taken as a *transient* object. Then, the picture would be one of grammars that favour certain variants proliferating or going extinct as speakers interact with one another in different social and linguistic contexts. In turn, this would lead to certain replicators being replicated more frequently than others in the manner that Hull has in mind.

It is, however, more natural to think of the speaker as an interactor whose life cycle corresponds with that of the speaker (i.e., it comes into being when the speaker is born, and dies with the speaker). Within Hull's framework, a language may only change as children acquire grammars that differ in structure to those of their parents. Such models have indeed been widely studied: see e.g., [22, 23]. However, Croft [19] rejects these in favour of a 'usage-based' model [24] that is based on the observation that speakers closely track the frequencies with which speakers use linguistic variants throughout their lives [25, 26]. In order to identify a speaker as an interactor, it is necessary to remove from Hull's definition of selection the requirement that the population dynamics of the replicators is a consequence of birth and death of the interactors. Croft has suggested (see, e.g., [27, p. 94]) an alternative definition which can be formulated as follows:

- *selection* – the process by which an interactor's interaction with its environment *causes* the differential replication of the relevant replicators.

This new definition opens the door for a range of processes, such as speakers producing variants to flatter or impress the listener, or to identify themselves with a particular social group, to be considered as selection mechanisms independently of the life cycle of the interactor. The key point here is that whilst certain aspects of the relationship between interactor and replicator are completely different in this formulation of language change as an evolutionary dynamics – in particular, the replicators are not "embedded" within the interactors as DNA is within individuals of a species – the components that are key to the process being an evolutionary process are present. That is, there is a structure that is replicated (linguistic behaviour), and that replication is differential as a consequence of an interactor's interaction with the environment (here, other speakers and their linguistic behaviour, which is governed by their grammars).

## 3.2 A Mathematical Analysis of Selection: The Price Equation

In 1970, Price [5] (see also [28]) attempted a general formulation of selection that abstracted away from the specific genetic mechanisms of inheritance that are usually emphasised by population geneticists, and is therefore of interest in our examination of evolutionary dynamics in different contexts. It is an equation for the change in the mean value of some quantitative character (trait) after one generation of reproduction. This trait may be discrete, like eye colour, or continuous, like height. As Price remarks "it holds for any sort of dominance or epistasis, sexual or asexual reproduction, for random or nonrandom mating, for diploid, haploid or polyploid species and even for imaginary species with more than two sexes" [5, p. 520]. The equation was used to understand such effects as kin and group selection, that is, selection that apparently takes place at a level other than the gene. Recall that such possibilities motivated Hull's formal distinction between the replicator and the interactor.

### 3.2.1 Derivation of the Price Equation

The key feature of the Price equation is that evolutionary changes are expressed purely in terms of properties of replicators within a parent population. To this end, let us define this parent population: it contains $N$ replicators, labelled $i = 1, 2, \ldots, N$, and the value of the trait of interest associated with replicator $i$ is denoted $z_i$. When the next generation of offspring is created, one can in principle count the number of offspring each replicator $i$ has: call this number $w_i$. A specific offspring of replicator $i$, labelled $j = 1, 2, \ldots, w_i$, inherits the trait $z_i$, subject to a change $\delta z_{i,j}$, which may be due to mutation, recombination, or any other process that can cause altered replication. Note that the interactor enters here implicitly: it may in principle affect both $w_i$ and $\delta z_{i,j}$.

We are interested in the difference in the mean value of the trait when averaged over the parent and offspring generations. Introducing the overbar to denote an average over the individuals in the population, unprimed variables to represent the parent population, and primed variables to represent the offspring population, we may write this change as

$$\delta \bar{z} = \bar{z}' - \bar{z} = \frac{1}{N'} \sum_{i=1}^{N} \sum_{j=1}^{w_i} (z_i + \delta z_{i,j}) - \frac{1}{N} \sum_{i=1}^{N} z_i \; . \tag{3.1}$$

Our aim is to rewrite the right-hand side purely in terms of properties of the parent population (i.e., unprimed variables). First of all, we note that $N' = N\bar{w}$, where $\bar{w}$ is the mean number of offspring, averaged over each of the parents. Then,

$$\bar{z}' - \bar{z} = \frac{1}{N} \sum_{i=1}^{N} \left( \frac{w_i}{\bar{w}} - 1 \right) z_i + \frac{1}{N\bar{w}} \sum_{i=1}^{N} \sum_{j=1}^{w_i} \delta z_{i,j} \; . \tag{3.2}$$

The first term on the right-hand side has the form of a covariance (or correlation function):

$$\mathrm{Cov}(x, y) = \overline{(x - \bar{x})(y - \bar{y})} = \overline{xy} - \bar{x}\bar{y} \; , \tag{3.3}$$

in which $x$ and $y$ are two random variables. Now, the mean of $w_i/\bar{w}$ when averaged over individuals $i$ is just unity, and hence

$$\frac{1}{N} \sum_{i=1}^{N} \left( \frac{w_i}{\bar{w}} - 1 \right) z_i = \frac{\overline{wz}}{\bar{w}} - 1 \times \bar{z} = \mathrm{Cov}\left( \frac{w}{\bar{w}}, z \right) \; . \tag{3.4}$$

The second term on the right-hand side of (3.2) may also be written in a slightly more compact form if we introduce for each parent $i$ the mean change in the trait value $\delta z_i$ averaged over its offspring, i.e.,

$$\delta z_i = \frac{1}{w_i} \sum_{j=1}^{w_i} \delta z_{i,j} . \tag{3.5}$$

Then,

$$\frac{1}{N\bar{w}} \sum_{i=1}^{N} \sum_{j=1}^{w_i} \delta z_{i,j} = \frac{1}{N} \sum_{i=1}^{N} \frac{w_i}{\bar{w}} \delta z_i = \frac{\overline{w\delta z}}{\bar{w}} . \tag{3.6}$$

Hence, (3.2) can be written in compact form as

$$\delta\bar{z} = \mathrm{Cov}\left(\frac{w}{\bar{w}}, z\right) + \frac{\overline{w\delta z}}{\bar{w}} , \tag{3.7}$$

which is the Price equation [5].

The two terms that appear on the right-hand side of the Price equation (3.7) can be ascribed to *selection* and *variation in transmission*, respectively. One sees this by noting that if all replicators have the same number of offspring, the first term vanishes by definition; whereas if all replicators are replicated faithfully, all $\delta z_{i,j} = 0$, and the second term vanishes. The fact that the selection term can be written as a covariance demonstrates that if any deviation of the trait $z$ away from the mean (up or down) is correlated with a higher rate of reproduction, the mean will shift in that direction.

An interesting feature of the Price equation that is seldom commented upon is that all the quantities appearing on its right-hand side are given by the *actual* numbers of offspring and changes that occur on going from one generation to the next. In principle, one could examine two generations of individuals and, by counting the number of offspring and measuring the changes in trait values, determine the contributions to the change in $\bar{z}$ due to selection and variation. In a finite population, however, one may infer a spurious selective component to the dynamics due to fluctuations in the number of offspring that a replicator has. We will discuss these fluctuations and the changes one would expect from them in Sect. 3.3 onwards. In the meantime, we explore a couple of applications of the Price equation.

### 3.2.2 Applications of the Price Equation

#### 3.2.2.1 Survival of the Fittest

If we suppose that the number of offspring that an individual has is a trait that is inherited by offspring without variation, then we can ask how the *fitness* of the population, defined as the mean number of offspring, changes over time. Putting $z = w$ into (3.7), we find

$$\delta \bar{w} = \frac{\text{Var}(w)}{\bar{w}} \, , \tag{3.8}$$

i.e., the variance of the offspring numbers (normalised by the mean). Since offspring numbers and variances are both nonnegative, this expression implies that – in the absence of variation – the fitness of the population will always increase (or remain constant). How is this achieved? Individuals that have a large number of offspring will contribute far larger numbers of individuals to later generations than those that have a small number of offsping. The numbers of the latter species relative to the former thus decreases, and the more poorly reproducing species will be eliminated. This is *survival of the fittest*, mathematically expressed.

Equation (3.8) has been called the *Fundamental Theorem of Natural Selection*, since it shows that the fitness of a population is bound to increase, and was introduced by Fisher in 1930 [29]. However, this theorem was misunderstood due to its giving only *part* of the contribution to changes in fitness: as can be seen from the full Price equation (3.7), the second term could counteract the first, and even drive the mean fitness of the population down if $\delta w_{i,j}$ can be negative.

### 3.2.2.2  Replicators, Interactors and Kin Selection

The theory of kin selection [30] provides a nice illustration of the relationship between replicator and interactor, and the way that properties of the latter may influence the dynamics of the former. Kin selection is one of the available explanations for the existence of *altruism* [3], defined in this context as an act performed by an interactor that leads to a reduction in the number of offspring it has (or, more precisely, offspring of the associated replicators), but causes another interactor in the population to have a larger number of offspring. The Price equation (3.7) allows one to understand the circumstances under which a gene that prompts a small number of individuals to perform the altruistic act may increase in frequency within the population. As we will see, such an increase can be inhibited or reversed by the presence of individuals who reap the benefit without paying the cost.

The starting point is a population of replicators, each of which leaves a single offspring in the following generation (i.e., $w_i = 1$). With each replicator we associate an indicator $\tau_i$, which equals 1 if replicator $i$ is an instance of the altruistic gene, and 0 otherwise. The interactor associated with replicator $i$ performs $g_i$ altruistic acts, each of which causes its offspring number to decrease by a *cost c*. Meanwhile, replicator $i$ is the recipient of $r_i$ such acts, each leading to an increase in the offspring number by the *benefit b*. Therefore, the number of offspring a replicator has in the following generation is

$$w_i = 1 + br_i - cg_i \, . \tag{3.9}$$

Notice that this relationship expresses the causal relation between proliferation and extinction of *interactors* (mediated by the quantities $r_i$ and $g_i$) and the differential perpetuation of the *replicators* (given by $w_i$) emphasised in Hull's analysis of selection (Sect. 3.1).

The frequency of altruistic genes in the population is $f = \bar{\tau}$. Assuming that replication is faithful (i.e., genes do not mutate on replication), we can use the Price equation (3.7) to ascertain that

$$\delta f = \frac{b\,\mathrm{Cov}(r, \tau) - c\,\mathrm{Cov}(g, \tau)}{1 + b\bar{r} - c\bar{g}} \; . \tag{3.10}$$

Offspring numbers are by definition nonnegative, so the denominator of the previous expression must also be nonnegative (in practice, this constrains the allowed combinations of $r_i$ and $g_i$). Therefore, the altruistic gene grows in frequency only if the numerator of the previous expression is positive, i.e., when

$$\frac{\mathrm{Cov}(r, \tau)}{\mathrm{Cov}(g, \tau)}\, b > c \; , \tag{3.11}$$

which is a version of Hamilton's rule for kin selection [30].

By definition, performing the altruistic act is positively correlated with carrying the altruistic gene, so the covariance in the denominator of the previous expression is positive. The cost $c$ and benefit $b$ are also positive quantities. On the other hand, $\mathrm{Cov}(r, \tau)$ can be positive or negative: it is positive if the recipient of an altruistic act is more likely to carry the altruistic gene than a randomly selected individual, and negative if the recipients are less likely than average to be altruists. If altruism is to be successful, a necessary requirement is that the beneficiaries of altruistic acts are also altruists. In particular, if there is no correlation between the recipient of an altruistic act, and that recipient being an altruist, the frequency of the altruistic gene will not grow (except, possibly, through stochastic fluctuations – see next section). Handing out benefits at random is not a viable long-term strategy.

It can also be shown that $\mathrm{Cov}(r, \tau) \leq \mathrm{Cov}(g, \tau)$ because recipients of the altruistic act need not themselves be altruists, whereas the benefactors always are. Therefore, the benefit $b$ received must always be larger than the cost $c$ by a factor that depends on how good the altruists are at targeting the benefit at other altruists. The better they are at this, the smaller the benefit they need to confer per unit cost in order for a rare altruistic gene to grow in frequency.

The example that is often used to illustrate Hamilton's rule (see, e.g., [31]) is based on a population of diploid, sexually reproducing interactors in which the altruistic gene has a very low frequency, $f \approx 0$. Recipients of the altruistic act are chosen exclusively from kin (brothers or sisters) of the benefactor. If the altruistic gene is present at a low frequency in the population, then, with high probability, exactly one of each altruist's parents carries exactly one altruistic gene. If each individual acquires one of the two genes carried by each parent, with each gene chosen at random, the probability that a kin of an altruist is also an altruist is $\frac{1}{2}$. When $N$ is large, averages over the population will be close to their expectation values, and so one will find

$$\overline{r\tau} = \frac{1}{N} \sum_i r_i \tau_i = \frac{1}{2N} \sum_i g_i \tau_i = \frac{1}{2}\overline{g\tau} \tag{3.12}$$

since the probability that any recipient is an altruist is half the probability that the corresponding benefactor is an altruist. Therefore the condition for growth of a rare altruistic gene becomes

$$\frac{\frac{1}{2}\overline{g\tau} - f\bar{g}}{\overline{g\tau} - f\bar{g}} b > c \; . \tag{3.13}$$

Now, as $f \to 0$, $\overline{g\tau}$ and $\bar{g}$ are both proportional to the gene frequency $f$, and so for vanishingly small gene frequency the condition for its growth becomes

$$\frac{1}{2}b > c \; . \tag{3.14}$$

That is, the altruistic gene will propagate if the benefit received exceeds twice the cost. More generally, one needs $rb > c$, where $r$ is the probability that the recipient of the altruistic act carries the altruistic gene, given the benefactor's strategy for choosing recipients. The reason kin selection can promote altruism is that kin are more likely to be genetically similar than randomly chosen members of the population.

If the mechanism for choosing beneficiaries of an altruistic act is not a very reliable indicator of their carrying the altruistic gene, Hamilton's rule (3.11) shows that the benefit conferred must be increased to compensate. We remark, however, that the overall rate of growth of the altruistic gene for 'strategies' that lie along the line $b\mathrm{Cov}(r, \tau) = \mathrm{const}$ is smaller for strategies that entail conferring large benefits on nearly randomly chosen individuals from the population than those that confer a small benefit on individuals likely to be altruists. As we will see in the next section, replicators with a small growth rate are more likely to be eliminated by random fluctuations before they can become established.

## 3.3 Neutral Demographic Fluctuations: Genetic Drift

In the previous section we saw that the Price equation (3.7) allows us to work out the change in the population average of some trait $z$ in one generation, assuming we know all the offspring numbers $w_i$, and the changes in the trait $\delta z_{i,j}$ of each of individual $i$'s offspring. In practice, the process of reproduction is stochastic – that is, the offspring numbers $w_i$ (and potentially also the changes $\delta z_{i,j}$) are *random* variables. The probabilistic changes in replicator frequencies due to stochasticity in the birth and death dynamics are sometimes referred to as *demographic fluctuations*.

As we previously observed, these fluctuations may cause us to infer erroneous selective effects where in fact there is no systematic contribution to an interactor's survival coming from variation in the underlying replicators. If one were able to average the first term in (3.7) over multiple realisations of the dynamics, one would indeed find that such random fluctuations would cancel. A model in which all replicator types have the same offspring numbers is called *neutral*; in genetics the fluctuations that arise are referred to as *genetic drift*.

Since we will not in general be able to perform multiple realisations of a population dynamics process, it is important to be able to disentangle systematic and purely random effects from a single realisation of the process. To do this, one needs to know something about the statistics of the purely neutral population dynamics, which is the subject of this and the following two sections. We do not attempt a detailed survey of the huge literature on genetic drift and its applications here; rather, we will highlight a few key points with reference to specific models and applications.

### 3.3.1 Models of Purely Neutral Evolution

There are two simple, illustrative concrete models of this neutral evolution that are widely used in genetics. The oldest is the Wright–Fisher model [29, 32], which comprises discrete generations of $N$ replicators, a number that stays constant over time. Since different replicator types have by definition no effect on the survival of any associated interactors in this model, we do not need to make any explicit reference to the latter and indeed one could use the terms interchangeably.

The next generation is formed from the current generation by repeating the following steps $N$ times: (i) a *parent* replicator is randomly chosen from the current generation; (ii) it is replicated, to create an *offspring*; (iii) the parent is returned to the current generation; (iv) the offspring is deposited in the next generation. Although all replicators have the same distribution of offspring numbers in this model, there is a competitive element in that only $N$ replicators can be accommodated in each generation – perhaps because the resources available become exhausted if the population grossly exceeds $N$. This competition is the origin of fluctuations in the frequencies of different replicator types. As we will see, the effect of this competition is that (without any mutation), all but one of the replicator types will eventually go extinct.

To examine these fluctuations and their consequences, it is conventional to consider a population with two replicator types. We will also couch our discussion not in terms of the Wright–Fisher model, but a more recent and mathematically convenient model due to Moran [33] that has very similar dynamics. The difference is that this model has overlapping generations. In step (iv) above, the offspring is also placed into the "current" generation, displacing one of the replicators already present, chosen at random. The displaced individual may be the same as that chosen as the parent. In the Moran model, time is usually measured in terms of the number of individual sampling events, i.e., iterations of steps (i)–(iv). Then, roughly speaking, $N$ sampling events in the Moran model correspond to one generation of the Wright–Fisher model (actually, the time scales of the two models are related by a factor $N/2$ [12, 18]). The results we obtain for Moran-type models here will often also apply to the Wright–Fisher model under suitable rescaling of time.

Mathematically, the Moran model can be formalised as a Markov chain in which there are $N+1$ states. We can label these states with the integer $n = 0, 1, 2, \ldots, N$, which counts the number of replicators of a specified type which we shall label $A$. This number changes if the parent and offspring chosen in a given step of the dynamics are of different types. It increases by one if the parent is of type $A$ (this event occurs with probability $\frac{n}{N}$) *and* the displaced replicator is not (probability $1 - \frac{n}{N}$). Thus the total probability that $n \to n+1$ is $\frac{n}{N}(1 - \frac{n}{N})$. Likewise, $n$ decreases by one if the types of parent and offspring are exchanged, an event that occurs with the same overall probability. Hence, we can write down the transition probabilities for this model:

$$
P(n \to m) = \begin{cases} \frac{n}{N}(1 - \frac{n}{N}) & m = n \pm 1 \\ 1 - 2\frac{n}{N}(1 - \frac{n}{N}) & m = n \\ 0 & \text{otherwise} \end{cases} \tag{3.15}
$$

where $0 \le n, m \le N$. In the language of Markov chains [34], the states $0 < n < N$ are *transient*, which means that once left, there is some probability that they are never returned to. In turn, this means that as the number of time steps $t \to \infty$, the probability of being in one of these transient states vanishes. Therefore, as $t \to \infty$, one has either $n = 0$ or $n = N$. In words, replicators of type $A$ have either gone extinct, or taken over the whole population. In genetics, this latter state of affairs is called *fixation*. Therefore, we see explicitly that undirected demographic fluctuations can cause the proliferation of a replicator. One reason why this does not count as selection in Hull's definition (Sect. 3.1) is that the fact that one replicator had more offspring than another in one generation does not cause it to have more offspring than another in subsequent generations. That is, a parent's actual offspring number $w_i$ is not inherited by its offspring. If anything is inherited, it is the statistical distribution of offspring number, which is common to all replicators.

### 3.3.2 Fixation Probability

From (3.15), we see explicitly that if we average over all realisations of the demographic fluctuations (an averaging we will denote with angle brackets in the following), the mean number of replicators of a given type, $\langle n \rangle$, remains unchanged over time. If $n(t) = n$ with probability one, then

$$
\langle n(t+1) \rangle = \sum_m m P(n \to m) \tag{3.16}
$$

$$
= (n+1)x(1-x) + n[1 - 2x(1-x)] + (n-1)x(1-x) \tag{3.17}
$$

$$
= n(t), \tag{3.18}
$$

where $x = n/N$. Averaging over any distribution of $n(t)$, we find, $\langle n(t+1) \rangle = \langle n(t) \rangle$, i.e., that the mean number of replicators of type $A$ (and hence any type) is conserved, as claimed. Above, we argued that as $t \to \infty$, $n(t) = N$ or $0$. Denote the probability of the former event as $\phi$. Then, we have that

$$\lim_{t \to \infty} \langle n(t) \rangle = \phi \times N + (1 - \phi) \times 0 = \phi N \tag{3.19}$$

Since $\langle n(t) \rangle = \text{const}$, we have that the fixation probability $\phi$ is given by

$$\phi = \frac{\langle n(0) \rangle}{N} \tag{3.20}$$

for any initial condition. The fact that the fixation probability of a replicator type is proportional to its initial frequency in the population (or the expectation value thereof, if the initial condition is a distribution) is one of the basic properties of genetic drift.

### 3.3.3 Mean Fixation Time

Another important basic property of genetic drift is the mean time $T(n)$ until replicator type $A$, represented $n$ times in the initial population, goes extinct or to fixation, averaged over all possible realisations of the dynamics. This can also be worked out straightforwardly. The key point to realise is that the mean number of time steps to fixation or extinction from time $t = 0$ is, if $0 < n < N$, equal to one plus the mean number of time steps from time $t = 1$, averaged over the distribution of states at time $t = 1$. As an equation,

$$T(n) = 1 + x(1-x)T(n+1) + x(1-x)T(n-1) + [1 - 2x(1-x)]\,T(n)\,, \tag{3.21}$$

where again $x = n/N$. It is useful make a change of variable to $x$ everywhere, i.e., by putting $T(n) = \tilde{T}(x)$, and expand in a Taylor series

$$T(n \pm 1) = \tilde{T}(x) \pm \frac{1}{N}\tilde{T}'(x) + \frac{1}{2N^2}\tilde{T}''(x) + \cdots . \tag{3.22}$$

Substituting into (3.22), and truncating at second order, we find

$$-1 = \frac{1}{N^2}x(1-x)\tilde{T}''(x)\,. \tag{3.23}$$

This equation can be solved once appropriate boundary conditions have been imposed. We are interested in the mean time until only one of the two replicator types remains, but are not interested in which one it is. For this question, the appropriate boundary condition is $\tilde{T}(0) = \tilde{T}(1) = 0$, since the mean time to reach one of the two absorbing states (extinction or fixation) from $n = 0$ or $n = N$ is zero. The solution (3.23) with these boundary conditions, first noted in the context of genetic drift by Kimura and Ohta [35], is

$$\tilde{T}(x) = -N^2 \left[ x \ln x + (1 - x) \ln(1 - x) \right] . \tag{3.24}$$

This equation is valid when $N$ is large. The key thing to notice is that when the initial number of both replicator types is of order $N$ (as opposed to order 1), the mean time to fixation of one of them is of order $N^2$ updates. Since the time scales of the Wright–Fisher and Moran models differ by a factor of $N$, in the former model the mean number of generations until one of the types goes to fixation is of order $N$, rather than order $N^2$.

### 3.3.4 Experimental Observation of Genetic Drift: Effective Population Size

The population size $N$ plays an important part in the analysis of genetic drift. This should not be too surprising given that this is the only parameter in the Wright–Fisher and Moran models. It is intuitively clear that if $N$ is small, the probability that one of the two types is never replicated in some sequence of sampling events is much higher than if $N$ is large, and hence that its extinction or proliferation through purely stochastic effects alone is likely to occur more rapidly. Indeed, in genetics, the strength of demographic fluctuations is typically quantified in terms of size of the Wright–Fisher population that would show the fluctuations of the same magnitude. This size is referred to as an *effective population size* for the system. In actual fact, a range of different effective population sizes can be defined [36, 37]; however, the key point is that the effective size may differ from the actual size.

This has been observed in an experimental realisation of genetic drift, conducted with small, artificially manipulated populations of *Drosophila melanogaster* [38]. The Wright–Fisher dynamics were imposed by allowing the organisms to reproduce, and then sampling a fixed number of offspring organisms at random, and using these as the parents for the next generation. The Wright–Fisher model predicts that if there are $n$ copies of an allele in the parent generation, the number in the offspring generation, $m$, should be distributed as a binomial

$$P(m|n) = \binom{N}{m} \left(\frac{n}{N}\right)^m \left(1 - \frac{n}{N}\right)^{N-m} . \tag{3.25}$$

In particular, the variance of the allele frequency $y = m/N$ in the offspring generation is $x(1 - x)/N$, where $x = n/N$ is the corresponding frequency in the parent generation. By running multiple replicates of this artificial evolution experiment, one can measure this variance experimentally, and the predicted parabolic dependence on $x$ was indeed observed: see Fig. 3.1, which shows data tabulated in [38]. However, the amplitude of the parabola was not given by $1/N$ with $N$ equal

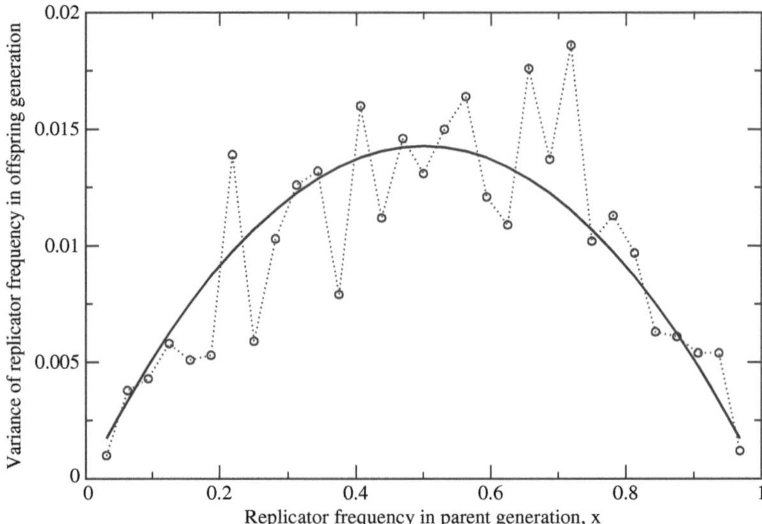

**Fig. 3.1** Variance in the frequency of the replicator type $A$ in the offspring generation as a function of its frequency in the parent generation as recorded for an experimental *Drosophila* population of 32 replicators. Points are plotted from data tabulated for Series I in [38]; the curve is a fit to $x(1-x)/N$ with $N$ an adjustable parameter approximately equal to 18

to the actual population size, but rather by a smaller, effective population size (a value of approximately 18 rather than 32). As we will see in Sect. 3.5, population subdivision – that is, situations in which parents are not sampled uniformly from the population, but are divided into subpopulations – generically has the effect of reducing the effective population size, a fact that could explain this experimental observation.

## 3.4 Immigration and Mutation in Neutral Models

Replication is only one of the processes involved in evolutionary dynamics: even in the absence of selection, changes can occur through mutation – a change in replicator type in the replication process – or immigration, the appearance of replicators from outside the immediate population. As we have seen above, when replication is faithful and the population is of finite size and closed to immigration, variability will always be lost after a sufficiently long time. Immigration and mutation are two processes that can reverse this effect, and allow steady states in which multiple replicator types can coexist.

From the point of view of simple models, immigration and mutation are equivalent processes. This is most simply understood if each new mutant or immigrant is of a different type to any that has been seen before. Then, whether by immigration or mutation, there is some probability at each time step of a new type appearing. We will refer to this type of mutation or immigration generically as *nonrecurrent immigration.* On the other hand, if mutation or immigration results in additional replicators of a fixed set of existing types, we call this *recurrent immigration.*

A neutral model with immigration has recently become prominent in community ecology [7, 8]. This model is controversial because typically it is assumed that species diversity is a consequence of selection rather than demographic fluctuations alone (see e.g. [39, 40] for prominent critiques). In particular, a common view is that a community can be represented as a set of ecological niches (essentially, a viable strategy for survival given the other species and the environment that is present), and that the fittest species occupying a niche will drive out all competitors in that niche through selection (as per the predictions of the Price equation). The neutral model, by contrast, does not make reference to multiple niches, and that as species compete for fixed resources within a single community, none is a priori fitter than any other.

The simplest version of this model has a single community of $N$ individuals, a number that remains fixed over time (as in the Wright–Fisher and Moran models). In each time step, a parent replicator (individual of a species) is selected, and its offspring displaces a randomly chosen replicator, as in the Moran model. The difference with the Moran model is that there is a probability $\nu$ that the parent is not sampled from within the local community (the population of $N$ replicators), but from the wider metacommunity. If $\nu$ is nonzero, ultimately one reaches a steady state in which multiple species coexist. Of particular interest is the abundance of different replicator types (species) in this steady state.

The mathematical analysis is different depending on whether immigration is recurrent or nonrecurrent – and different in interesting ways, so we shall examine both cases in more detail. Operationally, recurrent immigration can be realised by having a neutral metacommunity of very large size, so that on the time scales in which the population changes in the local community, the frequency of replicator type $A$ remains fixed at some value $\bar{x}$. We will assume the existence of one other replicator type, whose frequency is $1-\bar{x}$, so ultimately a steady state will be reached in which the two types coexist. On the other hand, when immigration is nonrecurrent, and every immigrant is of a new type, there will be a constant turnover of new species. Nevertheless, a type of steady state is reached in which the distribution of species abundances becomes time-independent, even though species labels keep changing.

### 3.4.1 Recurrent Immigration

Like the basic Moran model discussed in Sect. 3.3, its extension to include immigration can be handled within a Markov chain formulation. Our task is to work out how $n$, the number of replicators of type $A$, may change in one time step, given that the frequency of $A$ in the metacommunity is $\bar{x}$. As before, $n$ increases by one if the parent is of type $A$ (probability $\frac{n}{N}$), the displaced replicator is not (probability $1 - \frac{n}{N}$), *and* that the update that takes place in the time step is replication (probability $1 - \nu$). Alternatively, $n$ may increase by one through an immigration event (probability $v$), if the immigrant is of type $A$ (probability $\bar{x}$) and the displaced replicator is not (probability $1 - \frac{n}{N}$). Putting these probabilities together, and following the same logic for events that lead to $n$ decreasing by one, we find the transition probabilities for this model are

$$P(n \to m) = \begin{cases} \left(1 - \frac{n}{N}\right)\left[v\bar{x} + (1-v)\frac{n}{N}\right] & m = n+1 \\ \frac{n}{N}\left[v(1-\bar{x}) + (1-v)\left(1 - \frac{n}{N}\right)\right] & m = n-1 \\ 1 - 2(1-v)\frac{n}{N}\left(1 - \frac{n}{N}\right) + v\left[\frac{n}{N}(1-\bar{x}) + \left(1 - \frac{n}{N}\right)\bar{x}\right] & m = n \\ 0 & \text{otherwise} \end{cases}$$
$$(3.26)$$

where again $0 \le n, m \le N$. The probability $P(n,t)$ that there are $n$ replicators of type $A$ at time $t$ evolves via the master equation

$$P(n, t+1) = P(n-1, t)P(n-1 \to n) + P(n, t)P(n \to n)$$
$$+ P(n+1, t)P(n+1 \to n) . \qquad (3.27)$$

It is possible to solve for the steady state of this discrete-time master equation exactly (see, e.g., [41]). A simpler approach, however, is to take the limit of large community size $N$, and regard the frequency $x = n/N$ as a continuous variable, analogous to our treatment of Eq. (3.22) for the mean time to fixation in the Moran model ($v = 0$). Putting $\tilde{P}(x, t) = P(Nx, t)$, and expanding in powers of $1/N$, we find

$$P(n \pm 1) = \tilde{P}(x, t) \pm \frac{1}{N}\frac{\partial \tilde{P}(x, t)}{\partial x} + \frac{1}{2N^2}\frac{\partial^2 \tilde{P}(x, t)}{\partial x^2} + \cdots \qquad (3.28)$$

$$P(n-1 \to n) = (1-v)x(1-x) + v\bar{x}(1-x)$$
$$+ \frac{1}{N}\left[v\bar{x} + (1-v)(2x-1)\right] - \frac{1}{N^2}(1-v) \qquad (3.29)$$

$$P(n+1 \to n) = (1-v)x(1-x) + vx(1-\bar{x})$$
$$+ \frac{1}{N}\left[v(1-\bar{x}) + (1-v)(1-2x)\right] - \frac{1}{N^2}(1-v). \qquad (3.30)$$

Substituting into (3.27), and keeping only terms up to order $1/N^2$, we find

$$\tilde{P}(x, t+1) - \tilde{P}(x, t) = \frac{v}{N}\left[(x - \bar{x})\tilde{P}'(x, t) + \tilde{P}(x, t)\right]$$
$$+ \frac{1}{N^2}\left[x(1-x)\tilde{P}''(x, t) + 2(1-2x)\tilde{P}'(x, t) - 2\tilde{P}(x, t)\right]$$
$$+ \frac{v}{2N^2}\left[(\bar{x}(1-x) + x(1-\bar{x}) - 2x(1-x))\tilde{P}''(x, t)\right.$$
$$\left. + \left(2x - \bar{x} - \frac{1}{2}\right)\tilde{P}'(x, t) + \tilde{P}(x, t)\right] \qquad (3.31)$$

where a prime denotes differentiation with respect to the frequency $x$. This expression can be written more compactly as

$$\tilde{P}(x, t+1) - \tilde{P}(x, t) = \frac{\nu}{N}\left[(x-\bar{x})\tilde{P}(x,t)\right]' + \frac{1}{N^2}\left[x(1-x)\tilde{P}(x,t)\right]''$$
$$+ \frac{\nu}{2N^2}\left[\{\bar{x}(1-x) + x(1-\bar{x}) - 2x(1-x)\}\,\tilde{P}(x,t)\right]''$$
$$+ O(1/N^3)\,. \tag{3.32}$$

This equation tells us something very important about evolutionary dynamical processes that have both a systematic component (here, the migration/mutation process) and demographic fluctuations. The leading $\nu$-dependent term (which characterises the systematic component) is of order $1/N$, while the leading $\nu$-independent term (which characterises the random fluctuations that are present even when $\nu = 0$) is of order $1/N^2$. If we rescale time linearly with population size, i.e., put $\tau = t/N$, in the limit $N \to \infty$ we obtain the purely deterministic equation

$$\frac{\partial}{\partial \tau}\tilde{P}(x, \tau) = \nu\frac{\partial}{\partial x}(x - \bar{x})\tilde{P}(x, \tau)\,. \tag{3.33}$$

If initially a fraction $x_0$ of the replicators have a given type, their number after time $t$ is given by

$$x(t) = \bar{x} + (x_0 - \bar{x})e^{-\nu t}\,, \tag{3.34}$$

where $\bar{x}$ is the corresponding frequency in the metacommunity. That is, the population abundance in the local community simply approaches that of the metacommunity. Demographic fluctuations are completely irrelevant in this regime.

There is, however, a limit in which there is a balance between the immigration process (which tends to pull the relative abundance of different types towards that of the metacommunity) and the stochastic reproduction process (which, as we saw in the previous section, tends to eliminate diversity). This occurs if $\nu$ itself is of order $1/N$. Then, under a different rescaling of time, $\tau = t/N^2$, we find that as $N \to \infty$ Eq. (3.32) approaches

$$\frac{\partial}{\partial \tau}\tilde{P}(x, \tau) = \theta\frac{\partial}{\partial x}(x - \bar{x})\tilde{P}(x, \tau) + \frac{\partial^2}{\partial x^2}x(1-x)\tilde{P}(x, t)\,, \tag{3.35}$$

where the parameter $\theta$ is defined as

$$\theta = \lim_{N\to\infty} N\nu\,. \tag{3.36}$$

This statement is a purely mathematical one. In practice, one interprets it as meaning that if the immigration rate $\nu$ multiplied by the population size $N$ is of order 1, then both deterministic and stochastic effects are important. If it is much less than one, the evolution will be dominated by drift (demographic fluctuations) and if it is much larger than one, the evolution will be near-deterministic in character. This kind of thinking applies to other processes. Recall that at the end of Sect. 3.2 we discussed the rate of growth of a rare altruistic gene. If this rate $\delta f$ is of order $1/N$, then stochastic effects may bring the gene to extinction, even though *on average* it would tend to increase.

The full time-dependent solution for the probability distribution $\tilde{P}(x, \tau)$ governed by (3.35) is known (see, e.g., [11]). Here we shall mention only the steady-state solution: one can verify that the function

$$\tilde{P}(x) \propto x^{\theta \bar{x}-1}(1 - x)^{\theta(1-\bar{x})-1} \tag{3.37}$$

has zero time derivative if inserted in (3.35), a solution that has been known since the earliest studies of genetic drift [32]. For any value of $\theta > 0$, the mean of this distribution is $\bar{x}$. However, the distribution of $x$ can take different shapes as the parameter $\theta$ is varied. If $\theta$ is small, then the distribution has peaks near $x = 0$ and $x = 1$ – that is, at any given time, one of the replicator types is likely to be very abundant and the other rather rare, as shown in the leftmost plot of Fig. 3.2. In order to realise such a distribution, one must see over time transitions between each of the two types being in the majority. As $\theta$ is increased, one sees distributions that vanish at $x = 0$ and $x = 1$ and become strongly peaked around $\bar{x}$, as shown by Fig. 3.2. These distributions are consistent with our characterisation of small $\theta$ defining a fluctuation-dominated regime, and large $\theta$ an immigration-dominated regime.

### 3.4.2 Nonrecurrent Immigration

We now turn to the case where every immigration event introduces a new replicator type to the population. This is rather tricky to handle within the master-equation-

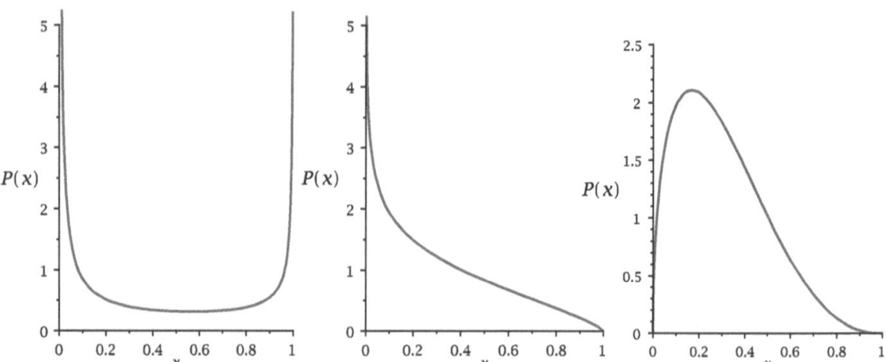

**Fig. 3.2** Stationary probability distribution of the $A$ replicator frequency $x$ with recurrent migration of both replicator types. The mean frequency of $A$ in the external metacommunity is $\bar{x} = 0.3$. The *three plots* show different immigration strengths $\theta$: from *left* to *right*, $\theta = 0.5$, 2.5 and 5.0

based approach that we have discussed so far, since one has birth and death of type labels, as well as replicators of a given type. A more tractable approach is obtained through an equivalent formalism where we consider the *ancestry* of a population going backward in time, rather than the descendants from some initial condition. Although notions of ancestry have long been used by geneticists to facilitate calculations, it was only relatively recently [42] that the backwards-time dynamics was formalised mathematically (see [43] for further discussion). A good overview of this approach can be found in various textbooks, see e.g., [12, 14, 44]. One of its most important features is its direct relevance to empirical applications: in genetics or ecology, one typically has observations about the diversity of a present-day population, but no knowledge of the initial condition. It is this empirical measure of diversity that can be tested against predictions from neutral theories.

### 3.4.2.1 Ancestral Formulation of Neutral Evolution

To understand this formalism, let us first consider the case where we have a sample of two replicators, and no immigration processes. We can construct *lineages* by considering the probability that, going back one time step, both replicators are distinct, or are copies of the same parent replicator. In the latter case, the lineages *coalesce*, reflecting the shared parentage of the two intervals. One may therefore represent the ancestry of the population in terms of a *genealogy*, a tree that shows the coalescence of lineages as common ancestors are found. An example is shown in Fig. 3.3 for a realisation of the dynamics within the Moran model.

At any given time step in the Moran model, coalescence of the two replicators in our sample occurs if it happens to include both the parent replicator that was selected for replication, and its offspring that was created. Since both were chosen from the population with replacement, the probability that our sample contains the parent individual is $2/N$; then, the probability that the other replicator is the offspring is $1/N$. Hence, this pair coalesces with probability $2/N^2$ per generation.

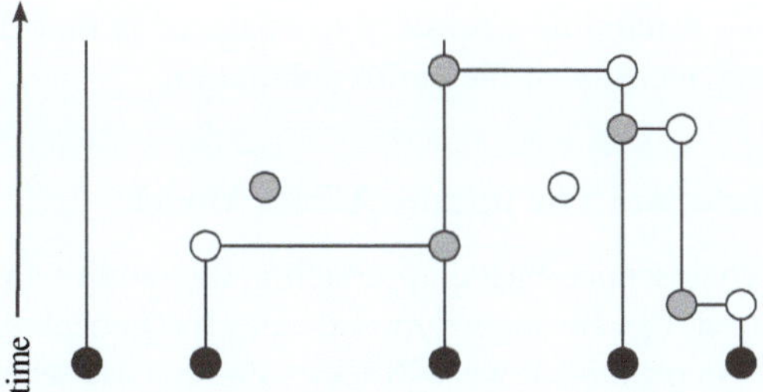

**Fig. 3.3** Ancestry of a sample of present-day replicators (shown *dark shaded*). As we go back one step in time, a parent and offspring were chosen from the population, shown *shaded* and *open* respectively. If the offspring individual happens to coincide with any of the ancestors of the present-day sample, its lineage coalesces or moves over to the parent, depending on whether the parent was also an ancestor of the sample or not

In a sample containing $n$ replicators, there are $\binom{n}{2}$ pairs that may coalesce in one time step. Therefore, the transition probability for a sample of $n$ distinct replicators to contact to a sample of $n-1$ distinct replicators when going back one time step is

$$Q(n \to n-1) = \binom{n}{2}\frac{2}{N^2} = \frac{n(n-1)}{N^2}. \tag{3.38}$$

As we look backwards in time, the distribution of the length of time spent in an a state with $n$ ancestors can be worked out fairly straightforwardly. Once the state has been entered, it is exited with probability $q_n = Q(n \to n-1)$ per time step. Otherwise, we remain in this state. Hence, we spend exactly $t$ time steps in state $n$ with probability

$$R(t) = (1-q_n)^{t-1}q_n \tag{3.39}$$

where $t \geq 1$ (because we always spend at least one time step in any given state). The mean time spent in state $n$ is then

$$\bar{t}_n = \sum_{t=1}^{\infty} R(t)t = q_n \sum_{t=0}^{\infty} t(1-q_n)^{t-1} = -q_n \frac{\mathrm{d}}{\mathrm{d}q_n} \sum_{t=0}^{\infty}(1-q_n)^t = -q_n \frac{\mathrm{d}}{\mathrm{d}q_n}\frac{1}{q_n} = \frac{1}{q_n}.$$
$$\tag{3.40}$$

If we start with a sample of $m$ replicators, then going back in time, we go through the states $n = m, m-1, m-2, \cdots, 2, 1$, i.e., until a single common ancestor of the entire population is found. The mean time to reach this common ancestor is obtained by summing up the mean time spent in each of the states $n = 2, 3, \ldots, m$. This is

$$\bar{T}_m = \sum_{n=2}^{m} \frac{N^2}{n(n-1)} = N^2 \sum_{n=2}^{m}\left[\frac{1}{n-1} - \frac{1}{n}\right] = N^2\left[1 - \frac{1}{m}\right]. \tag{3.41}$$

We see once again the characteristic scaling of all time scales with the square of the population size within the Moran model. It is also interesting to note that the mean time for the last two remaining ancestors to coalesce is half the mean age of the most recent common ancestor of the entire population.

### 3.4.2.2 Adding Mutation: The Infinite Alleles Model

The beauty of the coalescence-based approach is that additional processes, such as migration or mutation, can be superimposed onto the genealogical trees, and their statistics analysed. As promised, we will focus on the case where immigration into the local community (going forwards in time) brings with it a completely new replicator type (species). Recall that this happened with probability $\nu$, and displaced a randomly chosen replicator in the population.

We noted earlier that immigration and mutation processes are equivalent, in that both lead to the creation of new types. In the following, it will be most useful to think in terms of mutations (rather than migrations), since then one can think of the new replicator as being the offspring of an existing one in the population, even though its type is different. This model of mutation is called the *infinite alleles* model in the population genetics literature [12], because the possible number of alleles (replicator types) is infinite.

Within this model, we now have, going backwards in time, that a given pair of lineages coalesces with probability $2(1 - v)/N^2$ (as before, but incorporating the factor $1 - v$ to take into account the additional mutation process). Meanwhile, any one lineage acquires a mutation with probability $v/N$ per time step. One can think of the mutations "decorating" the genealogies, as shown in Fig. 3.4.

We are now interested in the number of ancestors at a given time in history whose types are represented in the present-day sample. Let us call these *contributing* ancestors. The requirement for this is that no mutations occur on any of the branches connecting a contributing ancestor to the present-day population. Looking backward in time, we can identify the contributing ancestors by terminating the genealogies whenever a mutation event occurs, i.e., by deleting the dotted lines in Fig. 3.4. From the figure, one can see that each event – coalescence or mutation – has the effect of reducing the number of contributing ancestors by one. When $n$ contributing ancestors are present, the rates of the coalescence and mutation processes are

$$Q_{\text{coal}}(n \to n - 1) = (1 - v)\frac{n(n - 1)}{N^2} \tag{3.42}$$

$$Q_{\text{mut}}(n \to n - 1) = v\frac{n}{N} \; . \tag{3.43}$$

We notice that mutation, by occurring at a rate of order $1/N$, is a much faster process than coalescence – this corresponds with our observations about the characteristic time scales of systematic and random processes in evolutionary dynamics. As previously, these processes occur on the same time scale if $Nv = \theta$ is of order 1. Making this substitution, we find that the total probability of an event that reduces the number of contributing ancestors by one is

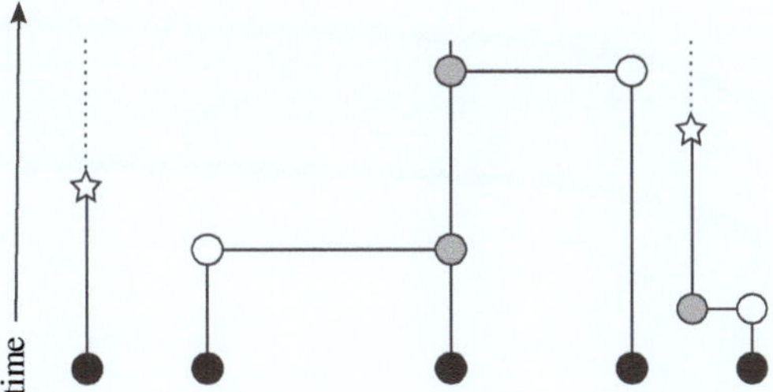

Fig. 3.4 An alternative realisation of the Moran process when mutation is allowed. Instead of a parent–offspring pair being chosen at a given time step, there is a probability $v$ that an individual mutated. If this happens to coincide with one of the lineages in the sample, it acquires a mutation, denoted by a *star* in the figure. The *dotted line* indicates that the type of the replicator attached to the lineage does not contribute to the present-day sample

$$Q(n \to n - 1) = Q_{\text{coal}}(n \to n - 1) + Q_{\text{mut}}(n \to n - 1) = \frac{n(n - 1 + \theta)}{N^2} \quad (3.44)$$

if we ignore terms of order $1/N^3$, as previously. We may now ask for the mean age of the oldest ancestor contributing to the present-day sample. This is obtained, as before, by summing $1/Q(n \to n - 1)$ over $n = 2$ to $m$, the size of the present-day sample:

$$\bar{T}_m = N^2 \sum_{n=2}^{m} \frac{1}{n(n - 1 + \theta)} \,. \quad (3.45)$$

Unfortunately, this expression does not have a convenient closed form. However, it can be readily computed and plotted. The behaviour of $\bar{T}_m/N^2$ as a function of $m$ for three different values of $\theta$ is shown in Fig. 3.5.

Other useful expressions can be found once we know the probability that, at any given time in the past, the next event (looking backward in time) is a mutation. Since both processes, coalescence and mutation, take place with a constant probability per unit time when $n$ is constant, the probability $M(n)$ that $n \to n - 1$ via a mutation, rather than a coalescence, is given by

$$M(n) = \frac{Q_{\text{mut}}(n \to n - 1)}{Q_{\text{coal}}(n \to n - 1) + Q_{\text{mut}}(n \to n - 1)} = \frac{\theta}{n - 1 + \theta} \quad (3.46)$$

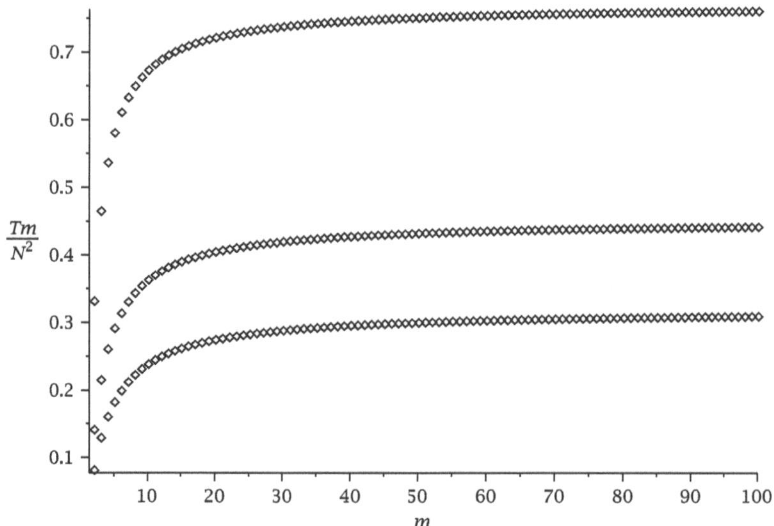

**Fig. 3.5** The age of the oldest ancestor, $\bar{T}_m/N^2$, as a function of the sample size $m$ for, from *top* to *bottom*, $\theta = 0.5$, 2.5 and 5.0

in the limit $N \to \infty$ where (3.36) holds. Now, given a sample of size $m$, the mean number of replicator types $k$ we may expect to see is equal to the mean number of mutation events that occurs in the history as $n$ descends from $m$ to 0. That is,

$$\bar{k} = \sum_{n=1}^{m} [M(n) \times 1 + (1 - M(n)) \times 0] = \sum_{n=1}^{m} \frac{\theta}{\theta + n - 1} . \tag{3.47}$$

Again, this expression has no convenient closed form, but again, it can be readily computed and plotted – see Fig. 3.6. One can also see that for large $m$ the summand behaves as $1/m$, and hence $S$ grows roughly logarithmically with the sample size. That is, asymptotically, one needs to roughly treble the sample size to see a new species.

It is possible to go even further and obtain the probability that, given a sample of size $m$, it contains $m_1$ replicators of one type, $m_2$ of a second, $m_3$ of a third, and so on. There are three parts to this computation. The first is to construct a standard ordering of the replicators that make up the sample. The second is to figure out how many of all possible replicator orderings are equivalent to the standard ordering, given that we know only that there are $m_i$ replicators of type $i$. The third part is, for a given standard ordering, to work out the probability that the ancestral coalescence and mutation events give rise to a sample with the desired abundances of replicator types.

We construct the standard ordering of replicators in the following way. First, we group them together by type, and then order the groups by the number of replicators

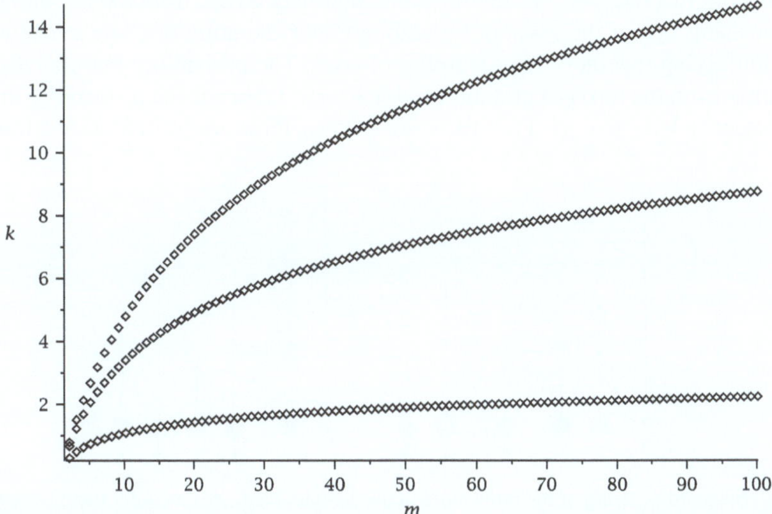

**Fig. 3.6** The mean number of replicator types, $\bar{k}$, as a function of the sample size $m$ for, from *bottom* to *top*, $\theta = 0.5, 2.5$ and $5.0$

they contain (largest first). Within each group, we order the replicators by their age (oldest first). To do this, we need to define an age ordering within a pair of lineages at a point of coalescence. This we achieve by saying that the offspring lineage ends at the time of coalescence, and the parent lineage continues. Within each group, there is one lineage that does not coalesce with any of the others in that group: this ancestor was created by a mutation (immigration) event. We define the age of a replicator through the length of the lineage that it is attached to. Figure 3.7 shows a standard ordering for one realisation of the Moran model dynamics.

How many distinct orderings are there when we consider all possible realisations? The replicators within the groups are indistinguishable (their age is not known to us), so there are

$$\frac{m!}{m_1!m_2!\cdots m_k!} \tag{3.48}$$

distinct age orderings of the replicators, where $k$ is the total number of types. Now, groups of different sizes can be distinguished; but groups of the same size cannot. Therefore, if there are $b_i$ groups of size $i$, this combinatorial factor must be further reduced by a factor $b_i!$ for each group size $i$. If there are $r$ different group sizes, we thus have that the total number of replicator orderings that can be distinguished by their group sizes is

$$\frac{m!}{m_1!m_2!\cdots m_k!} \frac{1}{b_1!b_2!\cdots b_r!} . \tag{3.49}$$

Finally, we need to identify the probability of the standard ordering with specified $m_i$. To work this out, we consider the construction of a tree, from top to bottom (i.e., in decreasing age of the events), by adding either mutation or coalescence events, and multiplying together the probability of each. The probability that the $j$th event (counting from the top) is a mutation is $\theta/(\theta + j - 1)$; hence the probability that it is a coalescence is $1/(\theta + j - 1)$. If there are $k$ types, there are by definition $k$ mutation

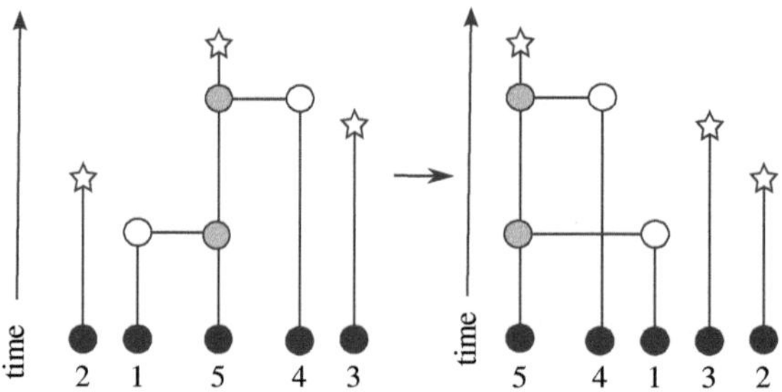

**Fig. 3.7** Standard ordering of the replicators in the sample. First, we can order them by the age at which they coalesced with a parent or mutated (whichever happened most recently). The *numerals* show the age order of the replicators, from youngest (*1*) to oldest (*5*). Then, we can order groups of replicators with a common ancestor by their size, largest first

events, and no matter how these are interleaved with the $m - k$ coalescence events, the probability of a particular ancestry within a standard ordering is

$$\frac{\theta^k}{\theta(\theta + 1) \cdots (\theta + m - 1)} . \tag{3.50}$$

Within a standard ordering, though, multiple ancestries are possible. When the $n$th replicator is added to a group (going from top to bottom down the tree), there are $n - 1$ parents to choose from to attach the offspring to. Each choice generates a topologically distinct tree. Hence the total probability of obtaining $k$ groups with $m_1, m_2, \ldots m_k$ replicators within each is

$$(m_1 - 1)!(m_2 - 2)! \cdots (m_k - 1)! \frac{\theta^k}{\theta(\theta + 1) \cdots (\theta + m - 1)} . \tag{3.51}$$

Multiplying this now by the number of distinct replicator orderings (3.49) we finally obtain the probability of seeing a sample with a particular distribution of group sizes within it as

$$S(m_1, m_2, \ldots, m_k | m) = \frac{m!}{m_1 m_2 \cdots m_k} \frac{1}{b_1! b_2! \cdots b_r!} \frac{\theta^k}{\theta(\theta + 1) \cdots (\theta + m - 1)} , \tag{3.52}$$

where we recall that $b_i$ is the number of groups of size $i$. It is possible to write this formula purely in terms of the $b_i$ and $m$ by noting that $k = \sum_i b_i$, and that each group size $i$ appears $b_i$ times in the denominator of the first term in the above product. That is, one can write

$$S(b_1, \ldots, b_r | m) = \frac{m!}{\theta(\theta + 1) \cdots (\theta + m - 1)} \prod_{i=1}^{r} \frac{1}{b_i!} \left(\frac{\theta}{i}\right)^{b_i} . \tag{3.53}$$

This expression is known as the Ewens sampling formula, as it was first written down by Ewens [45]. Shortly afterwards, it was proved by Karlin and McGregor [46].

An important feature of the Ewens sampling formula (3.53) is that the $\theta$-dependent part itself depends on the sample size $m$ and the number of types $k$, but not the abundances of each type. This in turn implies that if one is attempting to infer $\theta$ from a sample, no additional information about its value is provided by these abundances than is available from $m$ and $k$ alone. This statement can be made precise through information theory [47], but can be understood intuitively as follows. Suppose that the distribution of abundances conditioned on $m$ and $k$ did depend on $\theta$. For example, a small value of $\theta$ could imply that group sizes tend to be small or large, but not intermediate; whereas a large value of $\theta$ could imply the opposite for the same $m$ and $k$. Then, if one observed small and large groups, the inferred value of $\theta$ would be biased towards small values. If there is no dependence of the group sizes on $\theta$ for given $m$ and $k$, as here, then no such biasing takes place. It turns out

[12] that the maximum likelihood estimate of $\theta$ is given by solving (3.47) for $\theta$ with the observed number of types $k$ appearing on the left-hand side.

That fact that the distribution of group sizes is $\theta$-independent can also be used to design tests for neutral evolution in the presence of immigration of new types at a constant rate $\theta$ without needing to know the value of $\theta$. This is achieved by defining a *test statistic* that is some combination of the $b_i$, and computing its distribution, given $m$ and $k$, from (3.53). If the probability that the test statistic exceeds its observed value is less than some threshold (e.g., 1% or 5%), one has evidence against the neutral model. The choice of test statistic has some influence on the power of this technique, a discussion we defer to [12].

### 3.4.3 Applications in Ecology and Cultural Evolution

Variants of the simple model described above have been applied in the context of community ecology and cultural evolution. Hubbell [7] has discussed an ecological model in which a local community is in contact with a metacommunity that is of large (but finite) size. The local community receives immigrants from the metacommunity, as above, but speciation occurs within the metacommunity alone. This implies a species turnover, as before, but also that multiple individuals of the same species may migrate into the local community (as opposed to each arrival being of a new species).

Using this model, one can compute such properties as the species area relation (expected number of replicator types as a function of the community size), or the species abundance distribution (the number of individuals of each species, when ranked from most to least abundant). Initially these were obtained with computer simulation, and – by varying the speciation and immigration rates – one can obtain reasonable agreement with empirical data (see, e.g., [48, 49] for examples). Subsequently, these functions have been calculated analytically [41, 50], which obviates the need for repeating simulations with different parameter combinations.

More recently, the analogue of the sampling formula (3.53) has been obtained for this model by Etienne [51], and was used to show that this model provides a better fit to empirical data than the version considered above. The availability of this sampling formula also opens the door to obtaining maximum-likelihood estimates of the model parameters, which is a possibly stronger test of the neutral theory than fitting distributions. Certainly, one can compare the distributions obtained with the maximum-likelihood values of the parameters, and check that these two methods of analysing the data are consistent. Moreover, one can compare (as was done in [51]) the inferred value of a parameter, for example the immigration rate, with empirical measurements of related quantities. These successful applications of neutral models to ecological data remain however controversial [39, 40]. In part this is because some of the distributions obtained from the neutral theory differ from other candidate distributions to a sufficiently small degree that they are hard to distinguish empirically [50]. The development of more refined statistical tests may be one way to resolve this controversy.

Meanwhile, the simple model described in the previous section (albeit in a Wright–Fisher discrete generation formulation) has been applied to the cultural evolutionary process of baby naming. The idea here is that replicators are names for babies, and that a new baby is given a name that is sampled from an existing pool of $N$ named individuals; or a new name is invented with a probability $v$. Again, this model has been shown to fit certain empirical properties and distributions of baby names in the US rather well [52]. It has also very recently been shown that these dynamics are what one would expect from a model in which agents attempt to employ a Bayesian inference algorithm to estimate the frequency of a replicator within some population through contact with a limited sample [53]. A parameter in this inference model that corresponds to the degree of variability agents expect to see in this distribution turns out to fix the value of $\theta$. This looks to be a promising approach to better understanding the role of fluctuations in culturally evolving systems.

## 3.5 Population Subdivision

We now return to a point that came up in our discussion of pure drift in Sect. 3.3, namely that the scale of the fluctuations may be given by the reciprocal of a population size that differs from the actual number of replicators in the system. One way that this can occur is if there is population subdivision – that is, if offspring of a given replicator can displace individuals only in a restricted part of the population, for example, within a certain geographical distance of the parent. This effect has been discussed extensively in the literature, see e.g., [18, 37, 54]. We present only a simple demonstration here.

Let us consider a population that is subdivided into $L$ subpopulations. For brevity, we will call these *islands* – one also sees the word *demes* in the population genetics literature. Island $i$ hosts a population of a size $N_i$ which remains fixed over time but may vary from island to island. We now define the dynamics as follows: in each time step, we select the island $i$ that will receive the offspring replicator with probability $i$. A parent is randomly chosen from an island $j$ with probability $\mu_{ij}$. Then the parent is replicated, and displaces a randomly chosen individual in the target island $i$. Since a parent must come from somewhere, the probabilities $\mu_{ij}$ satisfy the sum rule $\mu_{ii} = 1 - \sum_{j \neq i} \mu_{ij}$.

One way to estimate the effective population size is to work out the corresponding dynamics of the backward-time lineages. Let us first ask for the probability that two lineages on the same island $i$ coalesce in a single time step. Both parent and offspring must have been chosen from island $i$; this event occurs with probability $f_i \mu_{ii}$. Then, as before, a randomly chosen pair of lineages will comprise parent and offspring with probability $2/N_i^2$. Therefore the probability of coalescence of a pair of lineages sampled from island $i$ is

$$c_i = \frac{2 f_i \mu_{ii}}{N_i^2} \,. \tag{3.54}$$

Now we ask for the probability $m_{ij}$ that a lineage sampled from island $i$ is a newly arrived immigrant from island $j$. The probability that in a replication event the parent is in $j$ and the offspring in $i$ is $f_i \mu_{ij}$; then the probability that the offspring is the individual in our sample is $1/N_i$. Hence

$$m_{ij} = \frac{f_i \mu_{ij}}{N_i} \, . \tag{3.55}$$

Again we see that these expressions are of the same order in $1/N_i$ if the between-island migration probabilities are $\mu_{ij} \sum 1/N_i$ for $i \neq j$. It is thus conventional to define rescaled migration probabilities

$$\theta_{ij} = \lim_{N_i \to \infty} N_i \mu_{ij} \tag{3.56}$$

analogous to (3.36). Then, to leading order in $1/N_i$, we have

$$c_i = \frac{2 f_i}{N_i^2} \tag{3.57}$$

$$m_{ij} = \frac{f_i \theta_{ij}}{N_i^2} \, . \tag{3.58}$$

There is a small probability that a lineage on island $i$ coalesces with a lineage on some other island $j$ in one time step. However, within this scaling, this probability vanishes faster than any $c_i$ or $m_{ij}$ as $N_i \to \infty$. One can therefore picture this backward-time coalescence dynamics in terms of lineages that "hop" from island $i$ to island $j$ with probability $m_{ij}$ per unit time, and coalescence between pairs of lineages on the same island happening with probability $c_i$ per unit time.

Eventually, as in the regular coalescence process, only one lineage – the common ancestor of the population – remains. Furthermore, the probability $Q_i$ that the lineage is found on island $i$ will become time independent. This steady-state distribution is given formally by the solution of the set of linear equations

$$\sum_{j \neq i} \left[ Q_i m_{ij} - Q_j m_{ji} \right] = 0 \, , \tag{3.59}$$

subject to the constraint $\sum_i Q_i = 1$. These island weights $Q_i$ turn out to be a central quantities in the analysis of neutral evolution in subdivided populations.

For example, if we go far forward in time, all replicators will eventually have the same type, and furthermore share a single common ancestor. Looking again backward in time, we find that this common ancestor has a probability $Q_i$ of having been on island $i$ at time $t = 0$. If, initially, a fraction $x_i$ of replicators on island $i$ are of a given type ($A$), the probability $\phi$ that $A$ becomes fixed in the population is

$$\phi = \sum_{i=1}^{L} Q_i x_i \, . \tag{3.60}$$

What is interesting here is that the network of migration pathways can lead to $Q_i$ being very much larger for some islands than others: a specific example will be given below. Thus, if the initial location of a replicator type is correlated with these high-weight islands, this type has a much larger chance of taking over the population than others, and one may potentially view this as a kind of selection. We will return to this point below.

The island weights $Q_i$ also enter into an estimate for an effective population size that is valid when migration is a very fast process relative to coalescence (that is, when the $\theta_{ij}$ parameters are much larger than one). We anticipate then that if $n$ lineages remain, the location of each one will be independent and be distributed according to $Q_i$. Thus, the probability that there are $n_i$ lineages on island $i$ will be given by the multinomial distribution

$$P(n_1, n_2, \ldots, n_L) = \frac{n!}{n_1! n_2! \cdots n_L!} Q_1^{n_1} Q_2^{n_2} \cdots Q_L^{n_L} . \tag{3.61}$$

We may then determine the mean coalescence rate within this state:

$$\bar{c} = \sum_{\{n_1, \ldots, n_L\}} P(n_1, n_2, \ldots, n_L) \sum_i \binom{n_i}{2} c_i \tag{3.62}$$

$$= \sum_i \sum_{n_i} \binom{n}{n_i} Q_i^{n_i} (1 - Q_i)^{n - n_i} \binom{n_i}{2} c_i \tag{3.63}$$

$$= \binom{n}{2} \sum_i Q_i^2 c_i . \tag{3.64}$$

In an undivided population of size $N_e$, the coalescence rate per pair is $2/N_e^2$. By setting this equal to the mean coalescence rate in this subdivided population, we arrive at an expression for its effective size $N_e$:

$$\frac{1}{N_e^2} = \frac{\bar{c}}{n(n-1)} = \sum_i \frac{Q_i^2 f_i}{N_i^2} . \tag{3.65}$$

A similar result was obtained using more rigorous methods for the Wright–Fisher model in the strong migration limit in [55].

It is worthwhile to identify the smallest and largest effective population sizes that are possible in the limit of strong migration. Unsurprisingly, the smallest effective population size is obtained if only the smallest island participates in the population dynamics. Then $N_e = N_j$, where $j$ labels the smallest island. More interesting is the opposite limit, in which the largest effective population size is obtained. This can be found by asking for an extremum of (3.65) subject to the constraint that $\sum_i Q_i = \sum_i f_i = 1$ (e.g., using the method of Lagrange multipliers [56]). One finds that the largest effective population size arises if $Q_i = f_i = N_i/N$, where $N$ is the total population size, $N = \sum_i N_i$. One way that this can be achieved is if parent and offspring islands are each chosen with a probability proportional to the size of the island (although the offspring need not be drawn from the entire population). Then, the effective population size equals that of the total population $N$, which is the case for any model where $Q_i = f_i = N_i/N$, not just that described here. Therefore we have that

$$\min_i \{N_i\} \leq N_e \leq \sum_i N_i \, , \qquad (3.66)$$

i.e., that when migration is very rapid, the effective population size lies somewhere between that of the smallest island and the total population.

When the time scales of migration and coalescence are more similar, there is no known general expression for the effective population size, although various approximate formulæ have been proposed [37, 54, 57]. One difficulty that arises is that the coalescence rate may not be constant over time when migration is slow, and hence the usefulness of a single parameter in characterising the entire history of a sample may be limited [58].

### 3.5.1 Voter-Type Models on Heterogeneous Networks

Recently, there have been a number of studies in the physics literature of a specific type of migration dynamics represented by the *voter model* defined on *heterogeneous* network structures, that is, where the parameters $Q_i$, $f_i$ or $N_i$ differ wildly between islands. Many of these studies have recently been discussed in a comprehensive review of statistical physical models of social dynamics [59, Sect. B3]. They demonstrate in particular that the migration structure can lead to an effective population size that differs considerably from its actual size. In the voter model proper, each island contains one individual. The dynamics are that an island $i$ is chosen at random, and then an individual occupying a randomly chosen neighbour of island $i$ on the network is replicated, with the offspring displacing the existing resident of island $i$.

We can use the above analysis to examine a version of this model in which the number of individuals per island is large, rather than equal to one. For simplicity, we will take all $L$ islands to be of equal size, $N_i = N/L$. As in the voter model, we will choose the target island for each replication event uniformly, $f_i = 1/L$. If island $i$ can receive immigrants from $k_i$ neighbouring islands, we have that the immigration probability $\mu_{ij}$ is proportional to $1/k_i$.

A special feature of this model is that the terms in (3.59) cancel term-by-term (technically, the migration process satisfies detailed balance, as can be shown by applying a Kolmogorov criterion [60]). That is,

$$Q_i \mu_{ij} = Q_j \mu_{ji} \qquad (3.67)$$

after cancelling common factors. Using the fact that $\mu_{ij} \propto 1/k_i$, we find that the normalised stationary distribution of the common ancestor for this model is

$$Q_i = \frac{k_i}{L\bar{k}} \, , \tag{3.68}$$

where $\bar{k}$ is the mean number of neighbours per island. Substituting into (3.65) we find that for this model, the effective population size is given by

$$N_e^2 = \frac{N^2 \bar{k}^2}{\overline{k^2}} \, . \tag{3.69}$$

Having determined the effective population size, we can now use it in expressions obtained for the Moran model without population subdivision to estimate properties of interest. For example, from (3.24) we have that the mean time until a variant goes extinct or takes over the whole population scales as the square of the (effective) population size. That is, we have

$$\frac{\overline{T}_{\mathrm{sub}}}{\overline{T}} \sim \frac{\bar{k}^2}{\overline{k^2}} \tag{3.70}$$

where $\overline{T}_{\mathrm{sub}}$ and $\overline{T}$ are the characteristic time scales of the subdivided and undivided populations respectively.

The implications of this expression can be understood if we write it in a slightly different form

$$\frac{\overline{T}_{\mathrm{sub}}}{\overline{T}} \sim \frac{1}{1 + \mathrm{Var}(k/\bar{k})} \, . \tag{3.71}$$

If all islands have similar connectivity, the variance in $k$ relative to its mean is small, and hence the characteristic time scales due to demographic fluctuations in the subdivided population are basically the same as in the undivided population. On the other hand, if some islands receive immigrants from many sources, but others are poorly connected, the variance in $k$ may be large, and the time scale on which one variant takes over the whole population is dramatically reduced.

A simple example that illustrates the effects of heterogeneity is the "star" network, which has a single central island that is connected to all the others whereas these peripheral islands are connected only to the central island, as shown in Fig. 3.8. Thus $k = 1$ for $L - 1$ peripheral islands, and $k = L - 1$ for the single central island, and one has

$$\bar{k} = 2 - \frac{1}{L} \quad \text{and} \quad \overline{k^2} = L - 1 \tag{3.72}$$

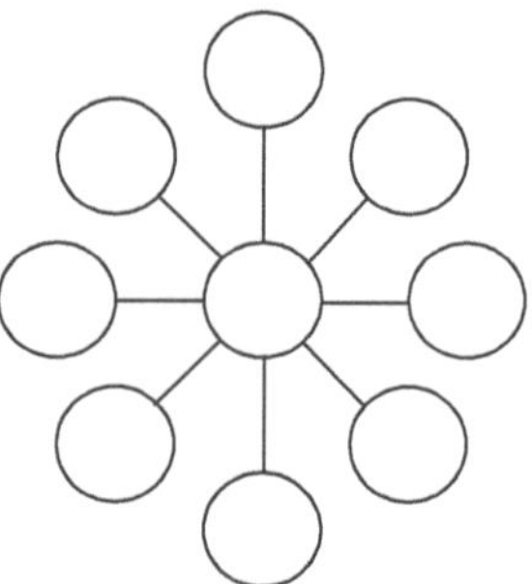

**Fig. 3.8** The "star" network structure, where peripheral islands are connected only to a single central islands. Migration can take place in both directions along any given pathway

and hence

$$\frac{\overline{T}_{\text{sub}}}{\overline{\overline{T}}} \sim \frac{4}{L} \tag{3.73}$$

as $L \to \infty$. That is, the characteristic time scale increases asymptotically linearly, rather than quadratically, with the number of islands in this case.

Notice that because $Q_i \propto k_i$ in this model, a replicator type that is created (e.g., by mutation) on the central island is $L - 1$ times as likely to go to fixation as a replicator type created elsewhere. This is an extreme example of the phenomenon noted above, that different replicators have different chances of survival depending on where they are created, despite the fact that the population dynamics of each replicator type is identical. This is a systematic effect: the first term in the Price equation (3.7) would be nonzero for these replicator frequencies, suggesting that this effect could be regarded as a form of selection. However, this selective advantage is not transmitted by replication of the replicators: if a replicator on the central island leaves an offspring on one of the other islands, the expected mean number of offspring *decreases*. Therefore it does not satisfy Hull's definition of selection (see Sect. 3.1).

Some clarity is attained if we return to the distinction between Fisher's fundamental theorem of natural selection (3.8) and the full Price equation (3.7) with the trait $z_i = w_i$, the expected number of offspring. Looking at Fisher's fundamental theorem, we would conclude from the fact that there is a variance in the offspring numbers, the mean fitness of the population must increase. However, we know that it does not, since the distribution of offspring numbers remains constant from one generation to the next. This apparent increase is exactly counterbalanced by the second term in the Price equation (3.7), which takes into account the change in expected offspring number that occurs as offspring are born onto a different island from their parents. Systematic changes in replicator frequency due to this mechanism have a different character to conventional natural selection of replicators, a fact that was recognised in an application of this type of model to the cultural evolutionary process of language change [61].

### 3.5.2 Application to Theories for Language Change

In [9], Baxter et al. introduced a formal mathematical model for language change that was based on Croft's evolutionary theory [19]. It was set up as an agent-based model in which tokens of linguistic utterances are replicated, and exposure of speakers to these replicators affects their experience of language and knowledge of the way in which it is used (a term referred to as grammar in [9, 19]. As discussed in Sect. 3.1, in this picture, speakers and their grammars are identified as the interactors. An interesting point here is that the mathematical model, which was defined purely in terms of linguistic interactions between agents, turns out to be equivalent to neutral evolution in a subdivided population, in which the interactors turn out to be equivalent to the islands central to the previous discussion.

In this model, the migration rates $\mu_{ij}$ derive from two factors, the frequency with which two speakers interact, and the importance that agent $i$ as a listener ascribes to agent $j$'s utterances. The first of these factors is necessarily symmetric: when agent $i$ meets agent $j$ then agent $j$ also meets agent $i$. However, the importance weights can be either symmetric or asymmetric. When they are symmetric, it is found that the effective population size for this model (to be interpreted as an effective population size of the replicators) is independent of the network structure connecting agents [57, 61]: this provides another example where the island (interactor) weights $Q_i$ are all equal. Asymmetric speaker importance weights admit the possibility of unequal interactor weights $Q_i$. The corresponding dominance that individual interactors with high weights $Q_i$ have on the linguistic behaviour of the wider community of the type described at the end of the last section has its origins in certain agents being preferentially listened to over others. It is perhaps legitimate to consider this act as a form of selection, acting at the level of interactors: Baxter et al. therefore call this form of selection *weighted interactor selection* to distinguish it from a selection process operating on replicators in the manner prescribed by Hull. They also introduced a term *neutral interactor selection* to handle the case where interaction frequencies vary between speakers, but importance weights are symmetric.

More usefully, this distinction between different types of selection processes directly relates to specific theories for language change, a connection discussed in [57, 61]. It was argued that theories implying purely neutral evolution or neutral interactor selection were not compatible with data for new-dialect formation [62, 63]. The essential problem is that the effective population size implied by these theories is so large that the resulting fixation time implied by (3.24) vastly exceeds what was actually observed. These conclusions do however rely on the appropriateness of the underlying model [9] and in particular the implicit assumptions that were made about psychology and human linguistic behaviour. Perhaps again, as with other applications of neutral theory to cultural evolution (such as [52]), there is a need to develop more sophisticated statistical methods, such as those based on the Ewens sampling formula (3.53) and its extensions to Hubbell's neutral theory of community ecology [51]. Again, the formal connections between specific models of induction and learning and neutral evolutionary dynamics [53] provide one possibility to illuminate the psychological basis of a neutral evolutionary dynamics as a null model for cultural change.

## 3.6 Summary and Outlook

In this chapter we have taken a broad-brush look at evolutionary dynamics in the contexts of genetics, community ecology and culture. As we pointed out in the introduction, change by replication has been advocated as a generic mechanism for explaining the dynamics of and diversity within populations of different kinds since the earliest days of the theory.

The central question that arises whenever evolution is operating is: what, if anything, is selection selecting for? Contributions to the understanding of this question have come from both philosophical considerations and the application of mathematical models, both as a conceptual tool to determine what evolution logically allows and as a means to analyse specific empirical data sets. As illustrative examples, we have discussed the utility of Hull's general analysis of selection and the notions of the interactor and replicator as a means to identify where the key components of evolution – replication, variation and selection – are operating in different contexts (Sect. 3.1).

In particular, we have shown how the Price equation (3.7) allows one to distinguish between the selective component of a change, and the component due to the generation of variation in the replication process. It also allows one to recognise the effect of selection on the interactor and how this may cause different replicator types to propagate or go extinct, as we saw through the specific example of kin selection. However, there still remains the possibility that any given change could have been due to fluctuations alone – survival of the luckiest, rather than survival of the fittest. A detailed understanding of these fluctuations, such as that expressed by the Ewens sampling formula (3.53), is pivotal in deciding whether this is the case. As such, neutral theories – which have been advocated separately for genetics [6], ecology [7] and cultural change [52] – serve as null models in quantitative analyses of evolution.

Being able to identify the target of selection is important in making predictions about evolving systems. For example, in the cultural context of language, we have recently argued that vital information about the nature of the cognitive processes that are involved in language could be revealed by analysing historical data about language change as a stochastic, evolutionary dynamics [10]. In turn, this may shed light on another important evolutionary question: namely why (apparently) only humans have yet evolved the spectacular communicative abilities afforded by language [64]. Thus the continued development of both conceptual and mathematical models, properly integrated with empirical data, has the potential to unravel scientific mysteries, among them what it means to be human.

# References


1. C.R. Darwin, *On the Origin of Species by Means of Natural Selection, or The Preservation of Favoured Races in the Struggle for Life*, 1st edn. (John Murray, London, 1859). http://darwin-online.org.uk/content/frameset?itemID=F373&viewtype=text&pageseq=1
2. C.R. Darwin, *The Descent of Man, and Selection in Relation to Sex*, 1st edn. (John Murray, London, 1871). http://darwin-online.org.uk/content/frameset?itemID=F937.1&viewtype=text&pageseq=1
3. E. Sober, D.S. Wilson, *Unto Others: The Evolution and Psychology of Unselfish Behavior* (Harvard University Press, Cambridge, MA, 1998)
4. D.L. Hull, *Science as a Process: An Evolutionary Account of the Social and Conceptual Development of Science* (University of Chicago Press, Chicago, IL, 1988)
5. G.R. Price, Nature **227**, 520 (1970)
6. M. Kimura, *The Neutral Theory of Molecular Evolution* (Cambridge University Press, Cambridge, 1985)
7. S.P. Hubbell, *The Unified Neutral Theory of Biodiversity and Biogeography* (Princeton University Press, Princeton, NJ, 2001)
8. D. Alonso, R.S. Etienne, A.J. McKane, Trends Ecol. Evol. **21**, 451 (2006)
9. G.J. Baxter, R.A. Blythe, W. Croft, A.J. McKane, Phys. Rev. E **73**, 46118 (2006)
10. D.J. Hruschka, M.H. Christiansen, R.A. Blythe, W. Croft, P. Heggarty, S.S. Mufwene, J.B. Pierrehumbert, S. Poplack, Trends Cogn. Sci. **13**, 464 (2009)
11. J.F. Crow, M. Kimura, *An Introduction to Population Genetics Theory* (Harper and Row, New York, NY, 1970)
12. W.J. Ewens, *Mathematical Population Genetics: Theoretical Introduction* (Springer, New York, NY, 2004)
13. N.H. Barton, D.E.G. Briggs, J.A. Eisen, D.B. Goldstein, N.H. Patel, *Evolution* (Cold Spring Harbor Laboratory Press, Cold Spring Harbor, NY, 2007)
14. J. Wakeley, *Coalescent Theory: An Introduction* (Roberts, Greenwood Village, CO, 2008)
15. L. Peliti. Introduction to the statistical theory of Darwinian evolution (1997). ArXiv:cond-mat/9712027v1 [cond-mat.dis-nn]
16. E. Baake, W. Gabriel, in *Annual Reviews of Computational Physics*, vol. 7, ed. by D. Stauffer (World Scientific, Singapore, 2000), pp. 203–264
17. B. Drossel, Adv. Phys. **50**, 209 (2001)
18. R.A. Blythe, A.J. McKane, J. Stat. Mech. P07018 (2007)
19. W. Croft, *Explaining Language Change: An Evolutionary Approach* (Longman, Harlow, 2000)
20. W.F. Doolittle, Science **284**, 2124 (1999)
21. R.J. Putnam, *Community Ecology* (Chapman and Hall, London, 1994)
22. P. Niyogi, R.C. Berwick, Linguist. Philos. **20**, 697 (1997)
23. M.A. Nowak, D.C. Krakauer, Proc. Natl. Acad. Sci. USA **96**, 8028 (1999)
24. M. Tomasello, *Constructing a Language: A Usage-Based Theory of Language Acquisition* (Harvard University Press, Cambridge, MA, 2003)
25. J.L. Bybee, *Phonology and Language Use* (Cambridge University Press, Cambridge, 2001)
26. J.B. Pierrehumbert, Lang. Speech **46**, 115 (2003)
27. W. Croft, in *Competing Models of Linguistic Change*, ed. by O.N. Tomsen (John Benjamins, Amsterdam, 2006)
28. A. Gardner, Curr. Biol. **18**, R198 (2008)
29. R.A. Fisher, *The Genetical Theory of Natural Selection* (Clarendon, Oxford, 1930)
30. W.D. Hamilton, J. Theor. Biol. **7**, 1 (1964)
31. A. Grafen, in *Behavioural Ecology: An Evolutionary Approach*, ed. by J.R. Krebs, N.B. Davies (Blackwell, Oxford, 1984)
32. S. Wright, Genetics **16**, 97 (1931)

33. P.A.P. Moran, Proc. Camb. Philos. Soc. **54**, 60 (1958)
34. R. Durrett, *Essentials of Stochastic Processes* (Springer, New York, NY, 1999)
35. M. Kimura, T. Ohta, Genetics **61**, 763 (1969)
36. J.L. Wang, A. Caballero, Heredity **82**, 212 (1999)
37. F. Rousset, *Genetic Structure and Selection in Subdivided Populations* (Princeton University Press, Princeton, NJ, 2004)
38. P. Buri, Evolution **10**, 367 (1956)
39. B.J. McGill, Nature **422**, 881 (2003)
40. B.J. McGill, B.A. Maurer, M.D. Weiser, Ecology **87**, 1411 (2006)
41. A.J. McKane, D. Alonso, R.V. Solé, Theor. Popul. Biol. **65**, 67 (2004)
42. J.F.C. Kingman, J. Appl. Prob. **19**(Essays), 27 (1982)
43. S. Tavaré, Theor. Popul. Biol. **26**, 119 (1984)
44. J. Hein, M.H. Schierup, C. Wiuf, *Gene Genealogies, Variation and Evolution: A Primer in Coalescent Theory* (Oxford University Press, Oxford, 2005)
45. W.J. Ewens, Theor. Popul. Biol. **3**, 87 (1972)
46. S. Karlin, J. McGregor, Theor. Popul. Biol. **3**, 113 (1972)
47. T.M. Cover, J.A. Thomas, *Elements of Information Theory*, 2nd edn. (Wiley, Hoboken, NJ, 2006)
48. G. Bell, Science **293**, 2413 (2001)
49. R. Condit, N. Pitman, E.G. Leigh, Jr., J. Chave, J. Terborgh, R.B. Foster, P. Núñez, S. Aguilar, R. Valencia, G. Villa, H.C. Muller-Landau, E. Losos, S.P. Hubbell, Science **295**, 666 (2002)
50. I. Volkov, J.R. Banavar, S.P. Hubbell, A. Maritan, Nature **424**, 1035 (2003)
51. R.S. Etienne, Ecol. Lett. **8**, 253 (2005)
52. M.W. Hahn, R.A. Bentley, Proc. R. Soc. Lond. B Biol. Sci. **270**, S120 (2003)
53. F. Reali, T.L. Griffiths, Proc. R. Soc. Lond. B Biol. Sci. **277**, 429 (2009)
54. B. Charlesworth, D. Charlesworth, N.H. Barton, Annu. Rev. Ecol. Evol. Syst. **34**, 99 (2003)
55. N. Takahata, Genetics **129**, 585 (1991)
56. M.L. Boas, *Mathematical Methods in the Physical Sciences* (Wiley, New York, NY, 1983)
57. G.J. Baxter, R.A. Blythe, A.J. McKane, Phys. Rev. Lett. **101**, 258701 (2008)
58. P. Sjödin, I. Kaj, S. Krone, M. Lascoux, M. Nordborg, Genetics **169**, 1061 (2005)
59. C. Castellano, S. Fortunato, V. Loreto, Rev. Mod. Phys. **81**, 591 (2009)
60. F.P. Kelly, *Reversibility and Stochastic Networks* (Wiley, Chichester, 1979)
61. G. Baxter, R. Blythe, W. Croft, A. McKane, Lang. Var. Change **21**, 257 (2009)
62. E. Gordon, L. Campbell, J. Hay, M. Maclagan, A. Sudbury, P. Trudgill, *New Zealand English: Its Origins and Evolution* (Cambridge University Press, Cambridge, 2004)
63. P. Trudgill, *New-Dialect Formation: The Inevitability of Colonial Englishes* (Edinburgh University Press, Edinburgh, 2004)
64. S. Kirby, M.H. Christiansen, *Language Evolution* (Oxford University Press, Oxford, 2003)

# Chapter 4
# A Simple General Model of Evolutionary Dynamics

Stefan Thurner

**Abstract** Evolution is a process in which some variations that emerge within a population (of, e.g., biological species or industrial goods) get selected, survive, and proliferate, whereas others vanish. Survival probability, proliferation, or production rates are associated with the "fitness" of a particular variation. We argue that the notion of fitness is an a posteriori concept in the sense that one can assign higher fitness to species or goods that survive but one can generally not derive or predict fitness per se. Whereas proliferation rates can be measured, fitness landscapes, that is, the inter-dependence of proliferation rates, cannot. For this reason we think that in a *physical* theory of evolution such notions should be avoided. Here we review a recent quantitative formulation of evolutionary dynamics that provides a framework for the co-evolution of species and their fitness landscapes (Thurner et al., 2010, Physica A **389**, 747; Thurner et al., 2010, New J. Phys. **12**, 075029; Klimek et al., 2009, Phys. Rev. E 82, 011901 (2010). The corresponding model leads to a generic evolutionary dynamics characterized by phases of relative stability in terms of diversity, followed by phases of massive restructuring. These dynamical modes can be interpreted as punctuated equilibria in biology, or Schumpeterian business cycles (Schumpeter, 1939, *Business Cycles*, McGraw-Hill, London) in economics. We show that phase transitions that separate phases of high and low diversity can be approximated surprisingly well by mean-field methods. We demonstrate that the mathematical framework is suited to understand *systemic properties* of evolutionary systems, such as their proneness to collapse, or their potential for diversification. The framework suggests that evolutionary processes are naturally linked to self-organized criticality and to properties of production matrices, such as their eigenvalue spectra. Even though the model is phrased in general terms it is also practical in the sense that it's predictions can be used to understand a series of experimental data ranging from the fossil record to macroeconomic indices.

---

S. Thurner (✉)
Science of Complex Systems, Medical University of Vienna, A-1090 Wien, Austria; Santa Fe Institute, Santa Fe, NM 87501, USA; IIASA, A-2361 Laxenburg, Austria
e-mail: stefan.thurner@meduniwien.ac.at

H. Meyer-Ortmanns, S. Thurner (eds.), *Principles of Evolution*,
The Frontiers Collection, DOI 10.1007/978-3-642-18137-5_4,
© Springer-Verlag Berlin Heidelberg 2011

## 4.1 Introduction

Evolutionary dynamics is at the core of countless biological, chemical, technical, social, and economic systems. An evolutionary description of dynamical systems describes the appearance of new elements within the system together with their interactions with already existing elements. These interactions often influence the rates of production (reproduction), not only of the new elements but also of those which are already present. Reproduction rates in biology are related to biological offspring; in economic systems they correspond to production rates of goods and services. In biology the set of all reproduction rates is often associated with *fitness*; in economics production rates correspond to *utility* or *competitive advantage*. Any element, a biological species or an industrial good, whose (re)production rate falls below a critical value, will vanish over time. This again can lead to a re-adjustment of (re)production rates of existing elements. The key aspect of evolutionary systems is that the set of existing elements and their corresponding reproduction rates (fitness landscapes) *co-evolve*. A mathematical formulation of dynamical systems where sets of elements co-evolve together with their (re)production rates is accompanied with tremendous difficulties. Even if the dynamics of abundance of elements could be exactly described – e.g., with differential equations – the boundary conditions of the systems could not be fixed and the equations cannot be solved. Boundary conditions of evolutionary systems constantly change as they evolve; they are coupled to the dynamics of the system itself. These systems are therefore hardly accessible with traditional mathematical methods.

In the present understanding of evolution the concept of fitness has been of central importance. Usually the relative abundance of species (with respect to other species) is described by replicator equations (first-order differential equations) and variants such as Lotka–Volterra equations [1–3]. Their mutual influence is quantified by a rate specifying how the abundance of one species changes when in contact with another. In biology this rate is called Malthusian fitness, in chemistry it is the reaction rate, and in economics it is related to production functions. Similarly, proliferation rates can be introduced for technological, financial, or even historical contexts. In the following we refer to all of them as fitness.

The fitness of an element within an evolutionary system is a function that generally depends on a large number of parameters, quantifying the reproductive state of the element and the states of its entire environment. Environment can mean the abundance of other elements or species [4] or the *ecosystem* [5]. Owing to the high dimensionality of parameter space, one talks about fitness landscapes (utility functions in economics), which are obviously hard – generally even impossible – to measure. The dependence of proliferation rates on all possibly relevant environmental conditions and parameters is beyond experimental control and the concept of fitness encounters severe limits of predictive applicability. This is particularly true when systems become large, which is often the case for biological or economic systems [6]. In general neither the set of species or environmental factors nor the corresponding inter-species and environmental interactions will be known with sufficient resolution. Random events may change the course of a system in unpredictable ways. For example, although the mutation rate of some particular virus may be known extremely well, some mutant could be harmless while another might wipe out entire species that had so far proliferated well. This lack of predictability means that (reproductive) fitness has to be viewed as an a posteriori concept, meaning that proliferation rates can be used to characterize fitness relative to other species only once these rates have been measured. Because of their co-evolving nature mentioned above, these rates can in general not be extrapolated into the future. It is not fruitful to predict future fitness of species from their present fitness. Instead, one has to understand how species and their fitness landscapes co-construct each other, how they co-evolve.

Similarly, the concept of *niches*, for example in ecologies or markets, is an a posteriori concept. It can in general not be prestated or deduced which proliferation rates of specific species will result from a set of conditions found in some spatially limited region. Niches – i.e., spaces where some species (goods) reproduce (sell) well – can in general not be foreseen, but are only identified once they have been occupied. Consequently, a *physical*, fully causal, and useful description of evolution has to consider abandoning the concept of fitness as the central concept and to look for alternatives that operate with the information actually available at a given time.

Of course it cannot be the aim of a theory of evolution to predict future populations in detail. This would be as nonsensical as to predict trajectories of gas particles in statistical physics. However, what should be possible to predict and understand within a sensible theory of evolution is a series of *systemic* facts of evolutionary systems, such as their characteristic time-series behavior, their punctuated equilibria, rates of crashes and booms, and their dependence of diversity on interaction patterns. To make progress in understanding the phenomenology of evolution, that is, in identifying principles that guide evolutionary dynamics, in recent decades a series of quantitative models have been suggested.

Inspired by special cases of Lotka–Volterra replicators, researchers have proposed important concepts, such as the hypercycle [7], replicator dynamics [8], molecular quasispecies [9], the Turing gas [10], and autocatalytic networks; for an overview see, for example, [11, 12] and references therein. A nonlinear version of autocatalytic networks has been studied in [13]. Nonlinear autocatalytic networks have been shown to be closely related to bit-string formulations and random grammars [12, 14]. Here species are represented as bit-strings with randomly assigned fitness values. More recently it has been noticed that nonlinear autocatalytic systems have a phase structure equivalent to thermodynamical systems (van der Waals gas) [15]. One phase corresponds to a highly populated, diverse phase, whereas the other is characterized by limited levels of diversification. In [16] a similar transition was explored for a destructive dynamics. A series of more specific evolutionary models have been presented recently [17–24]. In these publications explicit assumptions were made about several key elements: how new species come into being, how they interact, and under which conditions they vanish. Each of these models focuses on particular aspects of evolution. For example, Arthur [17] focuses on technological evolution with integrated circuits as species, whose fitness is examined by how well they execute certain pre-specified computational tasks. Jain and Krishna [18] consider ecological systems and demonstrate nicely the interplay between interaction topology and abundance of species. In most of these models detailed assumptions about the mechanisms involved were made. In Jain and Krishna's model, for example, species are actively removed and added to the system and innovations are externally enforced, not endogenously produced. In [17] the output of randomly assembled circuits is compared to a pre-specified list of desired computational tasks (such as bitwise addition). Although these assumptions are most reasonable within their specific contexts, it is relatively hard to identify those features which are universally valid across different evolutionary contexts, for example biological evolution, technical and industrial innovation [25, 26], economics [27], finance, sociodynamics and opinion formation [28–30], ecological dynamics (e.g., food-webs [31]), and maybe even history. To arrive at a general evolutionary description with as few ad hoc specifications as possible, a minimum set of general principles must be identified that are abstract enough to be applicable in the various evolutionary contexts and which – at the same time – must be specific enough to make useful quantitative and falsifiable predictions about systemic properties of evolutionary systems.

Motivated by statistical physics, we believe that it should be possible to formulate evolutionary dynamical systems by focusing on microscopic interactions of agents (determining (re)production rates on a microscale) and then to derive macroscopic – systemic – properties and laws. If successful, the macroscopic properties are determined by a few manageable parameters only. Further, there is room for hope that a variety of different specific microscopic interaction mechanisms may lead to the same class of macroscopic properties. In physics this led to the concept of *universality classes*. Here systems may differ in their microscopic details of interaction but nevertheless show the same physical properties on an aggregated level. Naturally, any physical model for evolution must include the basic dynamic elements of evolution; in evolutionary biology these are called replication, competition, mutation, and recombination. These names have their counterparts in other disciplines.

In the following we review a model in the above spirit [32–34] that is able to endogenously explain several systemic facts about evolutionary dynamics – such as intrinsic booms of diversity and large extinction events punctuated equilibria, as well as a series of statistical characteristics of evolutionary time series. In particular, many of these time series are characterized by power laws (for a financial context see, e.g., [21, 35]) and exhibit nontrivial correlations, see Fig. 4.1.

In Sect. 4.2 we present the generic version of the model and in Sect. 4.3 discuss its resulting dynamics. In Sect. 4.4 we sketch several ways to make the model more specific and realistic. Sect. 4.5 offers an understanding of diversity dynamics in terms of self-organized critical systems or in terms of a dynamical system driven by its largest eigenvalues. In Sect. 4.6 we formulate the model as a variational principle, which links evolutionary processes to the powerful physical concept that dynamics results as a consequence of minimization of properly chosen quantities. A stochastic variant of the model is easily solved within the mean-field approximation. To test the range of validity of the approximation we compare its results with those from numerical simulations. Finally, in Sect. 4.7 we demonstrate that model predictions allow us to understand several empirical facts of evolutionary time series.

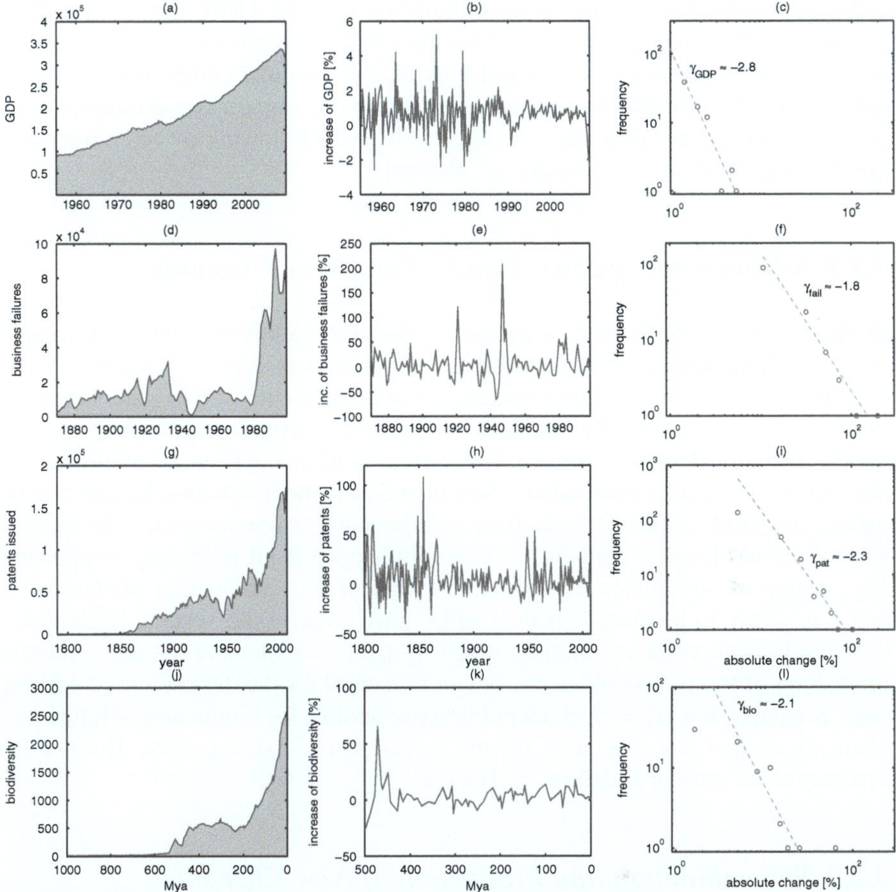

**Fig. 4.1** Several evolutionary time series from economics to biodiversity. **a** GDP of the UK starting 1950 [36]. **b** Percentage increase of GDP from (**a**). **c** Histogram for (**b**); a least *squares* fit to a power law yields a slope of $\approx -2.8$. For other countries [37] we found the slopes: ESP $-3.5$, FIN $-3.1$, FRA $-2.3$, NLD $-2.2$, and SWE $-3.3$, and for the US [38] $-3.3$. **d–f** Number of business failures in the conterminous United States from 1870 onward, data from [38]. **e** Annual percentage change for (**d**); **f** histogram for (**e**): power-law exponent $\approx -1.8$. **g** Total number of patents issued on inventions in the United States from 1790 to 2007, data from [38]. **h** Annual percentage increase of patents, starting 1800. **i** Histogram of absolute values of (**h**): power-law exponent $\approx -2.3$. **j** Biodiversity over time [39]. **k** Percentage change in biodiversity. **l** Histogram of absolute values of (**k**): power-law exponent $\approx -2.1$

## 4.2 A General Model for Evolution Dynamics

New things such as species, goods and services, ideas, or new chemical compounds often come into being as (re)combinations or substitutions of existing things. Water is the combination of hydrogen and oxygen, the iPod is a combination of electrical parts, Wikipedia is a combination of the internet and the concept of an encyclopedia, and a mutation emerges through a combination of some DNA and a $\gamma$-ray. New species, goods, compounds, etc. "act on the world" in three ways: (i) they can be used to produce yet other new things (think of, e.g., modular components), (ii) they can have a negative effect on existing things by suppressing (or outperforming) their (re)production, or (iii) they have no effect at all. In the following we refer to species, goods, compounds etc. generically as "elements".

### 4.2.1 A Notion for Species, Goods, Things, or Elements

In the following simple model *all* possibly thinkable elements (things) are components of a time-dependent $N$-dimensional vector $\boldsymbol{\sigma}(t) \equiv (\sigma_1(t), \cdots \sigma_N(t))$. $N$ can be very large, even infinite. For simplicity the components of this state vector are binary. $\sigma_i(t) = 1$ means that element $i$ is present (it exists) at time $t$; $\sigma_k(t) = 0$, means element $k$ does not exist at $t$, either because it has not been assembled or created yet, or it has been eliminated.[1] New elements come into being through combinations of already existing elements or "components". An innovation – the creation or production of a new element $i$ – can only happen if all necessary components (e.g., parts) are simultaneously available (exist). If the combination of elements $j$ and $k$ is a way to produce $i$, both $j$ and $k$ must exist in the system. Technically, $\boldsymbol{\sigma}(t)$ can be seen as the *availability status*: if $\sigma_i(t) = 1$, element $i$ is accessible for production of future new elements, or can it be used for the destruction of existing ones. Similarly, if $\sigma_i(t) = 0$, element $i$ is not accessible for production of future new elements, neither can it be used for the destruction of existing ones. The *product diversity* of the system is defined as $D(t) = \frac{1}{N} \sum_{i=1}^{N} \sigma_i(t)$.

### 4.2.2 Recombination and Production of New Elements

Whether an element $k$ can be produced from components $i$ and $j$ is encoded in a so-called *production table*, $\alpha_{ijk}^{+}$. If it is *possible* to produce $k$ from $i$ and $j$ (i.e., $\alpha_{ijk}^{+} > 0$), we call this a *production*. An entry in the production table is in principle a real number and quantifies the *rate* at which an element is produced from its components, substrates, etc. Again, for simplicity an entry in $\alpha_{ijk}^{+}$ is assumed to be binary, 0 or 1. If elements $i$ and $j$ can produce $k$, $\alpha_{ijk}^{+} = 1$, elements $i$ and $j$ are called the *productive set* of $k$. If it is (physically, chemically, or biologically) impossible to produce $k$ from $i$ and $j$, then $\alpha_{ijk}^{+} = 0$. In this notation, a production process is written

$$\sigma_k(t + 1) = \alpha_{ijk}^{+} \sigma_i(t) \sigma_j(t) , \tag{4.1}$$

---

[1] In a more general setting the state vector $0 < \sigma_i(t) < 1$ could represent the relative abundance of species $i$ with respect to the abundances of the others.

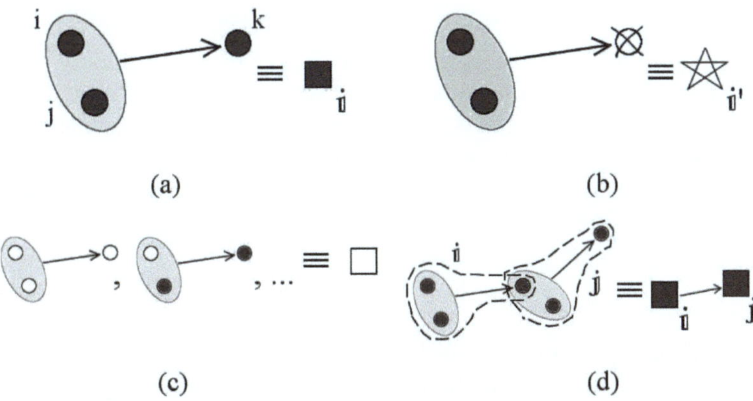

(a)                                                           (b)

(c)                                                           (d)

**Fig. 4.2** **a** Illustration of a (re)combination or production process. Elements $i$ and $j$ reside in a productive set. There exists a production rule $\alpha^+_{ijk} = 1$. Thus product $k$ becomes produced in the next time step. This *active production* is represented as a *full square* and indexed by $i$ (*boldface* indices indicate productions). **b** Same as (**a**) for a destruction process. Products $i'$ and $j'$ replace $k'$ via the destruction rule $\alpha^-_{i'j'k'} = 1$. **c** Examples of nonactive productions (in total 6 are possible). Nonactive productions are symbolized as *open squares*. **d** Definition of a link in the active production network: if an element – produced by a production $i$ – is part of the productive set of another active production $j$, production $j$ gets a directed link from production $i$. From [33]

regardless of whether $\sigma_k(t) = 0$ or 1. If a production is *actually* producing $k$, (i.e., if simultaneously $\sigma_i(t) = \sigma_j(t) = \sigma_k(t) = \alpha^+_{ijk} = 1$), we call it an *active production*, see Fig. 4.2a.

Note that the production table $\alpha^+_{ijk}$ exists whether it is known or accessible to mankind or not. For example the laws of chemical reactions exist whether one knows them or not. In general the details of the production table will not be known. Further, a particular element can often be produced through more than one production. In our (binary) notation the number of ways to produce element $k$ at time $t$ simply becomes $N^{\text{prod}}_k(t) = \sum_{ij} \alpha^+_{ijk} \sigma_i(t) \sigma_j(t)$.

### 4.2.3 Selection, Competition, Destruction

If a new element serves a function, a purpose, or a need that hitherto has been satisfied by another element, the new and the old elements are in competition. The element that can be produced in a more efficient way, that is more robust, etc., will sooner or later dominate the other (through a larger (re)production rate) and possibly drive it out of the system. We incorporate this competition mechanism into the model by allowing that a combination of two existing elements can lead to a destructive influence on another element. The combination of elements $i'$ and $j'$ produces an element $l$ which then drives product $k'$ out of a niche or the market. We say: the combination of $i'$ and $j'$ has a destructive influence on $k'$. We capture all possible destructive combinations in a *destruction rule table* $\alpha^-_{i'j'k'}$, see Fig. 4.2b.

If we have $\alpha^-_{i'j'k'} = 1$, elements $i'$ and $j'$ replace element $k'$. We call $\{i', j'\}$ the *destructive set* for $k'$. Note that in this way we do not have to explicitly produce the actually competing element $l$. In the absence of a destruction process, $\alpha^-_{i'j'k'} = 0$. As before, an *active destruction* happens only if all elements are actually present at a point in time $\sigma_{i'}(t) = \sigma_{j'}(t) = \sigma_{k'}(t) = \alpha^-_{i'j'k'} = 1$. The elementary dynamical update for a destructive process reads (for an element that is present at time $t$, i.e., $\sigma_k(t) = 1$),

$$\sigma_k(t + 1) = 1 - \alpha^-_{ijk}\sigma_i(t)\sigma_j(t) \ . \tag{4.2}$$

In general, at any given time, element $k$ can be in competition with more than one potential substitute – in our notation – by exactly $N_k^{\text{destr}}(t) = \sum_{ij} \alpha^-_{ijk}\sigma_i(t)\sigma_j(t)$ destructive sets.

Imagine the $N$ thinkable elements represented by circles in a plane, see Fig. 4.3a. If they exist they are plotted as full circles, if they are not produced, they are open circles. All existing elements have at least one productive set (pair of two existing circles); at time $t$ there exist $N_i^{\text{prod}}(t)$ such sets. Many circles will in the same way assemble to form destructive sets, the exact number for node $i$ being $N_i^{\text{destr}}(t)$. Now, draw a circle around each productive and destructive set and connect each set with the element it is producing/destroying. The graph that is produced in this way is seen in Fig. 4.3a. In general every element will be connected to several productive and destructive sets.

Given this notation for states $\boldsymbol{\sigma}$ and constructive/destructive interactions $(\alpha^+, \alpha^-)$, one now has to specify a dynamics for the system. There are many ways to do so; the details of the implementation depend on the system studied. As one possible (generic) example here we specify the following: if there exist more production processes associated with a particular element than there exist destructive

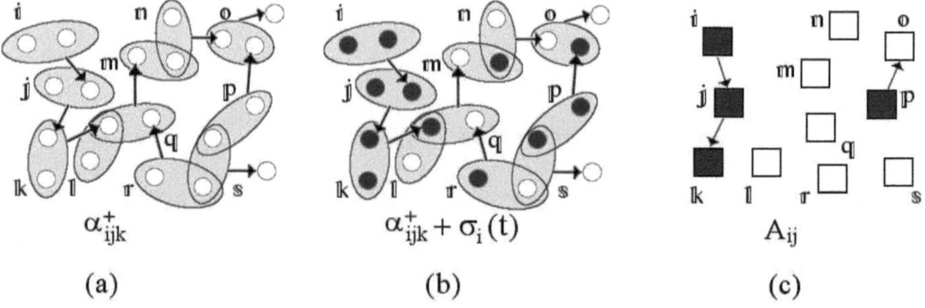



**Fig. 4.3** Comparison between static production rules $\alpha^+$ (**a**) and active production networks $A(t)$ (**c**). Production rules are all possible combinations to produce all thinkable species in a biological system, compounds in a chemical system, or goods in an economic system. They correspond to all nonzero entries in $\alpha^+$. In (**b**) the actual state of the system $\boldsymbol{\sigma}(t)$ (which elements exist) is superimposed on $\alpha^+$. Representing (**b**) not in terms of elements but in terms of productions (recombinations), we arrive at the active production or recombination network (**c**), using the definition for links in Fig. 4.2d. The same can be done for the active destruction networks. From [33]

processes acting on it, the element will be produced. Inversely, if there are more destructive than productive sets associated with an element, it will not be produced, or it will be destroyed if it already exists. If the number of productive and destructive sets for $i$ are the same, the state of $i$ will not change, i.e., $\sigma_i(t+1) = \sigma_i(t)$. More quantitatively this reads

$$
\begin{aligned}
&\text{if} \quad N_i^{\text{prod}}(t) > N_i^{\text{destr}}(t) \rightarrow \sigma_i(t+1) = 1 \;, \\
&\text{if} \quad N_i^{\text{prod}}(t) < N_i^{\text{destr}}(t) \rightarrow \sigma_i(t+1) = 0 \;, \\
&\text{if} \quad N_i^{\text{prod}}(t) = N_i^{\text{destr}}(t) \rightarrow \sigma_i(t+1) = \sigma_i(t) \;.
\end{aligned}
\tag{4.3}
$$

Note that a production or destruction is only active if both goods in its production or destruction set are currently available. Thus changes in the availability status of an element possibly induce changes in the status of the active production or destruction network. We discuss several alternatives to this dynamics in Sect. 4.4.


### 4.2.4 The Active Production or Recombination Network

It is essential to distinguish the production *rules* encoded as tensors $\alpha^{\pm}$ and the *active production networks* $A(t)$ (matrices). $\alpha^{\pm}$ is a collection of static rules of all potential ways to produce all thinkable elements. These rules exist regardless of whether elements exist. The production network $A(t)$ captures the set of actual active (!) productions taking place at any given time $t$. It maps the state of the evolutionary system $\sigma(t)$ (existing elements) with its rules $\alpha^{+}$ onto the set of active productions. The production network can be constructed in the following way: A *production* is defined as a pair $(i, j)$ that produces an element $k$, and is nothing but a nonzero entry in $\alpha^{+}$. There are $Nr^{+}$ productions in the system, where $r^{+}$ is the (average) number of productions per element. Nonexisting elements are open circles; the symbol used for a production is a square. A production is called an *active production* if the production set and the produced element (node) all exist ($\sigma_i(t) = \sigma_j(t) = \sigma_k(t) = 1$). An active production is shown in Fig. 4.2a symbolized as a filled square. In (c) we show some examples of nonactive productions (open square). We label active productions by boldface indices, $i \in \{1, \ldots Nr^{+}\}$. These constitute the nodes of the *active production network*. A directed link from active production node $i$ to node $j$ is defined if the *node* produced by production $i$ is in the productive set of production $j$, see Fig. 4.2d. It is then seen as an entry in an adjacency matrix and denoted as $A_{ij} = 1$. This definition is illustrated in Fig. 4.2c. For an example of how to construct the active production network $A(t)$ from $\sigma(t)$ and $\alpha^{+}$, see Fig. 4.3. In Fig. 4.3a we show a section of the static $\alpha^{+}$, in (b) we superimpose the knowledge of which of these nodes actually exist at time $t$. In (c) all productions are shown as squares (active ones full, nonactive ones empty). The links between the active productions constitute the active production (or recombination) network. In this way we map the production rule tensor $\alpha^{+}$ onto a production (adjacency) matrix $A(t)$. It is defined on the space of all productions and links two active productions if one production produces an element that serves as an input for the other production. The active destruction network is obtained in the same way. Active production networks can be viewed as representations of a multitude of evolutionary systems such as evolving ecologies (e.g., food webs) in biology [31], or as time-varying Leontief input–output matrices in economics [40].

To detect *dominant* links in the active production network, we introduce the following thresholding method: remove all links from the active production network which exist less than a pre-specified percentage of times $h$ within a moving time window of length $T$. For example, if $h = 95$ and $T = 100$, the network at time $t$, $A(t)$, only contains links that have existed more than 95 times within the time window $[t - T, t]$.

### 4.2.5 Spontaneous Creations, Innovations, Ideas, and Disasters

From time to time spontaneous ideas, inventions or mutations take place without the explicit need of the production network. Also from time to time elements disappear from the system, for example through exogenous events. To model these events we introduce a probability $p$ with which a given existing element is spontaneously annihilated, or a nonexisting element is spontaneously invented. This stochastic component plays the role of a driving force.

### 4.2.6 Formal Summary of the Model

Let $N$ be arbitrarily large, then the "phase space" of the system is given by

$$\Gamma \equiv \{0, 1\}^N = \{\boldsymbol{\sigma} \mid \sigma_i \in \{0, 1\}, 1 \leq i \leq N\} . \tag{4.4}$$

Further, define the quadratic forms

$$\Delta_i(\boldsymbol{\sigma}) \equiv \sum_{j,k}^{N} \alpha_{ijk}\sigma_j\sigma_k , \quad (1 \leq i \leq N) , \tag{4.5}$$

where the coefficients $\alpha_{ijk}$ take the values 0 and $\pm 1$. Here $\alpha_{ijk} \equiv \alpha_{ijk}^{+} - \alpha_{ijk}^{-}$. Both $\alpha^{+}$ and $\alpha^{-}$ take values from $\{0, 1\}$. Select a positive value $p \leq 1$. The dynamical update is then given by the map $F : \Gamma \to \Gamma$ via the difference equation

$$\boldsymbol{\sigma}(t + 1) = F(\boldsymbol{\sigma}(t)) , \tag{4.6}$$

where the map is of the form $F = \boldsymbol{\Psi} \circ \boldsymbol{\Phi}$, with $\boldsymbol{\Psi}$ a deterministic part and $\boldsymbol{\Phi}$ the stochastic part, which are defined as follows:

$$\boldsymbol{\Phi}(\boldsymbol{\sigma}) \equiv \mathbf{x} = (x_1, \cdots , x_N) , \tag{4.7}$$

where for all $i$, $x_i = 1(0)$ when $\Delta_i(\sigma) > 0(< 0)$, and $x_i = \sigma_i$, when $\Delta_i(\sigma) = 0$. The stochastic part is given by

$$\Psi \equiv \mathbf{x} = (x_1, \cdots, x_N) \,, \tag{4.8}$$

where $x_i = 1 - \sigma_i$ with probability $p$, and $x_i = \sigma_i$ with probability $1 - p$. This setup can be used as a starting point to compute several model properties analytically, see Sect. 4.6.

### 4.2.7  An Evolutionary Algorithm

The above model can be implemented in the following simple algorithm. Consider being at time step $t$; the update to $t + 1$ happens in three steps:

- Pick an element $i$ at random (random sequential update).
- Sum all productive and destructive influences on $i$, i.e., compute $\Delta_i(t)$ from (4.5). If $\Delta_i(t) > (<)0 \rightarrow \sigma_i(t + 1) = 1(0)$. For $\Delta_i(t) = 0$ do not change, $\sigma_i(t + 1) = \sigma_i(t)$.
- With probability $p$ switch the state of $\sigma_i(t + 1)$, i.e., if $\sigma_i(t + 1) = 1(0)$ set it to $\sigma_i(t + 1) = 0(1)$.
- Pick next element until all elements have been updated once, then go to next time step.

As initial conditions (at $t = 0$) we chose a fraction of randomly chosen initial elements to exist; typically we set $D(0) \sim 0.05$–$0.2$.

### 4.2.8  Random Interactions

In principle it is possible to empirically assess production or destruction networks in the real world (ecologies, economies, etc.), however, in practice this is unrealistic and would involve tremendous effort. For a *systemic* understanding of characteristics of evolutionary dynamics a detailed knowledge of these networks might, however, not be necessary – a *statistical* characterization of these networks will already provide some understanding of important key facts. The simplest implementation of a production/destruction network is to use random networks, i.e., to model $\alpha^{\pm}$ as random tensors. These tensors can then be described by a single number $r^+$ and $r^-$, which are the constructive/destructive *rule densities*. In other words, the probability that any given entry in $\alpha^+$ equals 1 is $P(\alpha^+_{ijk} = 1) = r^+ \binom{N}{2}^{-1}$, or – equivalently – each product has on average $r^{\pm}$ incoming productive/destructive links from productive/destructive sets. Further, which elements form which productive sets is also randomly assigned, i.e., the probability that a given product belongs to a given productive/destructive set is $2r^{\pm}/N$ (for $r^{\pm} \ll N$).

Real production networks may be highly structured and the assumption that production networks are purely random is unrealistic to some degree. For this reason

                                                    

**Table 4.1** Summary of model parameters

| Variable | | |
|---|---|---|
| $\sigma_i(t)$ | State of element $i$. exists / does not exist | Dynamic |
| $D(t)$ | Diversity at time $t$ | Dynamic |
| $A(t)$ | Active production network | Dynamic |
| Parameter | | |
| $\alpha^{\pm}$ | Productive/destructive interaction topology | Fixed |
| $r^{\pm}$ | Rule densities | Fixed |
| $p$ | Spontaneous-innovation parameter | Fixed |

we look at the effect of network topology by using alternative topologies such as scale-free versions of production/destruction tables in Sect. 4.4. Note that $\alpha^{\pm}$ is fixed in time throughout the simulation. The model parameters and variables are listed in Table 4.1.

## 4.3 Predictions of the Model

The model can be easily implemented in a computer simulation. Figure 4.4 shows the trajectories of 100 elements. Time progresses from left to right. Each column shows the state of each of the elements $i = 1, \ldots, N$ at any given time. If $i$ exists at $t$, $\sigma_i(t) = 1$ it is represented as a white cell; a black cell at position $(i, t)$ indicates $\sigma_i(t) = 0$. It is immediately visible that there exist two distinct dynamical modes in the system, a quasi-stationary phase, where the set of existing elements practically does not change over time, and a phase of massive restructuring. To extract the time series of diversity $D(t)$ we sum the number of all white cells within one column at time $t$ and divide by $N$ (Fig. 4.4b). Again it is seen that the plateaus of constant product diversity (punctuated equilibria) are separated

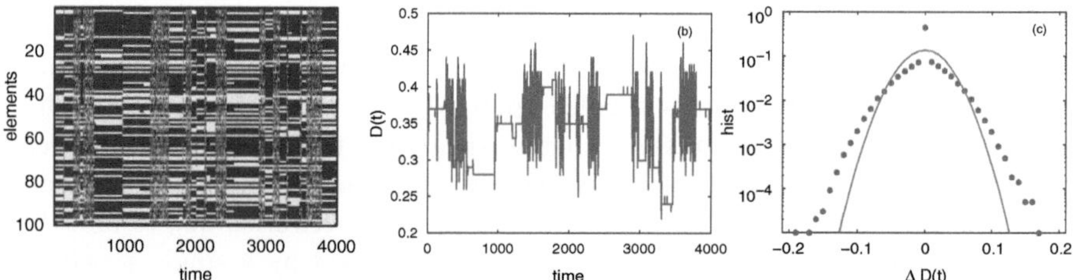

**Fig. 4.4** **a** Individual trajectories of elements (existing, *white*; nonexisting, *black*) for the parameter setting $r^{+} = 10$, $r^{-} = 15$, $p = 2 \times 10^{-4}$, $N = 10^2$ and an initial diversity of 20 randomly chosen elements. **b** Diversity $D(t)$ from the simulation in (**a**), which shows a punctuated equilibrium pattern. The system jumps between phases of relatively few changes – the plateaus – and chaotic restructuring phases. The length of the chaotic phases is distributed as a power law, which is identical with the fluctuation lifetime distributions shown in Fig. 4.6. **c** Histogram over the percentage-increments of diversity. The *line* is a Gaussian distribution with the same mean and variance

by restructuring periods, characterized by large fluctuations of the products. Note that depending on parameter settings, stationary diversity levels may differ by up to 50%. These fluctuations are strongly non-Gaussian, as can be inferred from Fig. 4.4c, where the histogram of the percentage changes in the diversity time series, $\Delta D(t) \equiv (D(t) - D(t-1))/D(t-1)$, is shown. If the extreme tails of the histogram are fitted to power laws the resulting exponents are in the range of those observed in Fig. 4.1 (not shown). For the chosen parameter setting the dynamics of the system does not reach a frozen state; stationary phases and chaotic ones continue to follow each other.

However, the dynamic changes with altering the "innovative rate" $p$, see Fig. 4.5. For a rate of $p = 0.01$ (i.e., in a system of $N = 100$ there is about one spontaneous innovation or destruction per time step), we observe extended restructuring processes, almost never leading to plateaus (a,b). For $p = 10^{-4}$ (one spontaneous innovation/destruction every 100 time steps), the situation is as described above, (c,d). When $p$ gets too small ($p \ll N^{-2}$), the system is eventually not driven from a stationary state and freezes (e,f). In the next section we discuss alternatives to the driving process, where innovation rates $p$ are replaced by, for example, *species lifetimes* or alternative competition/selection mechanisms. These alternatives also drive the system dynamically away from frozen states.

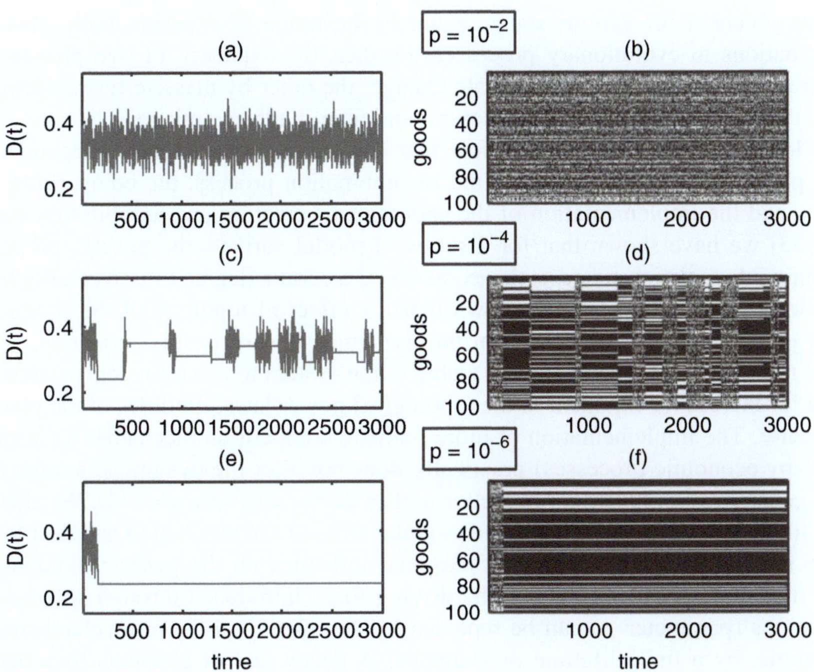

**Fig. 4.5** Time series of diversity (*left*) and element trajectories (*right*) for various values of $p$. For high innovation/mutation/spontaneity rates the system does not settle into plateaus, $p = 0.01$ (**a,b**); for intermediate levels $p = 10^{-4}$ plateaus form (**c,d**), and for low levels $p = 10^{-6}$ the system freezes (**e,f**)

Under which topological circumstances is the generic dynamics maintained for a fixed $p$ in the range of $10^3$–$10^5$? If there are too many destructive influences compared to constructive ones ($r^+ \ll r^-$), the system will evolve toward a state of low diversity in which innovations are mostly suppressed. If there are many more constructive than destructive interactions ($r^- \ll r^+$), that is, comparably little competition, the system is expected to saturate in a highly diverse state. Between these two extreme cases we find sustained dynamics as described above. This regime is indeed very broad, no fine-tuning of $r^+$ and $r^-$ is needed.

Elements tend to populate locations in element space that are locally characterized by high densities of productive rules and low densities of destructive influences. If the system remains in such a *basin of attraction* this results in diversity plateaus. Perturbations can force the population of elements out from these basins and a phase of restructuring can follow until another basin of attraction is found.

## 4.4 Model Variants

A central result of the model is that it endogenously produces fluctuations in the population of elements which resemble punctuated equilibria. The power-law exponents obtained for various statistics are in the range of experimentally observed fluctuations in evolutionary processes. Further, the existence of two phases (one characterized by relatively moderate change, the other by massive restructuring) is a feature commonly observed in natural and man-made evolutionary systems.

However, it is important to show that these characteristics are independent of the particular implementation of the recombination process, the competition process, and the implementation of the spontaneous creation and annihilation events. In [33] we have shown that for a series of model variants the generic properties remain intact, which suggests the existence of a certain degree of universality of the model. In particular we have shown that the effect of topology of the production and destruction rule tables $\alpha^\pm$ on the dynamics is surprisingly moderate. Scale-free network topologies proved to stabilize the system to a certain degree, meaning that for increasing exponent in the topological power laws, lifetimes of the plateaus increase. The implementation of more realistic competition mechanisms (inspired, e.g., by economic processes) practically does not alter the dynamics, neither does the generalization to larger productive and/or destructive sets, see [34]. We checked the influences of an asymmetry and modular structure in production and destruction rule tables and again found only marginal influence on the generic dynamics of the model. We have shown that the driving force introduced through spontaneous creations (parameter $p$) can be replaced by completely different mechanisms, for example, by a finite lifetime of elements. A decay rate of elements then acts as a stochastic driving force, keeping the system from frozen states, that is, even for $p = 0$ the system does not freeze. An important generalization of the recombination mechanism is that an element that *can* be produced *will not* necessarily be produced. To incorporate this we say that if an element can be produced, it will actually be produced with a probability $q$. This means that if an element should be produced or destroyed according to (4.3), this happens only with probability $q$. This variant is formally almost exactly the same as driving the system with the innovation parameter $p$; the two scenarios differ only marginally. Finally, we verified that the qualitative behavior of the model does not change if we employ a parallel update or a sequential update in deterministic order. The only impact of these changes is that for parallel updating the system needs longer to reach frozen states.

## 4.5  Understanding Evolutionary Dynamics

In the way the model is set up there are two ways of understanding evolutionary processes. First, a direct correspondence to self-organized critical (SOC) sandpile models is apparent; second, the diversity dynamics of evolutionary systems can be understood on the basis of the eigenvalues of the active production networks.

### 4.5.1  Evolutionary Dynamics as a Self-Organized Critical System

To see the similarities to self-organized critical systems [41] we proceed in the following way: Set $p = 0$ and let the system reach a frozen state, which we define as one or fewer changes in the state vector $\sigma$ occurring over five iterations. We then flip the state of one randomly chosen component $\sigma_i$. This perturbation may or may not trigger successive updates in diversity. In Fig. 4.6a the cluster-size (total number of elements that get updated as a consequence of this perturbation) distribution is shown. The observed power laws reflect typical features of self-organized criticality: one spontaneous event may trigger an avalanche of restructuring in the system; the observed power laws demonstrate that large events of macroscopic size are by no means rare events. We show the distribution of fluctuation-lifetimes in Fig. 4.6b, that is, the number of iterations that the system needs to arrive at a new frozen state after the perturbation. The distribution of lifetimes also follows a power law, which again points to the existence of self-organized criticality in the sense of [41].

### 4.5.2  Eigenvalues and Keystone Productions

Given the model setup it is an obvious step to analyze the topology of the (time-varying) active production networks to clarify how active production topology is related to the characteristics of the dynamical system. We compute the maximum real eigenvalue (EV) of the active production network $A(t)$. In Fig. 4.7a, we show the system diversity versus the maximum real eigenvalue of the adjacency matrix $A(t)$ associated with the active production network. The latter has been constructed as described in Sect. 4.2.4 without using thresholding, that is, $h = 1$ and $T = 1$. There is a high correlation of about $\rho \sim 0.85$; the observed slope is $\sim 16$.

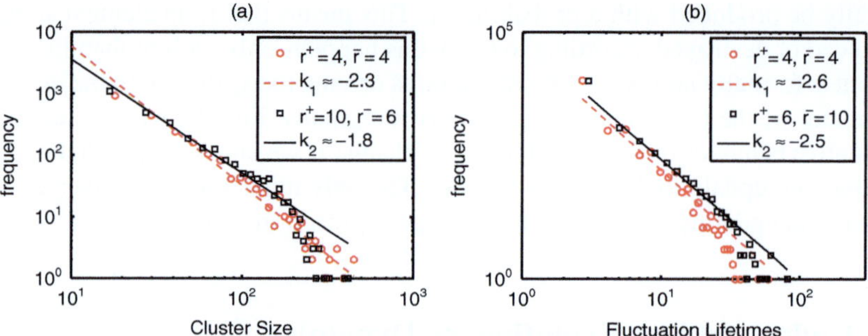

**Fig. 4.6** Cluster-size distribution (**a**) and fluctuation-lifetime distribution (**b**) of a system of size $N = 1,000$ for different rule densities $r^+$ and $r^-$. The slopes for power-law fits for $r^+ = r^- = 4$ are $k_1 = -2.3$ for the sizes, and $k_1 = -2.6$ for fluctuation durations, while for $r^+ = 10$ and $r^- = 6$, we get $k_2 = -1.8$ for sizes, and $k_2 = -2.5$ for the lifetimes. Given that these parameter settings correspond to highly different scenarios, the similarity of the power-law exponents indicates some degree of robustness. From [33]

For the chaotic phases (see Fig. 4.4) the maximal eigenvalue is mostly zero, which indicates that the active production networks is a directed acyclic graph. When a maximal eigenvalue of one is found, this indicates the existence of one or more simple cycles [42]. On the plateaus of diversity we typically find values larger than one, which is a sign of a larger number of interconnected cycles in $A$. In this sense the plateaus are characterized by a relatively long-lasting high level of "cooperation", whereas in the chaotic phase cooperation between cyclically driven production paths is absent. This is the topological manifestation of a collective "organization" that emerges from the model. Note that these structures form from a purely random setup in $\alpha^\pm$, the update, and the initial conditions.

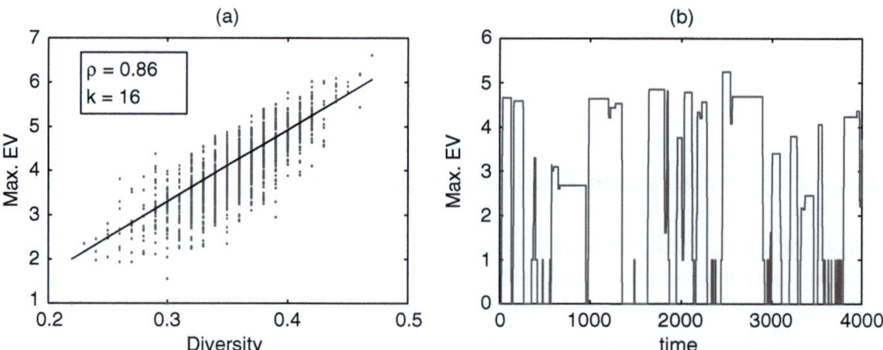

**Fig. 4.7** **a** At each time step we construct the active production network $A(t)$ for the run of Fig. 4.4, with $h = 1$ and $T = 1$. Its maximal eigenvalue is plotted vs. the system's diversity at that time $t$. Every point represents one time point. **b** Trajectory of the maximum EV when computed with thresholding using $T = 20$, $h = 0.95$ for the same run shown in Fig. 4.4. From [33]

The situation is very similar to what was found in a model of biological evolution by Jain and Krishna [18]. Unlike the dynamics of the active production/destruction network $A$, in their model, Jain and Krishna update their interaction matrix through an explicit selection mechanism. They could directly relate the topology of their dynamical interaction matrix to the diversity of the system. In particular they showed that highly populated phases in the system are associated with autocatalytic cycles and the presence of keystone species, that is, species building up these cycles. A drastic increase in diversity is associated with the spontaneous formation of such cycles, while the decline of species diversity is triggered by breaking a cycle. Even though we do not have such an explicit selection mechanism in the present model, the relation between topological structure of $A$ and the state of the dynamical system seems to be the same as in [18]. Note that in our framework the nodes are active productions and not species (or elements) as in their approach. To explicitly see the relation between eigenvalues, cycles, and product diversity we show the trajectory of the maximum eigenvalue of $A$ together with snapshots of the active production networks along the trajectory in Fig. 4.8.

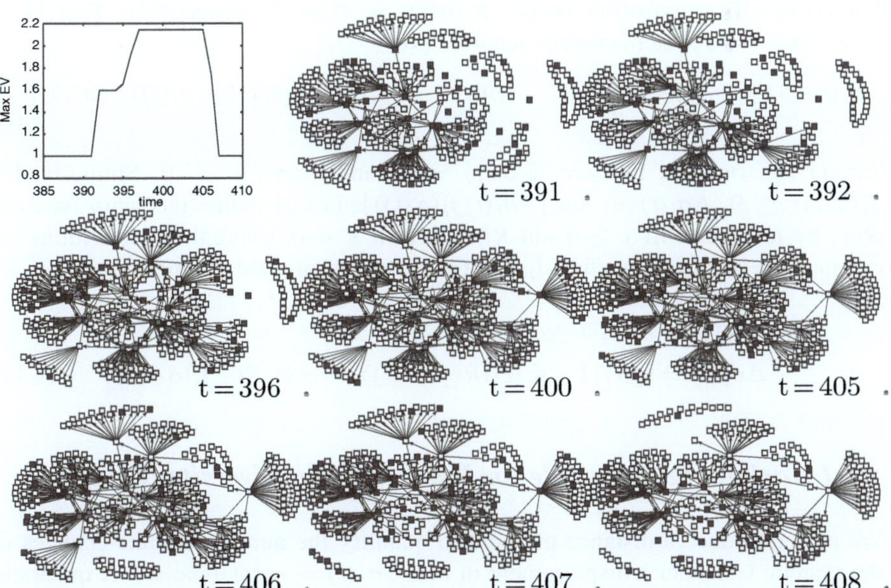



**Fig. 4.8**  Onset and breakdown of massive diversification. Trajectory of the maximum eigenvalue of $A$, for $T = 50$ and $h = 0.8$. The eigenvalue starts at around 1 builds up to about 2.1, then drops to 1 again. The active production networks along this trajectory are shown. At first the network contains simple cycles (maximal eigenvalue equals one). At each time step we show "keystone" productions as red nodes. A keystone production is defined as a node that – if removed from $A(t)$ – results in a reduction of the maximum eigenvalue of more than 10% (when compared to the maximum eigenvalue of initial $A$). Typically keystone nodes are components of cycles. Active productions are represented as blue squares. Note, active destruction graphs can be constructed in the same way but will always be much sparser than active production networks. From [33]

## 4.6 Toward a Unified Mathematical Framework

In the following we sketch how an analytical treatment of the proposed model can be achieved through methodology from spin systems in physics. With the formalism proposed in [34] we are able to obtain mean-field results that can be shown to provide a good approximation for understanding the phase diagram of evolutionary systems.

As before imagine that at each time each element $i$ experiences one of three scenarios, (i) creation $\sigma_i(t) = 0 \rightarrow \sigma_i(t+1) = 1$, (ii) annihilation $\sigma_i(t) = 1 \rightarrow \sigma_i(t+1) = 0$, or (iii) nothing $\sigma_i(t) = \sigma_i(t+1)$. Suppose the existence of a function $f_i(\boldsymbol{\sigma}(t)) : \{0, 1\}^N \rightarrow \mathbb{R}$, which indicates which of the transitions (i–iii) takes place. Let $f_i(\boldsymbol{\sigma}(t))$ indicate the following:

$$
\begin{aligned}
\text{(i)} \quad & f_i(\boldsymbol{\sigma}(t)) > 0 \Rightarrow \sigma_i(t+1) = 1 \,, \\
\text{(ii)} \quad & f_i(\boldsymbol{\sigma}(t)) < 0 \Rightarrow \sigma_i(t+1) = 0 \,, \\
\text{(iii)} \quad & f_i(\boldsymbol{\sigma}(t)) = 0 \Rightarrow \sigma_i(t+1) = \sigma_i(t) \,.
\end{aligned}
\tag{4.9}
$$

For (i) or (ii) a transition occurs if $\sigma_i(t) = 0$ or $1$, respectively. That is, if $f_i(\boldsymbol{\sigma}(t)) \geq 0$ the system evolves according to

$$
\sigma_i(t+1) = \sigma_i(t) + \Delta\sigma_i(t) \qquad \text{with} \qquad \Delta\sigma_i(t) = \mathrm{sgn}\left[(1 - \sigma_i(t))f_i(\boldsymbol{\sigma}(t))\right] \,.
\tag{4.10}
$$

$\Delta\sigma_i(t)$ can only be nonzero if $\sigma_i(t) = 0$ and $f_i(\boldsymbol{\sigma}(t)) > 0$. Similarly, for $f_i(\boldsymbol{\sigma}(t)) \leq 0$, $\Delta\sigma_i(t) = \mathrm{sgn}\left[-\sigma_i(t)f_i(\boldsymbol{\sigma}(t))\right]$. Let us define the ramp function $R(x)$ by $R(x) \equiv x$ iff $x \geq 0$ and $R(x) \equiv 0$ iff $x < 0$. Using these definitions we can map the indicator function $f_i$, Eq. (4.9), onto the update equation

$$
\begin{aligned}
\sigma_i(t+1) &= \sigma_i(t) + \Delta\sigma_i(t) \,, \\
\Delta\sigma_i(t) &= \mathrm{sgn}\left[(1 - \sigma_i(t))R(f_i(\boldsymbol{\sigma}(t))) - \sigma_i(t)R(-f_i(\boldsymbol{\sigma}(t)))\right] \,.
\end{aligned}
\tag{4.11}
$$

### 4.6.1 Variational Principle for Deterministic Diversity Dynamics

We next introduce a distance measure to quantify the number of state changes in the system. Consider a displacement of $\sigma_i(t)$, $\sigma_i'(t) = \sigma_i(t) + \delta\sigma_i(t)$. A quadratic distance measure is given by

$$
K_i(\sigma_i'(t), \sigma_i(t)) \equiv \frac{\mu}{2}\left[\sigma_i'(t) - \sigma_i(t)\right]^2 ,
\tag{4.12}
$$

with $\mu > 0$. Note the similarity to kinetic energy in classical mechanics. Analogously a *potential* $V_i$ is defined by

$$
V_i(\sigma_i'(t), \boldsymbol{\sigma}(t)) \equiv \left|(1 - \sigma_i'(t))R(f_i(\boldsymbol{\sigma}(t))) - \sigma_i'(t)R(-f_i(\boldsymbol{\sigma}(t)))\right| ,
\tag{4.13}
$$

which counts the number of possible interactions for the displaced state $\sigma_i'(t)$. Depending on $\sigma_i'(t)$, (4.13) will reduce to $V_i(\sigma_i'(t), \boldsymbol{\sigma}(t)) = |R(\pm f_i(\boldsymbol{\sigma}(t)))|$. Finally, we define the *balance function*,

$$B_i \equiv K_i + V_i \ . \tag{4.14}$$

While $K_i$ measures the *actual* activity in the system – it counts all state changes – the potential $V_i$ counts the *potential* activity in the newly obtained states. $B_i$ contains the full dynamical information of (4.9), which can now be expressed through a *variational principle*. Given $\boldsymbol{\sigma}(t)$, the solution $\sigma_i(t+1)$ of (4.11) is identical to the value of $\sigma_i'(t)$ for which $B_i$ assumes its minimum, that is,

$$\sigma_i(t+1) = \underset{\sigma_i'(t)}{\mathrm{argmin}} \left[ B_i\left(\sigma_i'(t), \boldsymbol{\sigma}(t)\right) \right] \ , \tag{4.15}$$

with $\underset{x}{\mathrm{argmin}} \left[ f(x) \right]$ denoting the value of $x$ for which $f(x)$ takes its minimum. This is proved in [34].

There exists a natural stochastic generalization of this diversity dynamics. In (4.11) a state transition $\sigma_i(t) \to \sigma_i(t+1)$ is *determined* by $\Delta\sigma_i(t) \in \{-1, 0, 1\}$. For the stochastic case we specify *transition probabilities* for this transition. From the variational principle (4.15), it follows that (4.11) always minimizes the balance function $B_i$. In the stochastic variant we assume that the lower $B_i$, the higher the probability of finding the system in the respective configuration $\sigma_i(t)$. In analogy to spin systems this probability is assumed to be a Boltzmann factor

$$p(\sigma_i(t)) \propto \mathrm{e}^{-\beta B_i(\boldsymbol{\sigma}(t))} \ , \tag{4.16}$$

with $\beta \equiv 1/T$ the inverse temperature. To obtain transition probabilities we demand detailed balance

$$\frac{p(\sigma_i(t) \to \hat{\sigma}_i(t))}{p(\hat{\sigma}_i(t) \to \sigma_i(t))} = \frac{p(\hat{\sigma}_i(t))}{p(\sigma_i(t))} = \mathrm{e}^{-\beta(\hat{B}_i - B_i)} \ , \tag{4.17}$$

with $\hat{B}_i \equiv B_i(\hat{\sigma}_i(t), \sigma(t)_{j \neq i})$. There are several ways to specify a transition probability such that (4.17) is satisfied. Here we use so-called Metropolis transition probabilities $p(\sigma_i(t) \to \hat{\sigma}_i(t)) = 1$ if $\hat{B}_i - B_i < 0$ and $p(\sigma_i(t) \to \hat{\sigma}_i(t)) = \exp[-\beta(\hat{B}_i - B_i)]$ otherwise. The stochastic diversity dynamics is fully specified by setting $\sigma_i(t+1) = \hat{\sigma}_i(t)$.

### 4.6.2 Mean-Field Approximation

Denote the expectation value of $\sigma_i(t)$ by $q_i(t) = \langle \sigma_i(t) \rangle$ and assume that the probability distribution factorizes, $p(\boldsymbol{\sigma}(t)) = \prod_i p_i(\sigma_i(t))$, with $p_i(\sigma_i(t)) =$

$(1 - q_i(t))\delta_{\sigma_i(t),0} + q_i(t)\delta_{\sigma_i(t),1}$. Note that this is a strong assumption. In this mean-field approximation the Boltzmann–Gibbs entropy $s$ for element $i$ is

$$s(\sigma_i(t)) = -\langle \ln p_i(\sigma_i(t)) \rangle \equiv s(q_i(t)) \,,$$
$$s(q_i(t)) = -\big(1 - q_i(t)\big) \ln(1 - q_i(t)) - q_i(t) \ln q_i(t) \,. \tag{4.18}$$

The "free energy" functional $\phi(\sigma_i(t))$ for the system becomes

$$\phi(q_i(t)) = \langle B_i \rangle_{p(\sigma(t))} - \frac{s(q_i(t))}{\beta} \,. \tag{4.19}$$

The asymptotic state of species $i$, $q_i(t \to \infty) \equiv q_i$, is then given by a minimum in free energy. The necessary condition for this, $\partial \phi(q_i)/\partial q_i = 0$, is

$$\frac{\partial \langle B_i \rangle}{\partial q_i} + \frac{1}{\beta} \ln\left(\frac{q_i}{1 - q_i}\right) = 0 \,,$$

and

$$q_i = \frac{1}{2}\left\{ \tanh\left[ -\frac{\beta}{2} \frac{\partial \langle B_i \rangle}{\partial q_i} \right] + 1 \right\} \,. \tag{4.20}$$

The self-consistent solution of (4.20) yields the asymptotic configuration.

Let us now calculate $\langle B_i \rangle$ for the stochastic scenario for random interaction topologies specified by rule densities $r^{\pm}$ and constructive/destructive set sizes $n^{\pm}$. This means that to produce (destroy) an element, $n^+$ ($n^-$) elements are necessary.[2] We start with an expression for the probability, such that $f_i(\sigma(t))$ is positive (negative), $p^{\pm}$. Define $p(k, r^+)$ as the probability that there are exactly $k$ active constructive interactions, that is, $p(k, r^+) \equiv \binom{r^+}{k} q^{n^+ k}(1 - q^{n^+})^{r^+ - k}$. Analogously, $q(l, r^-)$ is the probability that exactly $l$ out of $r^-$ destructive interactions are active. Then

$$p^+ = \sum_{k=1}^{r^+} p(k, r^+) \sum_{l=0}^{\min(k-1,r^-)} q(l, r^-) \,,$$
$$p^- = \sum_{l=1}^{r^-} q(l, r^-) \sum_{k=0}^{\min(l-1,r^+)} p(k, r^+) \,. \tag{4.21}$$

The average distance follows to be

$$\langle K_i \rangle_{p(\sigma)} = \frac{1}{2}\Big((1 - q_i)p^+ + q_i p^-\Big)^2 \,, \tag{4.22}$$

---

[2] For nonuniform productive/destructive set sizes in the system a power-set notation, as suggested in [34], might be practical.

and, abbreviating $f_i(\boldsymbol{\sigma}(t)) \equiv f_i$, the potential is

$$\langle V_i \rangle_{p(\sigma)} = |(1 - q_i)\, R(f_i) - q_i\, R(-f_i)| \ . \tag{4.23}$$

Taking the derivative with respect to $q_i$, we find the mean-field result

$$\frac{\partial \langle B_i \rangle}{\partial q_i} = -r^+ q^{n^+} + r^- q^{n^-} - \left[(1 - q)p^+ + qp^-\right](p^+ - p^-)\,, \tag{4.24}$$

with the self-consistent solution for $q$ to be found through

$$q = \frac{1}{2}\left\{ \tanh\left[\frac{\beta}{2}\left(r^+ q^{n^+} - r^- q^{n^-} + \left[(1 - q)p^+ + qp^-\right](p^+ - p^-)\right)\right] + 1 \right\} \ . \tag{4.25}$$

We compare the corresponding mean-field prediction to results of a Metropolis simulation of the asymptotic abundances in Fig. 4.9. These results are in close relation to the special cases reported in [15, 16].

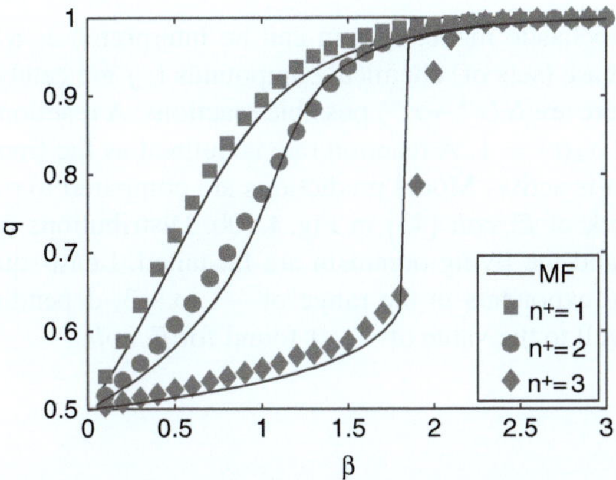


**Fig. 4.9** Asymptotic diversity $q$ as a function of inverse temperature $\beta$ for the mean-field approach (*lines*) and Metropolis simulations (*symbols*) of the stochastic dynamics for productive set sizes of $n^+ = 1, 2, 3$. The rule densities were $r^+ = 3$ and $r^- = 1$; the destructive set size was $n^- = 2$

## 4.7 Applicability to Specific Evolutionary Systems

In this last section we demonstrate in three examples that the presented general model can be easily specified to be applicable to data obtained from actual evolutionary systems.

### 4.7.1 Macroeconomic Instruments

If one models the diversity dynamics of an economy the number of active productions is a measure of the total productive output of an economy. The statistics of the model time series can be compared to the real-world equivalent, the gross domestic product (GDP) [33]. A production/destruction is active iff $\sigma_i(t) = \sigma_j(t) = \alpha_{ijk}^{\pm} = 1$ (for production/destruction set sizes of $n^{\pm} = 2$). In Fig. 4.10a we show a comparison of the actual distribution of percentage increments of the GDP of the UK and the number of active productions from the stochastic model for two different parameter settings. In one setting $\beta = 15$, $r^{\pm} = 5$, and $n^{\pm} = 2$ are used; the other has a denser interaction topology, $\beta = 15$, $r^+ = 8$, $r^- = 12$, $n^{\pm} = 2$. Both model and real-world GDP time series produce fat-tailed distributions, with power exponents in the range between $-2$ and $-4$. Exponents within this range are not only found in the GDP time series of other industrialized countries but also for a wide range of model parameters, see for example [33].

### 4.7.2 Chemical Reaction Networks

The presented stochastic model system can be interpreted as a chemical reaction network. In this case (sets of) chemical compounds $i$, $j$ are catalyzing or degrading compound $k$. There are $N(r^+ + r^-)$ possible reactions. A reaction is active if $\alpha_{ijk}^{\pm} = 1$, $\sigma_i(t) = 1$ and $\sigma_j(t) = 1$. A reaction rate is defined as the frequency with which a certain reaction is active. Model predictions are compared to reaction rates in the metabolic network of *E. coli* [43] in Fig. 4.10b. Distributions of reaction rates in both the model and the living organism are fat-tailed. Least-squares fits to model power laws yield exponents in the range of $-1$ to $-3$, depending on parameters. This compares well to the value of $\sim -1$ found for *E. coli*.

### 4.7.3 Lifetime Distributions of Species

In a macro-ecological setting one can compare the distribution of lifetimes of species from the model with the distribution of species lifetimes in the fossil record [44]. The lifetime of a given model species is the number of iterations for which it exists without interruption. The comparison is shown in Fig. 4.10c. Once more one finds power laws in the model with exponents between $-2$ and $-4$, depending on parameters. This matches well with paleontologic data, which suggest slopes between $-2$ and $-3$. Note that there is a strong dependence on the values used for the fit [21]. We work with an intermediate choice in Fig. 4.10c.

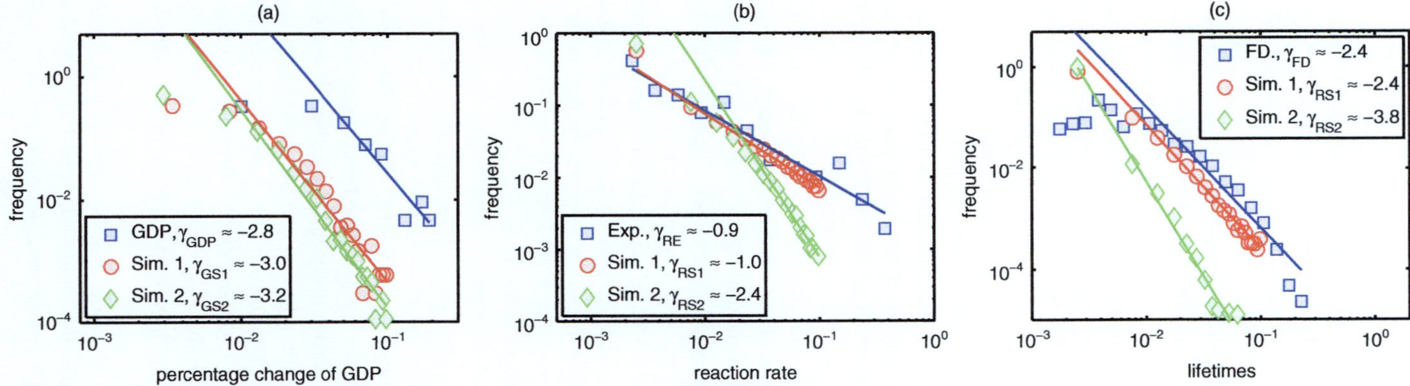


**Fig. 4.10** Comparison of some systemic observables of evolutionary systems with model predictions for two different parameter settings: Simulation 1 with $\beta = 15$, $r^{\pm} = 5$, $n^{\pm} = 2$ and Simulation 2 with $\beta = 15$, $r^{+} = 8$, $r^{-} = 12$, $n^{\pm} = 2$. Each distribution is normalized (sum over all data points equals one). **a** Percentage change of GDP of the UK since 1950 as compared to the model, see also Fig. 4.1. **b** Reaction rate distribution in the metabolic network of *E. coli* in comparison to the model. **c** Species lifetime distributions as found in the fossil record, again in agreement within the range the model is able to explain

## 4.8 Conclusions

We have presented a simple model for dynamical systems undergoing evolution that was presented in detail in [32–34]. We have explicitly taken into account that the existence or absence of species or elements in a system can strongly influence the rate of production or reproduction of existing elements. In other words, the proposed model fully captures the concept of a dynamically co-evolving fitness landscape.

The basic components of the model are a large – even infinite – space of goods, species, chemical compounds, etc. The abundances of these elements are captured in a state vector $\sigma(t)$. These elements can be combined to produce new elements, which, first, contribute to diversity, and second, might have a negative impact on the production rates of others. The possibilities of all combinations of elements are encoded in a rule table (tensor $\alpha$), which is time invariant and exists whether elements exist or not. The overlap of this rule table with the actual state of the system $\sigma(t)$ determines which elements will be produced and/or eliminated in the near future. These productions or destructions can be captured in the active production matrices $A(t)$, which vary over time and capture the state of creative and destructive activities in the system. Regardless of the details of the implementation of the actual production, competition, and update mechanisms, the model generically leads to a dynamics that is characterized by phases of relative stability in terms of diversity, followed by phases of massive restructuring. These dynamical modes can be interpreted as punctuated equilibria in biology, or Schumpeterian business cycles in economics. The presented framework suggests that *systemic properties* of evolutionary systems, such as their proneness to collapse or their potential for diversification, can be understood in terms of self-organized critical systems. Further we show that the actual active production networks $A$ (which in principle can be observed) determine the diversity of the system. In particular the importance of cyclic structures in the active production networks is demonstrated. These structures are related to the maximum eigenvalues of $A$.

We propose a mathematical framework that is inspired by statistical mechanics and express evolutionary system dynamics through a variational principle. This approach allows the phase diagram of evolution systems to be studied systematically, which was initially explored in [15, 16]. Within this framework mean-field approximations are easily obtained, which can be compared to Monte Carlo simulations of the model. We find surprisingly good overlap. Even though the model is phrased in general terms it is shown how it can be made rather specific such that its predictions can be used to understand series of experimental data ranging from the fossil record to macroeconomic instruments. We find that the model of constructive and destructive interactions reproduces stylized facts of man-made (economies) and natural evolutionary systems (metabolic networks, macro-ecology) across several orders of magnitude.

To some extent the presented model can be seen as a generalization of a series of previously introduced models, which are contained as special cases. A particular case of the mean-field approach (when destruction is off, $\alpha^- = 0$ for all entries) is identical to the random catalytic networks studied in [15]. As discussed in [33], $f_i$ in our stochastic model plays an identical role to that of the randomly assigned fitness values in the Bak–Sneppen model [19]. To recover the $NK$ model [14] as a special case of our model, associate each species with a bit-string, and assign a random fitness value to it. However, fitness in our framework emerges as a topological property of the entire system plus the set of existing species, whereas in $NK$ models fitness is basically a mapping of random numbers to bit-strings. In contrast to our (highly) nonlinear model, in the model of Solé and Manrubia [20] only linear interactions are allowed and new species are created not through endogenous recombinations, but by an explicit mutation mechanism.

Finally, we mention that our model systematically expands on the idea that the concept of fitness is an a posteriori concept. Fitness in the traditional sense can of course be reconstructed for every time step in our model. It is nothing but the co-evolving network of rates of the actually active (productive) processes at a given time, see [33]. It becomes clear that fitness cannot be used as a concept with much predictive power, even if a hypothetical computational entity, "Darwin's Demon" [34], knowing all mutual influences of all species at any given time, existed.

**Acknowledgments**  I am most grateful to Rudolf Hanel and Peter Klimek, with whom I have had the great pleasure of working on the topic over the past years.

# References


1. V. Volterra, ICES J. Mar. Sci. **3**, 3 (1928)
2. J.F. Crow, M. Kimura, *An Introduction to Population Genetics* (Burgess, Minneapolis, MN, 1970)
3. J. Hofbauer, K. Sigmund, *Evolutionary Games and Population Dynamics* (Cambridge University Press, Cambridge, 1998)
4. L. Van Valen, Evolutionary Theory **1**, 1 (1973)
5. R.V. Solé, J. Bascompte, *Self-organization in Complex Ecosystems*, Monographs in Population Biology, vol. 42 (Princeton University Press, Princeton, 2006)
6. S.A. Kauffman, S. Thurner, R. Hanel, *The evolving web of future wealth* (scientificamerican.com), http://www.scientificamerican.com/article.cfm?id=the-evolving-web-of-future-wealth. Accessed May 2008
7. M. Eigen, P. Schuster, *The Hypercycle* (Springer, Berlin, Heidelberg, 1979)
8. P. Schuster, K. Sigmund, J. Theor. Biol. **100**, 533 (1983)
9. M. Eigen, J. McCaskill, P. Schuster, Adv. Chem. Phys. **75**, 149 (1989)
10. W. Fontana, in *Artificial Life II*, ed. by C.G. Langton, C. Taylor, J.D. Farmer, S. Rasmussen (Addison-Wesley, Redwood City, CA, 1992), pp. 159–210
11. J.D. Farmer, S.A. Kauffman, N.H. Packard, Physica D **22**, 50 (1986)
12. S.A. Kauffman, *Origins of Order: Self-organization and Selection in Evolution,* (Oxford University Press, New York, NY, 1993)
13. P.F. Stadler, W. Fontana, J.H. Miller, Physica D **63**, 378 (1993)
14. S.A. Kauffman, J. Theor. Biol. **22**, 437 (1969)
15. R. Hanel, S.A. Kauffman, S. Thurner, Phys. Rev. E **72**, 036117 (2005)
16. R. Hanel, S.A. Kauffman, S. Thurner, Phys. Rev. E **76**, 036110 (2007)
17. B. Arthur, Complexity **11**, 23 (2006)
18. S. Jain, S. Krishna, Phys. Rev. Lett. **81**, 5684 (1998). S. Jain, S. Krishna, Proc. Natl. Acad. Sci. USA **99**, 2055 (2002)

19. P. Bak, K. Sneppen, Phys. Rev. Lett. **71**, 4083 (1993)
20. R.V. Solé, S.C. Manrubia, Phys. Rev. E **54**, R42 (1996)
21. M.E.J. Newman, R.G. Palmer, *Modeling Extinction* (Oxford University Press, New York, NY, 2003)
22. P. Dittrich, P. Speroni di Fenizio, Bull. Math. Biol. **69**, 1199 (2007)
23. M. Aldana, S. Coppersmith, L.P. Kadanoff, in *Perspectives and Problems in Nonlinear Science: A Celebratory Volume in Honor of Lawrence Sirovich*, ed. by E. Kaplan, J.E. Marsden, K.R. Sreenivasan (Springer, New York, 2002), p. 23
24. P. Klimek, S. Thurner, R. Hanel, J. Theor. Biol. **256**, 142 (2009)
25. P. Romer, J. Pol. Econ. **98**, 71 (1990)
26. B. Arthur, *The Nature of Technology: What It Is and How It Evolves* (Free Press, Simon & Schuster, New York, 2009)
27. J.A. Schumpeter, *Capitalism, Socialism and Democracy* (Harper, New York, 1947)
28. V. Sood, S. Redner, Phys. Rev. Lett. **94**, 178701 (2005)
29. S. Galam, J. Math. Psychol. **30**, 426 (1986)
30. R. Lambiotte, S. Thurner, R. Hanel, Phys. Rev. E **76**, 046101 (2007)
31. R. Hanel, S. Thurner, Eur. Phys. J. B **62**, 327 (2008)
32. S. Thurner, R. Hanel, P. Klimek, Physica A **389**, 747 (2010)
33. S. Thurner, P. Klimek, R. Hanel, New J. Phys. **12**, 075029 (2010); arXiv:0909.3482v1 [physics.soc-ph]
34. P. Klimek, S. Thurner, R. Hanel, Phys. Rev. E **82**, 011901 (2010)
35. J.D. Farmer, J. Geanakoplos, arXiv:0803.2996v1 [q-fin.GN] (2008)
36. Office for National Statistics, http://www.statistics.gov.uk/StatBase/TSDdownload1.asp. Accessed October 2010
37. Groningen Growth and Development Centre, http://www.ggdc.net/databases/hna.htm. Accessed October 2010
38. U.S. Bureau of the Census, *Historical Statistics of the United States, Colonial Times to 1970, Bicentennial Edition, Parts 1 & 2* (Washington, DC, 1975); More recent data: U.S. Census Bureau, http://factfinder.census.gov. Accessed October 2010
39. http://en.wikipedia.org/wiki/Biodiversity. Accessed October 2010
40. W.W. Leontief, *Input–Output Economics* (Oxford University Press, New York, 1986)
41. P. Bak, C. Tang, K. Wiesenfeld, Phys. Rev. Lett. **59** , 381 (1987)
42. R.J. Bagley, J.D. Farmer, W. Fontana, in *Artificial Life II*, ed. by C.G. Langton, C. Taylor, J.D. Farmer, S. Rasmussen (Addison-Wesley, Redwood City, CA, 1992), pp. 141–158
43. M. Emmerling, M. Dauner, A. Ponti, J. Fiaux, M. Hochuli, T. Szyperski, K. Wüthrich, J. E. Bailey, U. Sauer, J. Bacteriol. **184**, 152 (2002)
44. J.J. Sepkoski, Jr., Milwaukee Public Museum Contrib. Biol. Geol. **83**, 156 (1992)

# Chapter 5
# Can We Recognize an Innovation? Perspective from an Evolving Network Model

Sanjay Jain and Sandeep Krishna

**Abstract** "Innovations" are central to the evolution of societies and the evolution of life. But what constitutes an innovation? We can often agree after the event, when its consequences and impact over a long term are known, whether something was an innovation, and whether it was a "big" innovation or a "minor" one. But can we recognize an innovation "on the fly" as it appears? Successful entrepreneurs often can. Is it possible to formalize that intuition? We discuss this question in the setting of a mathematical model of evolving networks. The model exhibits self-organization, growth, stasis, and collapse of a complex system with many interacting components, reminiscent of real-world phenomena. A notion of "innovation" is formulated in terms of graph-theoretic constructs and other dynamical variables of the model. A new node in the graph gives rise to an innovation, provided it links up "appropriately" with existing nodes; in this view innovation necessarily depends upon the existing context. We show that innovations, as defined by us, play a major role in the birth, growth, and destruction of organizational structures. Furthermore, innovations can be categorized in terms of their graph-theoretic structure as they appear. Different structural classes of innovation have potentially different qualitative consequences for the future evolution of the system, some minor and some major. Possible general lessons from this specific model are briefly discussed.

## 5.1 Introduction

In everyday language, the noun innovation stands for something new that brings about a change; it has a positive connotation. Innovations occur in all branches of human activity – in the world of ideas, in social organization, in technology. Innovations may arise by conscious and purposeful activity, or serendipitously; in either case, innovations by humans are a consequence of cognitive processes. However, the word innovation does not always refer to a product of cognitive activity. In

S. Jain (✉)
Department of Physics and Astrophysics, University of Delhi, Delhi 110 007, India
e-mail: jain_physics@yahoo.co.in

H. Meyer-Ortmanns, S. Thurner (eds.), *Principles of Evolution*,
The Frontiers Collection, DOI 10.1007/978-3-642-18137-5_5,
© Springer-Verlag Berlin Heidelberg 2011

biology, we say, for example, that photosynthesis, multicellularity, and the eye were evolutionary innovations. These were products not of any cognitive activity, but of biological evolution. It nevertheless seems fair to regard them as innovations; these novelties certainly transformed the way organisms made a living. The notion of innovation seems to presuppose a context provided by a complex evolutionary dynamics; for example, in everyday language the formation of the earth, or even the first star, is not normally referred to as an innovation.

Innovations are a crucial driving force in chemical, biological and social systems, and it is useful to have an analytical framework to describe them. This subject has a long history in the social sciences (see, e.g., [1, 2]). Here we adopt a somewhat different approach. We give a mathematical example of a complex system that seems to be rich enough to exhibit what one might intuitively call innovation, and yet simple enough for the notion of innovation to be mathematically defined and its consequences analytically studied. The virtue of such a stylized example is that it might stimulate further discussion about innovation, and possibly help clarify the notion in more realistic situations.

Innovations can have "constructive" and "destructive" consequences at the same time. The advent of the automobile (widely regarded as a positive development) was certainly traumatic for the horse-drawn carriage industry and several other industries that depended upon it. When aerobic organisms appeared on the earth, their more efficient energy metabolism similarly caused a large extinction of several anaerobic species [3]. The latter example has a double irony. Over the first 2 billion years of life on earth, there was not much oxygen in the earth's environment. Anaerobic creatures (which did not use free oxygen for their metabolism) survived, adapted, innovated new mechanisms (e.g., photosynthesis) in this environment, and spread all over the earth. Oxygen in the earth's environment was largely a by-product of photosynthetic anaerobic life, a consequence of anaerobic life's "success". However, once oxygen was present in the environment in a substantial quantity, it set the stage for another innovation, the emergence of aerobic organisms which used this oxygen. Because of their greater metabolic efficiency, the aerobic organisms out-competed and decimated the anaerobic ones. In a very real sense, therefore, anaerobic organisms were victims of their own success. Innovation has this dynamic relationship with "context": what constitutes "successful" innovation depends upon the context, and successful innovation then alters the context. Our mathematical example exhibits this dynamic and explicitly illustrates the two-faced nature of innovation. We show that the ups and downs of our evolutionary system as a whole are also crucially related to innovation.

## 5.2 A Framework for Modeling Innovation: Graph Theory and Dynamical Systems

Systems characterized by complex networks are often represented in terms of a graph consisting of nodes and links. The nodes represent the basic components of the system, and links between them their mutual interactions. A graph representation is quite flexible and can represent a large variety of situations [4]. For a society, nodes can represent various agents, such as individuals, firms, and institutions, as well as goods and processes. Links between nodes can represent various kinds of interactions, such as kinship or communication links between individuals, inclusion links (e.g., a directed link from a node representing an individual to a node representing a firm implying that the individual is a member of the firm), production links (from a firm to a good that it produces), links that specify the technological web (for every process node, incoming links from all the goods it needs as input and outgoing links to every good it produces), etc. In an ecological setting, nodes can represent biological species, and links their predator–prey or other interactions. In cellular biology, nodes might represent molecules such as metabolites and proteins as well as genes, and links their biochemical interactions.

A graph representation is useful for describing several kinds of innovation. Often, an innovation is a new good, process, firm, or institution. This is easily represented by inserting a new node in the graph, together with its links to existing nodes. Of course, not every such insertion can be called an innovation; other conditions have to be imposed. The existing structure of the graph provides one aspect of the "context" in which a prospective innovation is to be judged, reflecting its "location" or relationship with other entities. In this formulation it is clear that innovations such as the ones mentioned above are necessarily a change in the graph structure. Thus a useful modeling framework for innovations is one where graphs are not static but change with time. In real systems graphs are always evolving: new nodes and links constantly appear, and old ones often disappear as individuals, firms, and institutions die; goods lose their utility; species become extinct; or any of these nodes lose some of their former interactions. It is in such a scenario that certain kinds of structures and events appear that earn the nomenclature "innovation". We will be interested in a model where a graph evolves by the deletion of nodes and links as well as the insertion of new ones. Insertions will occasionally give rise to innovations. We will show that innovations fall in different categories that can be distinguished from each other by analyzing the instantaneous change in the graph structure caused by the insertion, locally as well as globally. We will argue that these different "structural" categories have different "dynamical" consequences for the "well-being" of other nodes and the evolution of the system as a whole in the short as well as long run.

In addition to an evolving graph, another ingredient seems to be required for modeling innovation in the present approach: a graph dependent dynamics of some variables associated with nodes or links. In a society, for example, there are flows of information, goods and money between individuals that depend upon their mutual linkages, which affect node attributes such as individual wealth, power, etc. The structure of an ecological food web affects the populations of its species. Thus, a change in the underlying graph structure has a direct impact on its "node variables". Deciding whether a particular graph change constitutes an innovation must necessarily involve an evaluation of how variables such as individual wealth and populations are affected by it. Changes in these variables in turn trigger further changes in the graph itself, sometimes leading to a cascade of changes in the graph and other variables. For example, the decline in wealth of a firm (node variable) may cause it to collapse; the removal of the corresponding node from the market (graph change) may cause a cascade of collapses. The invention of a new product (a new node in the graph) that causes the wealth of the firm inventing it to rise (change in a node variable) may be emulated by other firms, causing new linkages and further new products.

In order to "recognize an innovation on the fly" it thus seems reasonable to have a framework that has (a) a graph or graphs representing the network of interactions of the components of the system, (b) the possibility of graph evolution (the appearance and disappearance of nodes and links) and (c) a graph-dependent dynamics of node or link variables that in turn has a feedback upon the graph evolution. The example discussed below has these features. They are implemented in a simple framework that has only one type of node, one type of link, and only one type of node variable.

## 5.3 Definition of the Model System

The example is a mathematical model [5] motivated by the origin of life problem, in particular, the question of how complex molecular organizations could have emerged through prebiotic chemical evolution [6–10]. There are $s$ interacting molecular species in a "prebiotic pond", labeled by $i \in S \equiv \{1, 2, \ldots, s\}$. Their interactions are represented by the links of a directed graph, of which these species are nodes. The graph is defined by its $s \times s$ adjacency matrix $C = (c_{ij})$, with $c_{ij} = 1$ if there exists a link from node $j$ to node $i$ (chemically that means that species $j$ is a catalyst for the production of species $i$), and $c_{ij} = 0$ otherwise. $c_{ii}$ is assumed zero for all $i$: no species in the pond is self-replicating. Initially the graph is chosen randomly, each $c_{ij}$ for $i \neq j$ is chosen to be unity with a small probability $p$ and zero with probability $1 - p$. $p$ represents the "catalytic probability" that a given molecular species will catalyze the production of another randomly chosen one [11].

The pond sits by the side of a large body of water such as a sea or river, and periodically experiences tides or floods, which can flush out molecular species from the pond and bring in new ones, changing the network. We use a simple graph update rule in which exactly one node is removed from the graph (along with all its links) and one new node is added, whose links with the remaining $s - 1$ nodes are chosen randomly with the same probability $p$. We adopt the rule that the species with the least relative population (or, if several species share the least relative population, one of them chosen randomly) is removed. This is where selection enters the model: species with smaller populations are punished. This is an example of "extremal" selection [10] in that the least populated species is removed; the results of the model are robust to relaxing the extremality assumption [12, 13].

In order to determine which node will have the least population, we specify a population dynamics that depends upon the network. The dynamics of the relative populations, $x_i$ ($0 \leq x_i \leq 1$, $\sum_{i=1}^{s} x_i = 1$), is given by

$$\dot{x}_i = \sum_{j=1}^{s} c_{ij} x_j - x_i \sum_{k=1}^{s} \sum_{j=1}^{s} c_{kj} x_j. \tag{5.1}$$

This is a set of rate equations for catalyzed chemical reactions in a well stirred chemical reactor.[1] They implement the approximate fact that under certain simplifying assumptions a catalyst causes the population of whatever it catalyzes to grow at a rate proportional to its own (i.e., the catalyst's) population [5, 14]. Between successive tides or floods the set of species and hence the graph remains unchanged, and the model assumes that each $x_i$ reaches its attractor configuration $X_i$ under (5.1) before the next graph update. The species with the least $X_i$ is removed at the next graph update.

Starting from the initial random graph and random initial populations, the relative populations are evolved according to (5.1) until they reach the attractor $\mathbf{X}$, and then the graph is updated according to the above rules. The new incoming species is given a fixed relative population $x_0$; all $x_i$ are perturbed about their existing values (and rescaled to restore normalization). This process is iterated several times. Note that the model has two inbuilt time scales, the population dynamics relaxes on a fast time scale, and graph evolution on a slow time scale. The above model may be regarded as an evolutionary model in nonequilibrium statistical mechanics.

## 5.4 Time Evolution of the System

A sample run is depicted in Figs. 5.1 and 5.2. For concreteness, we will discuss this run in detail, describing the important events, processes, and the graph structures that arise, with an emphasis on the role of innovation. The same qualitative behavior is observed in hundreds of runs with the various parameter values. Quantitative estimates of average time scales, etc., as a function of the parameters $s$ and $p$ are discussed in [5, 14] and Appendix A. The robustness of the behavior to various changes of the model are discussed in [12, 13].

Broadly, Figs. 5.1 and 5.2 exhibit the following features: Initially, the graph is sparse and random (see Fig. 5.2a–d), and remains so until an autocatalytic set (ACS), defined below, arises by pure chance. On average the ACS arrives on a time scale $1/(p^2 s)$ in units of graph update time[2]; in the exhibited run it arrives at $n = 2854$ (Fig. 5.2e). In this initial regime, called the "random phase", the number of populated species, $s_1$, remains small. The appearance of the ACS transforms the population and network dynamics. The network self-organizes, its density of links increases (Fig. 5.1a), and the ACS expands (Fig. 5.2e–n) until it spans the entire network (as evidenced by $s_1$ becoming equal to $s$, at $n = 3880$, Figs. 5.1b and 5.2n). The ACS grows across the graph exponentially fast, on a time scale $1/p$ [5]. This growth is punctuated by occasional drops (e.g., Fig. 5.1b at $n = 3387$, see also

---

[1] See *Derivation of Equation (5.1)* in Appendix A.
[2] See *Time Scale for Appearance and Growth of the Dominant ACS* in Appendix A.

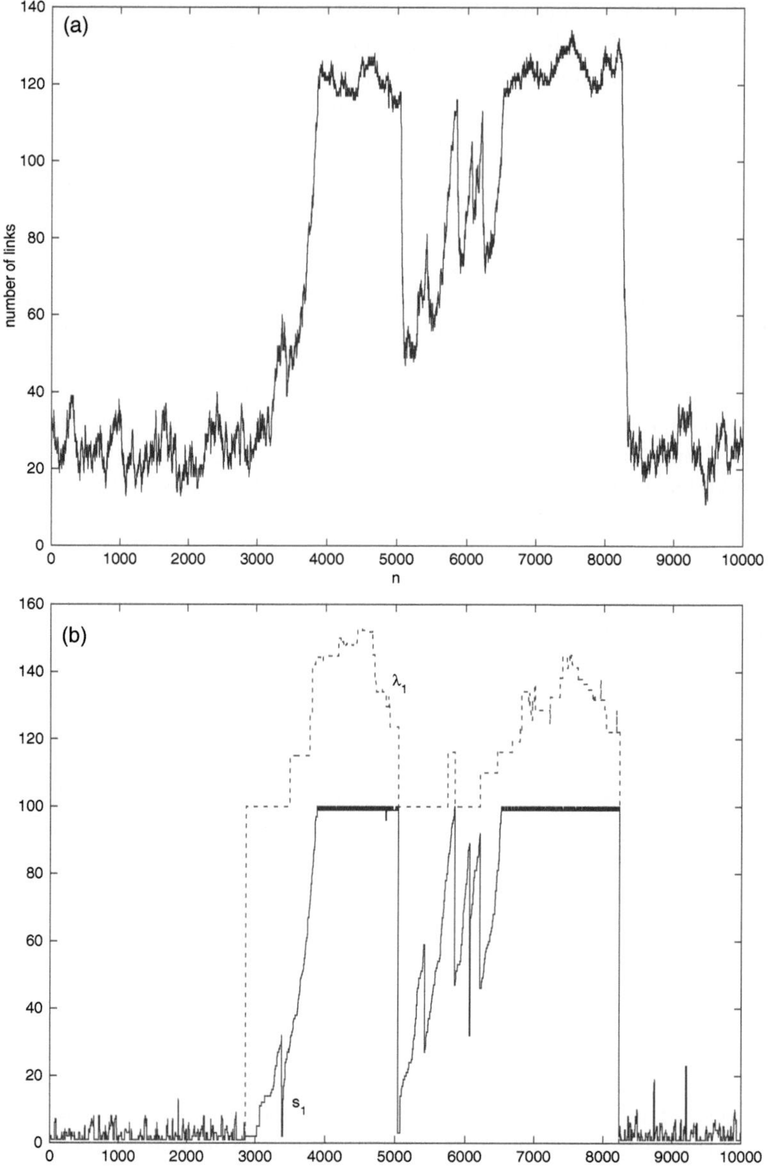

**Fig. 5.1** A run with parameter values $s = 100$ and $p = 0.0025$. The $x$-axis shows time, $n$ (= number of graph updates). **a** Number of links in the graph as a function of time. **b** *Continuous line*: $s_1$, the number of populated species in the attractor (= the number of nonzero components of $X_i$) as a function of time. *Dotted line*: $\lambda_1$, the largest eigenvalue of $C$ as a function of time. (The $\lambda_1$ values shown are 100 times the actual $\lambda_1$ values)

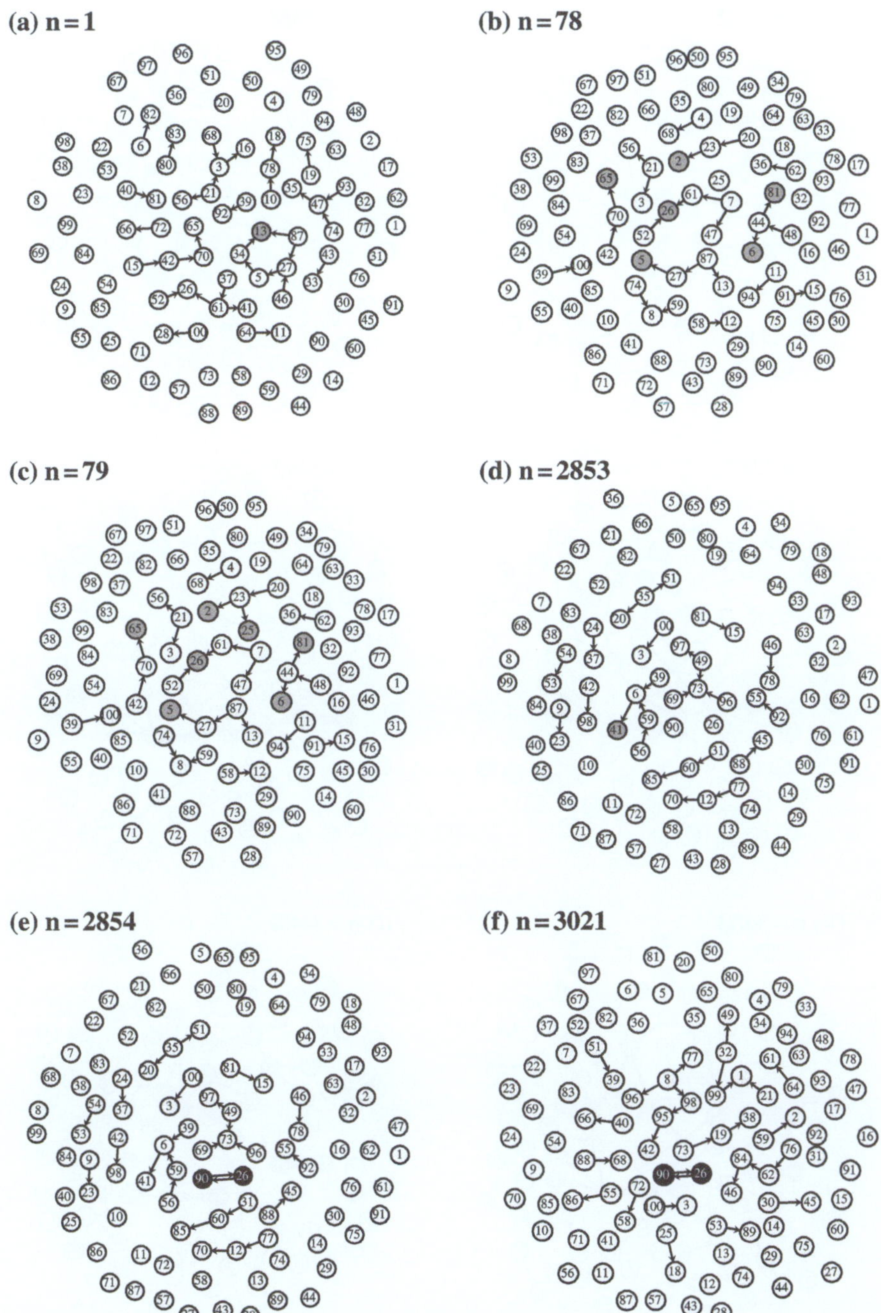

**Fig. 5.2** The structure of the evolving graph at various time instants for the run depicted in Fig. 5.1. Examples of several kinds of innovation and their consequences for the evolution of the system are shown (see text for details). Nodes with $X_i = 0$ are shown in *white*; according to the evolution rules all *white* nodes in a graph are equally likely to be picked for replacement at the next graph

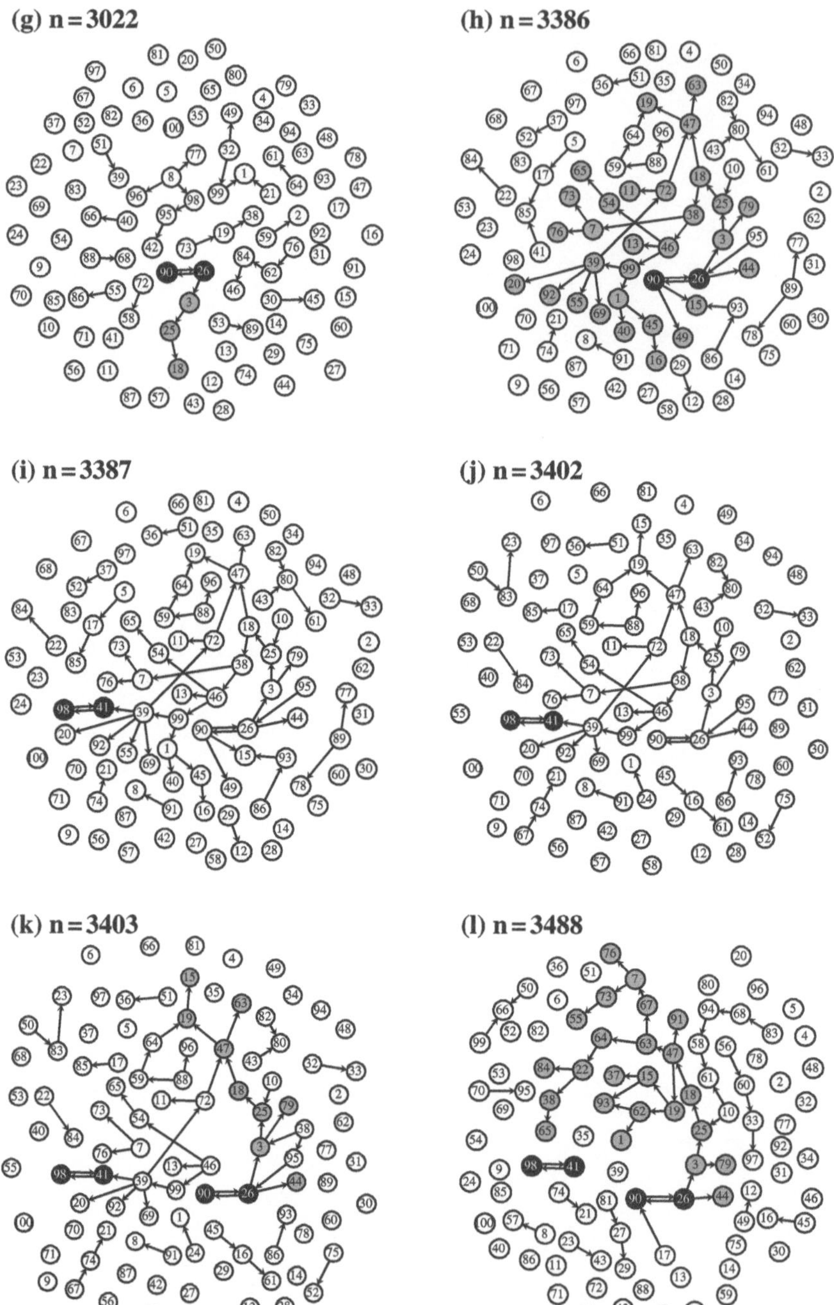

**Fig. 5.2** (continued) update. *Black* and *gray* nodes have $X_i > 0$. Thus the number of black and gray nodes in a graph equals $s_1$, plotted in Fig. 5.1b. *Black* nodes correspond to the core of the dominant ACS and *gray* nodes to its periphery. Only mutual links between the nodes are of significance, not their spatial location, which is arranged for visual convenience. The graphs are drawn using LEDA [15]

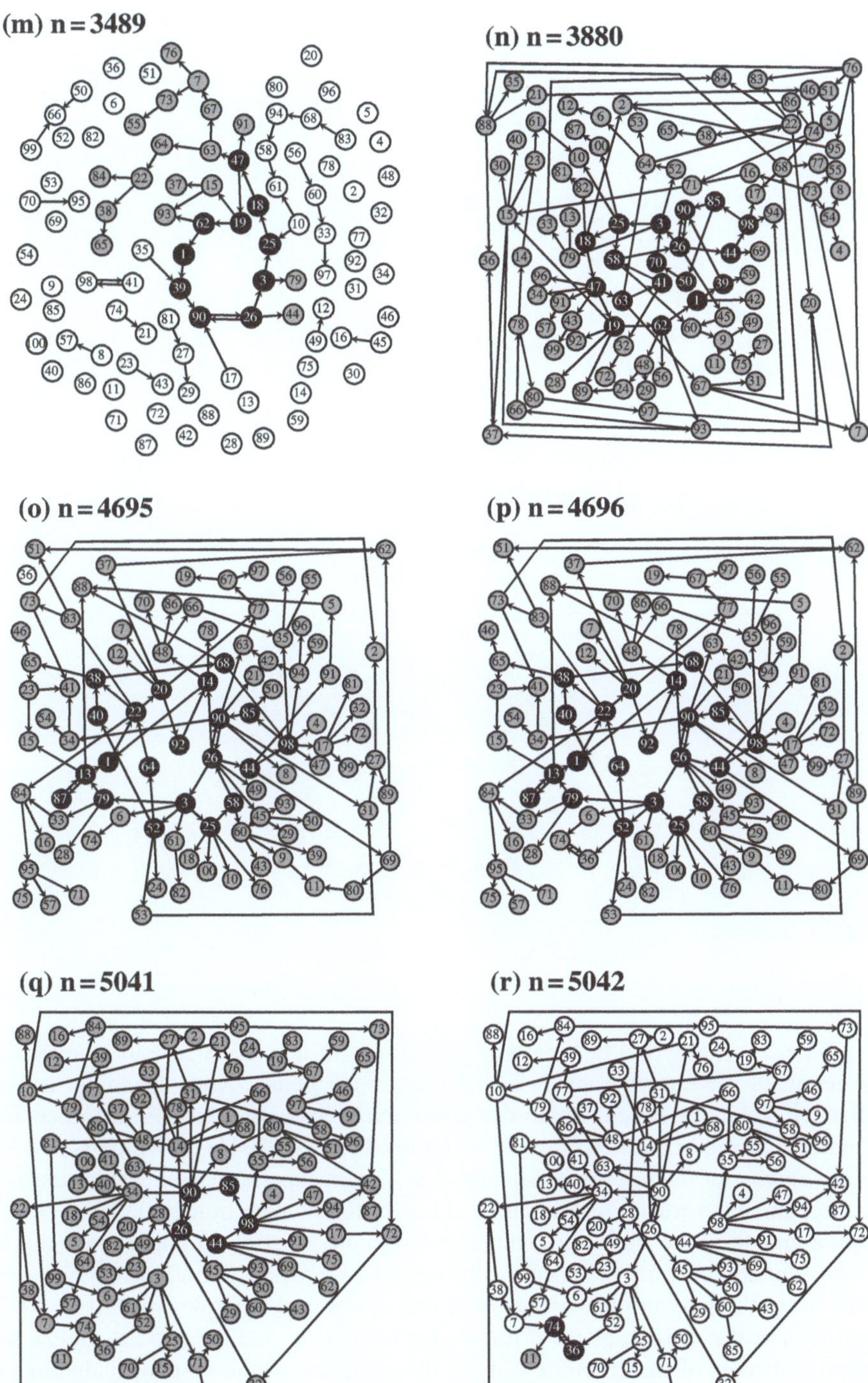



Fig. 5.2  (continued)

**(s) n = 8232**   **(t) n = 8233**

**(u) n = 10000**

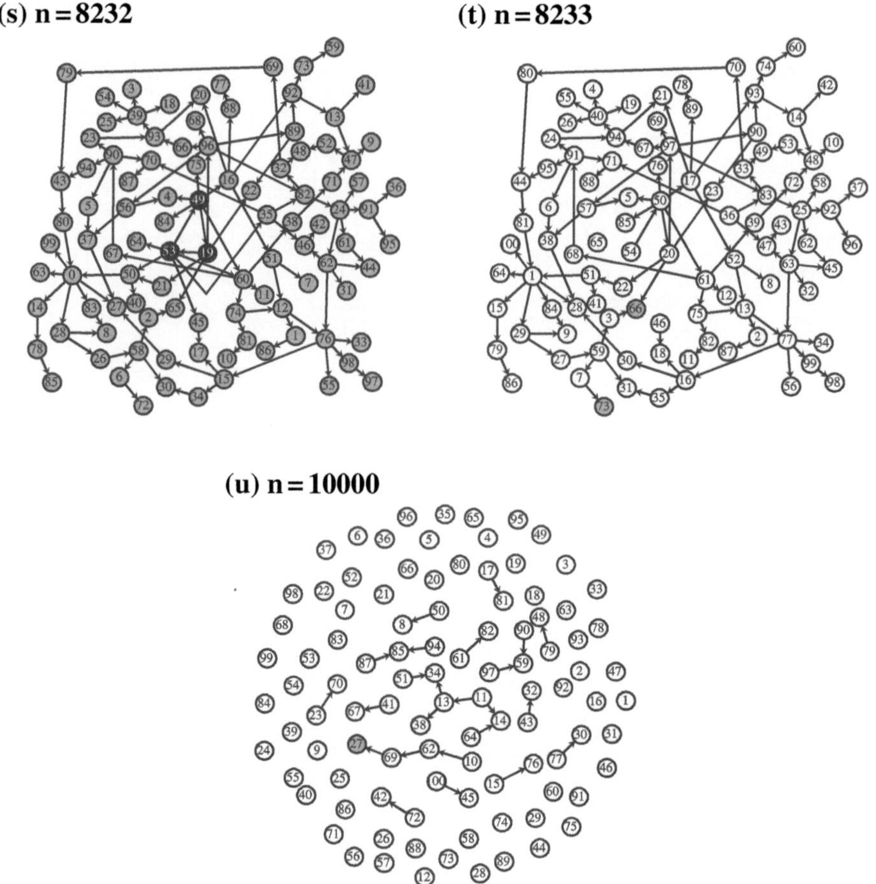

**Fig. 5.2** (continued)

Fig. 5.2h, i). The period between the appearance of a small ACS and its eventual spanning of the entire graph is called the *growth phase*. After spanning, a new effective dynamics arises, which can cause the previously robust ACS to become fragile, resulting in crashes (the first major one is at $n = 5041$), in which $s_1$ as well as the number of links drops drastically. The system experiences repeated rounds of crashes and recoveries (Fig. 5.2o–u, see [12, 16] for a longer time scale). The period after a growth phase and up to a major crash (more precisely, a major crash that is a *core-shift*, defined below) is called the *organized phase*. After a crash, the system ends up in the growth phase if an ACS still exists in the graph (as at $n = 5042$, Fig. 5.2r) or the random phase if it does not (as at $n = 8233$, Fig. 5.2t). Below we argue that most of the crucial events in the evolution of the system, including its self-organization and collapse, are caused by "innovation".

## 5.5 Innovation

The word "innovation" certainly connotes something new. In the present model, at each graph update a new structure enters the graph: the new node and its links with existing nodes. However, not every new thing qualifies as an innovation. In order for a novelty to bring about some change, it should confer some measure of at least temporary "success" to the new node. (A mutation must endow the organism in which it appears with some extra fitness, and a new product must have some sale, in order to qualify as an innovation.) In the present model, after a new node appears, the population dynamics takes the system to a new attractor of (5.1), which depends upon the mutual interactions of all the nodes. In the new attractor this node (denoted $k$) may go extinct, $X_k = 0$, or may be populated, $X_k > 0$. The only possible criterion of individual "success" in the present model is population. Thus we require the following minimal *performance criterion* for a new node $k$ to give rise to an innovation: $X_k$ should be greater than zero in the attractor that follows after that particular graph update. That is, the node should "survive" at least until the next graph update.

This is obviously a minimal requirement, a necessary condition, and one can argue that we should require of an innovation more than just this "minimal performance". A new node that brings about an innovation ought to transform the system or its future evolution in a more dramatic way than merely surviving until the next graph update. Note, however, that this minimal performance criterion nevertheless eliminates from consideration a large amount of novelty that is even less consequential. Out of the 9,999 new nodes that arise in the run of Fig. 5.1, as many as 8,929 have $X_k = 0$ in the next population attractor; only 1,070 have $X_k > 0$. Furthermore, the set of events with $X_k > 0$ can be systematically classified in the present model using a graph-theoretic description. Below we describe an exhaustive list of six categories of such events, each with a different level of impact on the system (see Fig. 5.3, discussed in detail below). One of these categories consists of nodes that disappear after a few graph updates leaving no subsequent trace on the system. Another category consists of nodes that have only an incremental impact. The remaining four categories cause (or can potentially cause) more drastic changes in the structure of the system, its population dynamics, and its future evolution.

In view of this classification it is possible to exclude one or more of these categories from the definition of innovation and keep only the more "consequential" ones. However, we have chosen to be more inclusive and will regard all the above categories of events as innovations. In other words we will regard the above "minimal" performance criterion as a "sufficient" one for innovation. Thus we will call the introduction of a new node $k$ and the graph structure so formed an *innovation* if $X_k > 0$ in the population attractor that immediately follows the event, i.e., if the node "survives" at least until the next graph update. This definition then includes both "small" and "big" innovations that can be recognized based on their graph-theoretic structure upon appearance.

As will be seen below, it turns out that a new node generates an innovation only if it links "appropriately" to "suitable" structures in the existing graph. Thus the above

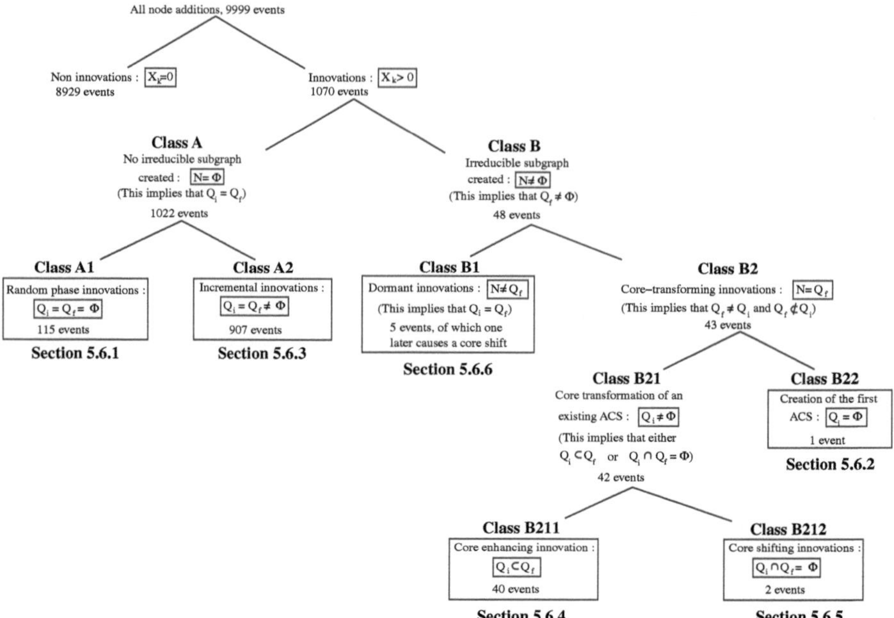

**Fig. 5.3** A hierarchy of innovations. Each node in this binary tree represents a class of node addition events. Each class has a name; the *small box* contains the mathematical definition of the class. All classes of events except the leaves of the tree are subdivided into two exhaustive and mutually exclusive subclasses represented by the two branches emanating downwards from the class. The number of events in each class pertains to the run of Fig. 5.1 with a total of 9,999 graph updates, between $n = 1$ (the initial graph) and $n = 10,000$. In that run, out of 9,999 node addition events, most (8,929 events) are not innovations. The rest (1,070 events), which are innovations, are classified according to their graph-theoretic structure. The classification is general; it is valid for all runs. $X_k$ is the relative population of the new node in the attractor of (5.1). $N$ stands for the new irreducible subgraph, if any, created by the new node. If the new node causes a new irreducible subgraph to be created, $N$ is the *maximal* irreducible subgraph that includes the new node. If not, $N = \emptyset$ (where $\emptyset$ stands for the empty set). $Q_i$ is the core of the graph just before the addition of the node and $Q_f$ the core just after the addition of the node. The six leaves of the innovation subtree are numbered (below the corresponding *box*) according to the subsection in which they are discussed in the main text. The graph-theoretic classes A, B, A1, B1, etc., are described in section 5.7 and Appendix B

definition makes the notion of innovation "context dependent". It also captures the idea that an innovation rests on new linkages between structures.

## 5.6 Six Categories of Innovation

### 5.6.1 A Short-Lived Innovation: Uncaring and Unviable Winners

There are situations where a node, say an agent in society or a species in an ecosystem, acquires the ability to live parasitically off another, without giving the system anything substantive in return. The parasite gains as long as the host survives, but often this situation doesn't last very long. The host dies, and eventually so does the parasite that is dependent on it. It is debatable whether the acquiring of such a parasitic ability should be termed an innovation, but from the local vantage point of the parasite, while the going is still good, it might seem like one.

Figure 5.2b, c show an example of an innovation of such a type that appears in the random phase of the model. Node 25 is the node that is replaced at $n = 78$ (see Fig. 5.2b, where node 25 is colored white, implying that $X_{25} = 0$ at $n = 78$.) The new node that replaces it (also numbered 25 in Fig. 5.2c) receives a link from node 23, thus putting it at the end of a chain of length 2 at $n = 79$. This is an innovation according to the above definition, for, in the attractor configuration corresponding to the graph of Fig. 5.2c, node 25 has a nonzero relative population; $X_{25} > 0$. This is because for any graph that does not contain a closed cycle, one can show that the attractor $\mathbf{X}$ of (5.1), for generic initial conditions, has the property that only those $X_i$ are nonzero whose nodes $i$ are the endpoints of the longest chains in the graph.[3] The $X_i$ for all other nodes is zero [17]. Since the longest chains in Fig. 5.2c are of length 2, node 25 is populated. (This explains why a node is gray or white in Fig. 5.2a–d.) Note that node 25 in Fig. 5.2c has become a parasite of node 23 in the sense that there is a link from 23 to 25 but none from 25 to any other node. This means that node 25 receives catalytic support for its own production from node 23, but does not give support to any other node in the system.

However, this innovation does not last long. Nodes 20 and 23, on which the well being of node 25 depends, are unprotected. Since they have the least possible value of $X_i$, namely, zero, they can be eliminated at subsequent graph updates, and their replacements in general do not feed into node 25. Sooner or later selection picks 23 for replacement, and then 25 also gets depopulated. By $n = 2853$ (Fig. 5.2d), node 25 and all others that were populated at $n = 79$ have joined the ranks of the unpopulated. Node 25 (and others of its ilk) are doomed because they are "uncaring winners": they do not feed into (i.e., do not catalyze) the nodes upon which their well being depends. That is why when there are no closed cycles, all structures are transitory; the graph remains random. Of the 1,070 innovations, 115 were of this type.

### 5.6.2 Birth of an Organization: Cooperation Begets Stability

At $n = 2853$ node 90 is an unpopulated node (Fig. 5.2d). It is eliminated at $n = 2854$ and the new node 90 forms a two-cycle with node 26 (Fig. 5.2e). This is the first time (and the only time in this run) an innovation forms a closed cycle in a graph that previously had no cycles. A closed cycle between two nodes is the simplest *cooperative* graph theoretical structure possible. Nodes 26 and 90 help each other's population grow; together they form a self-replicating system. Their populations

---

[3] See *The Attractor of Equation 5.1* in Appendix A.

grow much faster than other nodes in the graph; it turns out that in the attractor for this graph only nodes 26 and 90 are populated, with all other nodes having $X_i = 0$ ([17, 18]; see also Appendix A). Because node 90 is populated in the new attractor this constitutes an innovation. However, unlike the previous innovations, this one has a greater staying power, because nodes 26 and 90 do well *collectively*. At the next graph update *both* nodes 26 and 90 will be immune to removal since one of the other nodes with zero $X_i$ will be removed. Notice that nodes 26 and 90 do not depend on nodes that are part of the least fit set (those with the least value of $X_i$). The cycle has all the catalysts it needs for for the survival of each of its constituents. This property is true not just for cyclic subgraphs but for a more general cooperative structure, the *autocatalytic set* (ACS).

An ACS is a set of species that contains a catalyst for each species in the set [11, 19, 20]. In the context of the present model we can define an ACS to be a subgraph such that each of its nodes has at least one incoming link from a node of the same subgraph. While ACSs need not be cycles, they must contain at least one cycle. If the graph has (one or more) ACSs, one can show that the set of populated nodes ($X_i > 0$) must be an ACS, which we call the dominant ACS[4] [5, 17]. (In Fig. 5.2e–s, the subgraph of the gray and black nodes is the dominant ACS.) Therefore none of the nodes of the dominant ACS can be hit in the next graph update as long as there is any node outside it. In other words, the collective well-being of all the constituents of the dominant ACS, ensured by cooperation inherent within its structure, is responsible for the ACS's relative robustness and hence longevity.

In societies, this kind of event is akin to the birth of an organization wherein two or more agents improve upon their performance by entering into an explicit cooperation. A booming new township or industrial district perhaps can be analysed in terms of a closure of certain feedback loops. In prehistory, the appearance of tools that could be used to improve other tools may be regarded as events of this kind which probably unleashed a lot of artifact building. On the prebiotic earth one can speculate that the appearance of a small ACS might have triggered processes that eventually led to the emergence of life [14].

If there is no ACS in the graph then the largest eigenvalue of the adjacency matrix of the graph, $\lambda_1$ is zero. If there is an ACS then $\lambda_1 \geq 1$[5] [5, 17]. In Fig. 5.1b, $\lambda_1$ jumped from zero to one when the first ACS was created at $n = 2854$.

### 5.6.3 Expansion of the Organization at Its Periphery: Incremental Innovations

Consider Fig. 5.2f, g. Node 3, which is unpopulated at $n = 3021$, gets an incoming link from node 90 and an outgoing link to node 25 at $n = 3022$, which results in three nodes adding onto the dominant ACS. Node 3 is populated in the new attractor

---

[4] See *Dominant ACS of a Graph* in Appendix A.
[5] See *Graph-Theoretic Properties of ACSs* in Appendix A.

and hence this is an innovation. This innovation has expanded the *periphery* of the organization, defined below.

Every dominant ACS is a union of one or more *simple ACSs*, each of which have a substructure consisting of a *core* and *periphery*. For example, the dominant ACS in Fig. 5.1g has one simple ACS and in Fig. 5.1k it has two. For every simple ACS there exists a maximal subgraph, called the *core* of that ACS, from each of whose nodes there is a directed path to every node of that ACS. The rest of that ACS is termed its *periphery*. In Fig. 5.2e–s, the core is colored black, and the periphery gray. Thus in Fig. 5.2g, the two-cycle of nodes 26 and 90 is the core of the dominant ACS, and the chain of nodes 3, 25 and 18 along with the incoming link to node 3 from 26 constitutes its periphery. The core of a simple ACS is necessarily an *irreducible subgraph*. An irreducible subgraph is one that contains a directed path from each of its nodes to every other of its nodes [21]. When the dominant ACS consists of more than one simple ACS, its core is the union of their cores, and its periphery the union of their peripheries. Note that the periphery nodes by definition do not feed back into the core; in this sense they are parasites that draw sustenance from the core. The core, by virtue of its irreducible property (positive feedback loops within its structure, or cooperativity), is self-sustaining, and also supports the periphery. The $\lambda_1$ of the ACS is determined solely by the structure of its core [12, 16].

The innovation at $n = 3022$, one of 907 such innovations in this run, is an "incremental" one in the sense that it does not change the core (and hence does not change $\lambda_1$). However, such incremental innovations set the stage for major transformations later on. The ability of a core to tolerate parasites can be a boon or a bane, as we will see below.

### 5.6.4 Growth of the Core of the Organization: Parasites Become Symbionts

Another kind of innovation that occurs in the growth phase is illustrated in Fig. 5.2l, m. In Fig. 5.2l, the dominant ACS has two disjoint components. One component, consisting of nodes 41 and 98, is just a two-cycle without any periphery. The other component has a two-cycle (nodes 26 and 90) as its core that supports a large periphery. Node 39 in Fig. 5.2l is eliminated at $n = 3489$. The new node 39 (Fig. 5.2m) gets an incoming link from the periphery of the larger component of the dominant ACS and an outgoing link to the core of the same ACS. This results in expansion of the core, with several nodes getting added to it at once and $\lambda_1$ increasing. This example illustrates two distinct processes:

i. This innovation co-opts a portion of the parasitic periphery into the core. This strengthens cooperation: 26 contributes to the well-being of 90 (and hence to its own well-being) along two paths in Fig. 5.2m instead of only one in Fig. 5.2l. This is reflected in the increase of $\lambda_1$; $\lambda_1 = 1.15$ and 1 for Fig. 5.2m, l, respectively. The larger the periphery, the greater the probability of such core-enhancing innovations. This innovation is an example of how tolerance and

support of a parasitic periphery pays off for the ACS. Part of the parasitic periphery turns symbiont. Note that this innovation builds upon the structure generated
by previous incremental innovations. In Fig. 5.1b each rise in $\lambda_1$ indicates an
enlargement of the core [12, 16]. There are 40 such events in this run. As a result
of a series of such innovations that add to the core and periphery, the dominant
ACS eventually grows to span the entire graph at $n = 3880$, Fig. 5.2n, and the
system enters the *organized phase*.

ii. This example also highlights the competition between different ACSs. The two-
cycle of nodes 41 and 98 was populated in Fig. 5.2l, but is unpopulated in
Fig. 5.2m. Since the core of the other ACS becomes stronger than this two-cycle,
the latter is driven out of business.

### 5.6.5 Core-Shift 1: Takeover by a New Competitor

Interestingly, the same cycle of nodes 41 and 98 that is driven out of business at
$n = 3489$ had earlier (when it first arose at $n = 3387$) driven the two-cycle of nodes
26 and 90 out of business. Up to $n = 3,386$ (Fig. 5.2h), the latter two-cycle was
the only cycle in the graph. At $n = 3,387$ node 41 was replaced and formed a new
two-cycle with node 98 (Fig. 5.2i). Note that at $n = 3,387$ only the new two-cycle
is populated; all the nodes of the ACS that was dominant at the previous time step
(including its core) are unpopulated. We call such an event, where there is no overlap
between the old and the new cores, a *core shift* (a precise definition is given in [16]).
This innovation is an example of how a new competitor takes over.

Why does the new two-cycle drive the old one to extinction? The reason is that
the new two-cycle is downstream of the old one (node 41 has also acquired an
incoming link from node 39; thus there exists a directed path from the old cycle
to the new one, but none from the new to the old). Both two-cycles have the same
intrinsic strength, but the new two-cycle does better than the old because it draws
sustenance from the latter without feeding back. In general if the graph contains
two nonoverlapping irreducible subgraphs $A$ and $B$, let $\lambda_1(A)$ and $\lambda_1(B)$ be the
largest eigenvalues of the submatrices corresponding to $A$ and $B$. If $\lambda_1(A) > \lambda_1(B)$,
then $A$ wins (i.e., in the attractor of (5.1), nodes of $A$ and all nodes downstream
of $A$ are populated), and nodes of $B$ are populated if $B$ is downstream of $A$ and
unpopulated otherwise. When $\lambda_1(A) = \lambda_1(B)$, then if $A$ and $B$ are disconnected,
both are populated, and if one of them is downstream of the other, it wins and
the other is unpopulated [12, 16]. At $n = 3387$ the latter situation applies (the
$\lambda_1$ of both cycles is 1, but one is downstream of the other; the downstream cycle
wins at the expense of the upstream one). Examples of new competitors taking
over because their $\lambda_1$ is higher than that of the existing ACS are also seen in the
model.

In the displayed run, two core-shifts of this kind occurred. The first was at
$n = 3387$, which has been discussed above. One more occurred at $n = 6062$, which
was of an identical type with a new downstream two-cycle driving the old two-cycle
to extinction. Both these events resulted in a sharp drop in $s_1$ (Fig. 5.1b). A core-
shifting innovation is a traumatic event for the old core and its periphery. This is

reminiscent of the demise of the horse-drawn carriage industry upon the appearance of the automobile, or the decimation of anaerobic species upon the advent of aerobic ones.

At $n = 3403$ (Fig. 5.2k) an interesting event (that is not an innovation) happens. Node 38 is hit and the new node 38 has no incoming link. This cuts the connection that existed earlier (see Fig. 5.2j) between the cycle 98–41 and the cycle 26–90. The graph now has two disjoint ACSs with the same $\lambda_1$ (see Fig. 5.2k). As mentioned above, in such a situation both ACSs coexist; the cycle 26–90 and all nodes dependent on it once again become populated. Thus the old core has staged a "come-back" at $n = 3402$, leveling with its competitor. As we saw in the previous subsection, at $n = 3489$ the descendant of this organization strengthens its core and in fact drives its competitor out of business (this time permanently).

It is interesting that node 38, though unpopulated, still plays an important role in deciding the structure of the dominant ACS. It is purely a matter of chance that the core of the old ACS, the cycle 26–90, did not get hit before node 38. (All nodes with $X_i = 0$ have an equal probability of being replaced in the model.) If it had been destroyed between $n = 3387$ and 3402, then nothing interesting would have happened when node 38 was removed at $n = 3403$. In that case the new competitor would have won. Examples of that are also seen in the runs. In either case an ACS survives and expands until it spans the entire graph. It is worth noting that, while overall behavior like the growth of ACSs (including their average time scale of growth) is predictable, the details are shaped by historical accidents.

### 5.6.6 Core-Shift 2: Takeover by a Dormant Innovation

A different kind of innovation occurs at $n = 4696$. At the previous time step, node 36 is the least populated node (Fig. 5.2o). The new node 36 forms a two-cycle with node 74 (Fig. 5.2p). This two-cycle remains part of the periphery since it does not feed back into the core; this is an incremental innovation at this time since it does not enhance $\lambda_1$. However, because it generates a structure that is intrinsically self-sustaining (a two-cycle), this innocuous innovation is capable of having a dramatic impact in the future.

At $n = 5041$, Fig. 5.2q, the core has shrunk to five nodes (the reasons for this decline are briefly discussed later). The 36–74 cycle survives in the periphery of the ACS. Now it happens that node 85 is one of those with the least $X_i$ and gets picked for removal at $n = 5042$. Thus of the old core only the two-cycle 26–90 is left. But this is now upstream from another two-cycle 74–36 (see Fig. 5.2r). This is the same kind of structure as discussed above, with one cycle downstream from another. The downstream cycle and its periphery wins; the upstream cycle and all other nodes downstream from it except nodes 36, 74, and 11 are driven to extinction. This event is also a core-shift and is accompanied by a huge crash in the $s_1$ value (see Fig. 5.1b). This kind of an event is what we call a "takeover by a dormant innovation" [12]. The innovation 36–74 occurred at $n = 4696$. It lay dormant until $n = 5042$ when the old core had become sufficiently weakened that this dormant innovation could take over as the new core.

In this run 5 of the 1,070 innovations were dormant innovations. Of them only the one at $n = 4696$ later caused a core-shift of the type discussed above. The others remained as incremental innovations.

At $n = 8233$ a *complete crash* occurs. The core is a simple three-cycle (Fig. 5.2s) at $n = 8232$ and node 50 is hit, completely destroying the ACS. $\lambda_1$ drops to zero accompanied by a large crash in $s_1$. Within $O(s)$ time steps most nodes are hit and replaced and the graph has become random like the initial graph. The resemblance between the initial graph at $n = 1$ (Fig. 5.2a) and the graph at $n = 10,000$ (Fig. 5.2u) is evident. This event is not an innovation but rather the elimination of a *keystone species* [12].

## 5.7 Recognizing Innovations: A Structural Classification

The six categories of innovations discussed above occur in all the runs of the model and their qualitative effects are the same as described above. The above description was broadly chronological. We now describe these innovations structurally. Such a description allows each type of innovation to be recognized the moment it appears; one does not have to wait for its consequences to be played out. The structural recognition in fact allows us to predict qualitatively the kinds of impact it can have on the system. A mathematical classification of innovations is given in Appendix B; the description here is a plain-English account of that (with some loss of precision).

As is evident from the discussion above, positive feedback loops or cooperative structures in the graph crucially affect the dynamics. The character of an innovation will also depend upon its relationship with previously existing feedback loops and the new feedback loops it creates, if any. Structurally an *irreducible subgraph* captures the notion of feedback in a directed graph. By definition, since there exists a directed path (in both directions) between every pair of nodes belonging to an irreducible subgraph, each node exerts an influence on the other (albeit possibly through other intermediaries).

Thus the first major classification depends on whether the new node creates a new cycle and hence a new irreducible subgraph. One way of determining whether it does so is to identify the nodes downstream of the new node (namely those to which there is a directed path from this node) and those that are upstream (from which there is a directed path to this node). If the intersection of these two sets is empty the new node has not created any new irreducible subgraph, otherwise it has.

A. *Innovations in which the new node does not create any new cycles and hence no new irreducible subgraph is created.* These innovations will have a relatively minor impact on the system. There are two subclasses here, which depend upon the context: whether an irreducible subgraph already exists somewhere else in the graph or not.

   A1. *Before the innovation, the graph does not contain an irreducible subgraph.* Then the innovation is a shortlived one discussed in Sect. 5.6.1

(Fig. 5.2b, c). There is no ACS before or after the innovation. The largest eigenvalue $\lambda_1$ of the adjacency matrix of the graph being zero both before and after such an innovation is a necessary and sufficient condition for it to be in this class. Such an innovation is doomed to die when the first ACS arises in the graph for the reasons discussed in the previous section.

A2. *Before the innovation an irreducible subgraph already exists in the graph.* One can show that such an innovation simply adds to the periphery of the existing dominant ACS, leaving the core unchanged. Here the new node gets a nonzero $X_k$ because it receives an incoming link from one of the nodes of the existing dominant ACS; it has effectively latched on to the latter like a parasite. This is an incremental innovation (Sect. 5.6.3, Fig. 5.2f, g). It has a relatively minor impact on the system at the time it appears. Since it does not modify the core, the ratios of the $X_i$ values of the core nodes remain unchanged. However, it does eat up some resources (since $X_k > 0$) and causes an overall decline in the $X_i$ values of the core nodes. $\lambda_1$ is nonzero and does not change in such an innovation.

B. *Innovations that do create some new cycle.* Thus a new irreducible subgraph gets generated. Because these innovations create new feedback loops, they have a potentially greater impact. Their classification depends upon whether or not they modify the core and the extent of the modification caused; this is directly correlated with their immediate impact.

B1. *The new cycles do not modify the existing core.* If the new irreducible subgraph is disjoint from the existing core and its intrinsic $\lambda_1$ is less than that of the core, then the new irreducible subgraph will not modify the existing core but will become part of the periphery. Like incremental innovations, such innovations cause an overall decline in the $X_i$ values of the core nodes but do not disturb their ratios and the value of $\lambda_1$. However, they differ from incremental innovations in that the new irreducible subgraph has self-sustaining capabilities. Thus in the event of a later weakening of the core (through elimination of some core nodes), these innovations have the potential to cause a core-shift, wherein the irreducible graph generated in the innovation becomes the new core. At that point it would typically cause a major crash in the number of populated species, as the old core and all its periphery that is not supported by the new core would become depopulated. Such innovations are the dormant innovations (Sect. 5.6.6, Fig. 5.2o, p). Note that not all dormant innovations cause core-shifts. Most in fact play the same role as incremental innovations.

B2. *Innovations modify the existing core.* If the new node is part of the new core, the core has been modified. The classification of such innovations depends on the kind of core that exists before and the nature of the modification.

B21. *The existing core is nonempty,* i.e., an ACS already exists before the innovation in question arrives.

B211. *The innovation strengthens the existing core.* In this case the new node receives an incoming link from the existing dominant ACS and has an outgoing link to the existing core. The existing core nodes get additional positive feedback, and $\lambda_1$ increases. Such an event can cause some members of the parasitic periphery to be co-opted into the core. These are the core-enhancing innovations discussed in Sect. 5.6.4 (Fig. 5.2l, m).

B212. *The new irreducible subgraph is disjoint from the existing core and "stronger" than it.* "Stronger" means that the intrinsic $\lambda_1$ of the new irreducible graph is greater than or equal to the $\lambda_1$ of the existing core, and in the case of equality it is downstream from the existing core. Then it will destabilize the existing core and become the new core itself, causing a core-shift. The takeovers by new competitors, discussed in Sect. 5.6.5 (Fig. 5.2h, i) belong to this class.

B22. *The existing core is empty*, i.e., no ACS exists before the arrival of this innovation. Then the new irreducible graph is the core of the new ACS that is created at this time. This event is the beginning of a self-organizing phase of the system. This is the birth of an organization discussed in Sect. 5.6.2 (Fig. 5.2d, e). This is easily recognized graph theoretically as $\lambda_1$ jumps from zero to a positive value.

Note that the 'recognition' of the class of an innovation is contingent upon knowing graph-theoretic features like the core, periphery, $\lambda_1$, and being able to determine the irreducible graph created by the innovation. The above rules are an analytic classification of all innovations in the model, irrespective of values of the parameters $p$ and $s$. Note, however, that their relative frequencies depend upon the parameters. In particular, note that innovations of class A require the new node to have at least one link (an incoming one) and class B require at least two links (an incoming and an outgoing one). Thus as the connection probability $p$ declines, for fixed $s$, the latter innovations (the more consequential ones) become less likely.

## 5.8 Some Possible General Lessons

In this model, due to the simplicity of the population dynamics, it is possible to make an analytic connection between the graph structure produced by various innovations and their subsequent effect on the short- and long-term dynamics of the system. In addition, we are able to completely enumerate the different types of innovations and classify them purely on the basis of their graph structure. Identifying innovations and understanding their effects is much more complicated in real-world processes in both biological and social systems. Nevertheless, the close parallel between the qualitative categories of innovation we find in our model and real-world examples means that there may be some lessons to be learnt from this simple mathematical model.

One broad conclusion is that in order to guess what might be an innovation, we need an understanding of how the patterns of connectivity influence system dynamics and vice versa. The inventor of a new product or a venture capitalist asks: what inputs will be needed, and whose needs will the product connect to? Given these potential linkages in the context of other existing nodes and links, what flows will actually be generated along the new links? How will these new flows impact the generation of other new nodes and links and the death of existing ones and how that will feed back into the flows again? The detailed rules of this dynamics are system dependent, but presumably successful entrepreneurs have an intuitive understanding of this very dynamics.

In our model, as in real processes, there are innovations that have an immediate impact on the dynamics of the system (e.g., the creation of the first ACS and core-shifting innovations) and ones that have little or no immediate impact. Innovation in real processes analogous to the former are probably easier to identify because they cause the dynamics of the system to immediately change dramatically (in this model, triggering a new round of self-organized growth around a new ACS). Of the latter, the most interesting innovations are the ones that eventually do have a large impact on the dynamics: the dormant innovations. In this model dormant innovations sometimes lead to a dramatic change in the dynamics of the system at a later time. This suggests that in real-world processes too it might be important, when observing a sudden change in the dynamics, to examine innovations that occurred much before the change. Of course, in the model and in real processes, there are innovations that have nothing to do with any later change in the dynamics. In real processes it would be very difficult to distinguish such innovations from dormant innovations, which do cause a significant impact on the dynamics. The key feature distinguishing a dormant innovation from incremental innovations in this model is that a dormant innovation creates an irreducible structure that can later become the core of the graph.

This suggests that in real-world processes it might be useful to find an analogy of the core and periphery of the system and then focus on innovations or processes that alter the core or create structures that could become the core. In the present model, it is possible to define the core in a purely graph-theoretic manner. In real systems it might be necessary to define the core in terms of the dynamics. One possible generalization is based on the observation that removal of a core node causes the population growth rate to reduce (due to the reduction of $\lambda_1$) while the removal of a periphery node leaves $\lambda_1$ unchanged. This could be used as an algorithmic way of identifying core nodes or species in more complex mathematical models, or in real systems where such testing is possible.

## 5.9 Discussion

As in real systems, the model involves an interplay between the force of selection that weeds out underperforming nodes, the influx of novelty that brings in new nodes and links, and an internal (population) dynamics that depends upon the mutual interactions. In an environment of nonautocatalytic structures, a small ACS is very successful and drives the other nodes to the status of "have-nots" ($X_i = 0$). The latter are eliminated one by one, and if their replacements "latch on" to the ACS, they survive, else they suffer the same fate. The ACS "succeeds" spectacularly: eventually all the nodes join it. But this sets the stage for enhanced internal competition between the members of the ACS. Before the ACS spanned the graph, only have-nots, nodes outside the dominant ACS, were eliminated. After spanning the eliminated node must be one of the "haves", a member of the ACS (whichever has the least $X_i$). This internal competition weakens the core and enhances the probability of collapse due to core-transforming innovations or elimination of keystone species. Thus the ACS's very success creates the circumstances that bring about its destruction.[6] Both its success and a good part of its destruction are due to innovations (see also [12]).

It is of course true that we can describe the behavior of the system in terms of attractors of the dynamics as a function of the graph without recourse to the word "innovation". The advantage in introducing the notion of innovation as defined above is that it captures a useful property of the dynamics in terms of which many features can be readily described. Further, we hope that the examples discussed above make out a reasonable case that this notion of innovation is sufficiently close (as close as is possible in an idealized model such as this) to the real thing to help in discussions of the latter.

In the present model, the links of the new node are chosen randomly from a fixed probability distribution. This might be appropriate for the prebiotic chemical scenario for which the model was constructed, but is less appropriate for biological systems and even less for social systems. While there is always some stochasticity, in these systems the generation of novelty is conditioned by the existing context, and in social systems also by the intentionality of the actors. Thus the ensemble of choices from which the novelty is drawn also evolves with the system. This feedback from the recent history of system states to the ensemble of system perturbations, though not implemented in the present version of the model, certainly deserves future investigation.

## Appendix A: Definitions and Proofs

In this Appendix we collect some useful facts about the model. These and other properties can be found in [5, 13, 17, 18].

### *Derivation of Equation (5.1)*

Let $i \in \{1, \ldots, s\}$ denote a chemical (or molecular) species in a well-stirred chemical reactor. Molecules can react with one another in various ways; we focus on only

---

[6] For related discussion of discontinuous transitions in other systems, see [22–24].

one aspect of their interactions: catalysis. The catalytic interactions can be described by a directed graph with $s$ nodes. The nodes represent the $s$ species and the existence of a link from node $j$ to node $i$ means that species $j$ is a catalyst for the production of species $i$. In terms of the adjacency matrix, $C = (c_{ij})$ of this graph, $c_{ij}$ is set to unity if $j$ is a catalyst of $i$ and is set to zero otherwise. The operational meaning of catalysis is as follows:

Each species $i$ will have an associated nonnegative population $y_i$ in the reactor that changes with time. Let species $j$ catalyze the ligation of reactants $A$ and $B$ to form the species $i$, $A + B \xrightarrow{j} i$. Assuming that the rate of this catalyzed reaction is given by the Michaelis–Menten theory of enzyme catalysis, $\dot{y}_i = V_{\max} ab \frac{y_j}{K_M + y_j}$ [25], where $a, b$ are the reactant concentrations, and $V_{\max}$ and $K_M$ are constants that characterize the reaction. If the Michaelis constant $K_M$ is very large this can be approximated as $\dot{y}_i \propto y_j ab$. Combining the rates of the spontaneous and catalyzed reactions and also putting in a dilution flux $\phi$, the rate of growth of species $i$ is given by $\dot{y}_i = k(1 + \nu y_j)ab - \phi y_i$, where $k$ is the rate constant for the spontaneous reaction, and $\nu$ is the catalytic efficiency. Assuming the catalyzed reaction is much faster than the spontaneous reaction, and that the concentrations of the reactants are nonzero and fixed, the rate equation becomes $\dot{y}_i = K y_j - \phi y_i$, where $K$ is a constant. In general because species $i$ can have multiple catalysts, $\dot{y}_i = \sum_{j=1}^{s} K_{ij} y_j - \phi y_i$, with $K_{ij} \sim c_{ij}$. We make the further idealization $K_{ij} = c_{ij}$, giving

$$\dot{y}_i = \sum_{j=1}^{s} c_{ij} y_j - \phi y_i \ . \tag{5.2}$$

The relative population of species $i$ is by definition $x_i \equiv y_i / \sum_{j=1}^{s} y_j$. As $0 \leq x_i \leq 1$, $\sum_{i=1}^{s} x_i = 1$, $\mathbf{x} \equiv (x_1, \ldots, x_s)^T \in J$. Taking the time derivative of $x_i$ and using (5.2) it is easy to see that $\dot{x}_i$ is given by (5.1). Note that the $\phi$ term, present in (5.2), cancels out and is absent in (5.1).

## *The Attractor of Equation (5.1)*

A graph described by an adjacency matrix $C$ has an eigenvalue $\lambda_1(C)$ that is a real, non-negative number that is greater than or equal to the modulus of all other eigenvalues. This follows from the Perron–Frobenius theorem [21] and this eigenvalue is called the Perron–Frobenius eigenvalue of $C$.

*The attractor $\mathbf{X}$ of (5.1) is an eigenvector of $C$ with eigenvalue $\lambda_1(C)$.*

Since (5.1) does not depend on $\phi$, we can set $\phi = 0$ in (5.2) without loss of generality for studying the attractors of (5.1). For fixed $C$ the general solution of (5.2) is $\mathbf{y}(t) = e^{Ct}\mathbf{y}(0)$, where $\mathbf{y}$ denotes the $s$-dimensional column vector of populations. It is evident that if $\mathbf{y}^\lambda \equiv (y_1^\lambda, \ldots, y_s^\lambda)$ viewed as a column vector is a right eigenvector of $C$ with eigenvalue $\lambda$, then $\mathbf{x}^\lambda \equiv \mathbf{y}^\lambda / \sum_i y_i^\lambda$ is a fixed point of (5.1). Let $\lambda_1$ denote the eigenvalue of $C$ that has the largest real part; it is clear that $\mathbf{x}^{\lambda_1}$ is an attractor of

(5.1). By the theorem of Perron–Frobenius for nonnegative matrices [21], $\lambda_1$ is real and $\geq 0$ and there exists an eigenvector $\mathbf{x}^{\lambda_1}$ with $x_i \geq 0$. If $\lambda_1$ is nondegenerate, $\mathbf{x}^{\lambda_1}$ is the unique asymptotically stable attractor of (5.1), $\mathbf{x}^{\lambda_1} = (X_1, \ldots, X_s)$.

## *The Attractor of Equation (5.1) When There Are No Cycles*

*For any graph with no cycles, in the attractor only the nodes at the ends of the longest paths are nonzero. All other nodes are zero.*

Consider a graph consisting only of a linear chain of $r + 1$ nodes, with $r$ links, pointing from node 1 to node 2, node 2 to node 3, etc. Node 1 (to which there is no incoming link) has a constant population $y_1$ because the right-hand side (rhs) of (5.2) vanishes for $i = 1$ (taking $\phi = 0$). For node 2, we get $\dot{y}_2 = y_1$, hence $y_2(t) = y_2(0) + y_1 t \sim t$ for large $t$. Similarly, it can be seen that $y_k$ grows as $t^{k-1}$. In general, it is clear that for a graph with no cycles, $y_i \sim t^r$ for large $t$ (when $\phi = 0$), where $r$ is the length of the longest path terminating at node $i$. Thus, nodes with the largest $r$ dominate for sufficiently large $t$. Because the dynamics (5.1) does not depend upon the choice of $\phi$, $X_i = 0$ for all $i$ except the nodes at which the longest paths in the graph terminate.

## *Graph-Theoretic Properties of ACSs*

   i. *An ACS must contain a closed walk.*
  ii. *If a graph, C, has no closed walk then $\lambda_1(C) = 0$.*
 iii. *If a graph, C, has a closed walk then $\lambda_1(C) \geq 1$.*
     *Consequently:*
 iv. *If a graph C has no ACS then $\lambda_1(C) = 0$.*
  v. *If a graph C has an ACS then $\lambda_1(C) \geq 1$.*

   i. Let $A$ be the adjacency matrix of a graph that is an ACS. Then by definition, every row of $A$ has at least one nonzero entry. Construct $A'$ by removing, from each row of $A$, all nonzero entries except one that can be chosen arbitrarily. Thus $A'$ has exactly one nonzero entry in each row. Clearly the column vector $\mathbf{x} = (1, 1, \ldots, 1)^T$ is an eigenvector of $A'$ with eigenvalue 1 and hence $\lambda_1(A') \geq 1$. Proposition iii therefore implies that $A'$ contains a closed walk. Because the construction of $A'$ from $A$ involved only removal of some links, it follows that $A$ must also contain a closed walk.

  ii. If a graph has no closed walk then all walks are of finite length. Let the length of the longest walk of the graph be denoted $r$. If $C$ is the adjacency matrix of a graph then $(C^k)_{ij}$ equals the number of distinct walks of length $k$ from node $j$ to node $i$. Clearly $C^m = 0$ for $m > r$. Therefore all eigenvalues of $C^m$ are zero. If $\lambda_i$ are the eigenvalues of $C$ then $\lambda_i^k$ are the eigenvalues of $C^k$. Hence, all eigenvalues of $C$ are zero, which implies $\lambda_1 = 0$. This proof was supplied by V. S. Borkar.

iii. If a graph has a closed walk then there is some node $i$ that has at least one closed walk to itself, i.e., $(C^k)_{ii} \geq 1$, for infinitely many values of $k$. Because the trace of a matrix equals the sum of the eigenvalues of the matrix, we have $\sum_{i=1}^{s}(C^k)_{ii} = \sum_{i=1}^{s} \lambda_i^k$, where $\lambda_i$ are the eigenvalues of $C$. Thus, $\sum_{i=1}^{s} \lambda_i^k \geq 1$, for infinitely many values of $k$. This is only possible if one of the eigenvalues $\lambda_i$ has a modulus $\geq 1$. By the Perron–Frobenius theorem, $\lambda_1$ is the eigenvalue with the largest modulus, hence $\lambda_1 \geq 1$. This proof was supplied by R. Hariharan.
iv. and (v) follow from the above.

## *Dominant ACS of a Graph*

*If a graph has (one or more) ACSs, i.e., $\lambda_1 \geq 1$, then the subgraph corresponding to the set of nodes $i$ for which $X_i > 0$ is an ACS.*
Renumber the nodes of the graph so that $x_i > 0$ only for $i = 1, \ldots, k$. Let $C$ be the adjacency matrix of this graph. Since $\mathbf{X}$ is an eigenvector of the matrix $C$, with eigenvalue $\lambda_1$, we have $\sum_{j=1}^{s} c_{ij} X_j = \lambda_1 X_i \Rightarrow \sum_{j=1}^{k} c_{ij} X_j = \lambda_1 X_i$. Since $X_i > 0$ only for $i = 1, \ldots, k$ it follows that for each $i \in \{1, \ldots, k\}$ there exists a $j$ such that $c_{ij} > 0$. Hence the $k \times k$ submatrix $C' \equiv (c_{ij}), i, j = 1, \ldots, k$ has at least one nonzero entry in each row. Thus each node of the subgraph corresponding to this submatrix has an incoming link from one of the other nodes in the subgraph. Hence the subgraph is an ACS. We call this subgraph the *dominant ACS* of the graph.

## *Time Scales for Appearance and Growth of the Dominant ACS*

The probability for an ACS to be formed at some graph update in a graph that has no cycles can be closely approximated by the probability of a two-cycle (the simplest ACS with one-cycles being disallowed) forming by chance, which is $p^2 s$ (the probability that in the row and column corresponding to the replaced node in $C$ any matrix element and its transpose are both assigned unity). Thus, the "average time of appearance" of an ACS is $\tau_a = 1/p^2 s$, and the distribution of times of appearance is $P(n_a) = p^2 s (1 - p^2 s)^{n_a - 1}$. This approximation is better for small $p$.

Assuming that the possibility of a new node forming a second ACS is rare enough to neglect, and that the dominant ACS grows by adding a single node at a time, one can estimate the time required for it to span the entire graph. Let the dominant ACS consist of $s_1(n)$ nodes at time $n$. The probability that the new node gets an incoming link from the dominant ACS and hence joins it is $p s_1$. Thus in $\Delta n$ graph updates, the dominant ACS will grow, on average, by $\Delta s_1 = p s_1 \Delta n$ nodes. Therefore $s_1(n) = s_1(n_a)\exp((n - n_a)/\tau_g)$, where $\tau_g = 1/p$, $n_a$ is the time of appearance of the first ACS and $s_1(n_a)$ is the size of the first ACS. Thus $s_1$ is expected to grow exponentially with a characteristic timescale $\tau_g = 1/p$. The time taken from the appearance of the ACS to its spanning is $\tau_g \ln(s/s_1(n_a))$.

## Appendix B: Graph-Theoretic Classification of Innovations

In the main text we defined an innovation to be the new structure created by the addition of a new node, when the new node has a nonzero population in the new attractor. Here, we present a graph-theoretic hierarchical classification of innovations (see Fig. 5.3). At the bottom of this hierarchy we recover the six categories of innovations described in the main text.

Some notation follow: We need to distinguish between two graphs, one just before the new node is inserted, and one just after. We denote them by $C_i$ and $C_f$ respectively, and their cores by $Q_i$ and $Q_f$. Note that a graph update event consists of two parts – the deletion of a node and the addition of one. $C_i$ is the graph after the node is deleted and before the new node is inserted. The graph before the deletion will be denoted $C_0$; $Q_0$ will denote its core[7]. If a graph has no ACS, its core is the null set.

The links of the new node may be such that new cycles arise in the graph (that were absent in $C_i$ but are present in $C_f$). In this case the new node is part of a new irreducible subgraph that has arisen in the graph. $N$ will denote the maximal irreducible subgraph which includes the new node. If the new node does not create new cycles, $N = \emptyset$. If $N \neq \emptyset$, then $N$ will either be disjoint from $Q_f$ or will include $Q_f$ (it cannot partially overlap with $Q_f$ because of its maximal character). The structure of $N$ and its relationship with the core before and after the addition determines the nature of the innovation. With this notation all innovations can be grouped into two classes:

A.    Innovations that do not create new cycles, $N = \emptyset$. This implies $Q_f = Q_i$ because no new irreducible structure has appeared and therefore the core of the graph, if it exists, is unchanged.

B.    Innovations that do create new cycles, $N \neq \emptyset$. This implies $Q_f \neq \emptyset$ because if a new irreducible structure is created then the new graph has at least one ACS and therefore a nonempty core.

Class A can be further decomposed into two classes:

A1.    $Q_i = Q_f = \emptyset$. In other words, the graph has no cycles both before and after the innovation. This corresponds to shortlived innovations discussed in Sect. 5.6.1 (Fig. 5.2b, c).

A2.    $Q_i = Q_f \neq \emptyset$. In other words, the graph had an ACS before the innovation, and its core was not modified by the innovation. This corresponds to incremental innovations discussed in Sect. 5.6.3 (Fig. 5.2f, g).

---

[7] Most of the time the deleted node (being the one with the least relative population) is outside the dominant ACS of $C_0$ or in its periphery. Thus, in most cases the core is unchanged by the deletion: $Q_i = Q_0$. However, sometimes the deleted node belongs to $Q_0$. In that case $Q_i \neq Q_0$. In most such cases, $Q_i$ is a proper subset of $Q_0$. In very few (but important) cases, $Q_i \cap Q_0 = \emptyset$ (the null set). In these latter cases, the deleted node is a *keystone node* [16]; its removal results in a "core shift".

Class B of innovations can also be divided into two subclasses:

B1. $N \neq Q_f$. If the new irreducible structure is not the core of the new graph, then $N$ must be disjoint from $Q_f$. This can only be the case if the old core has not been modified by the innovation. Therefore $N \neq Q_f$ necessarily implies that $Q_f = Q_i$. This corresponds to dormant innovations discussed in Sect. 5.6.6 (Fig. 5.2o, p).

B2. $N = Q_f$, i.e., the innovation becomes the new core after the graph update. This is the situation where the core is transformed due to the innovation. The "core-transforming theorem" [12, 13, 18] states that an innovation of type B2 occurs whenever either of the following conditions are true:

(a)  $\lambda_1(N) > \lambda_1(Q_i)$ or
(b)  $\lambda_1(N) = \lambda_1(Q_i)$ and $N$ is downstream of $Q_i$.

Class B2 can be subdivided as follows:

B21. $Q_i \neq \emptyset$, i.e., the graph contained an ACS before the innovation. In this case an existing core is modified by the innovation.

B22. $Q_i = \emptyset$, i.e., the graph had no ACS before the innovation. Thus, this kind of innovation creates an ACS in the graph. It corresponds to the birth of a organization discussed in Sect. 5.6.2 (Fig. 5.2d, e).

Finally, class B21 can be subdivided:

B211. $Q_i \subset Q_f$. When the new core contains the old core as a subset we get an innovation that causes the growth of the core, discussed in Sect. 5.6.4 (Fig. 5.2l, m).

B212. $Q_i$ and $Q_f$ are disjoint (note that it is not possible for $Q_i$ and $Q_f$ to partially overlap, else they would form one big irreducible set which would then be the core of the new graph and $Q_i$ would be a subset of $Q_f$). This is an innovation where a core-shift is caused due to a takeover by a new competitor, discussed in Sect. 5.6.5 (Fig. 5.2h, i).

Note that each branching above is into mutually exclusive and exhaustive classes. This classification is completely general and applicable to all runs of the system. Figure 5.3 shows the hierarchy obtained using this classification.

**Acknowledgments** S.J. thanks John Padgett for discussions.

# References

1. J.A. Schumpeter, *The Theory of Economic Development* (Harvard University Press, Cambridge, MA, 1934)
2. E.M. Rogers, *The Diffusion of Innovations*, 4th edn. (Free Press, New York, NY, 1995)
3. P.G. Falkowski, Science **311**, 1724 (2006)

4. S. Bornholdt, H.G. Schuster (eds.), *Handbook of Graphs and Networks* (Wiley-VCH, Weinheim, 2003)
5. S. Jain, S. Krishna, Phys. Rev. Lett. **81**, 5684 (1998)
6. F. Dyson, *Origins of Life* (Cambridge University Press, Cambridge, 1985)
7. S.A. Kauffman, *The Origins of Order* (Oxford University Press, Oxford, 1993)
8. R.J. Bagley, J.D. Farmer, W. Fontana, in *Artificial Life II*, ed. by C.G. Langton, C. Taylor, J.D. Farmer, S. Rasmussen (Addison-Wesley, Redwood City, CA, 1991), pp. 141–158
9. W. Fontana, L. Buss, Bull. Math. Biol. **56**, 1 (1994)
10. P. Bak, K. Sneppen, Phys. Rev. Lett. **71**, 4083 (1993)
11. S.A. Kauffman, J. Cybernetics **1**, 71 (1971)
12. S. Jain, S. Krishna, Proc. Natl. Acad. Sci. USA **99**, 2055 (2002)
13. S. Krishna, Ph.D. Thesis (2003), arXiv:nlin/0403050v1 [nlin.AO]
14. S. Jain, S. Krishna, Proc. Natl. Acad. Sci. USA **98**, 543 (2001)
15. LEDA, The Library of Efficient Data Types and Algorithms (presently distributed by Algorithmic Solutions Software GmbH; http://www.algorithmic-solutions.com)
16. S. Jain, S. Krishna, Phys. Rev. E **65**, 026103 (2002)
17. S. Jain, S. Krishna, Comput. Phys. Commun. **121–122**, 116 (1999)
18. S. Jain, S. Krishna, in *Handbook of Graphs and Networks*, ed. by S. Bornholdt, H.G. Schuster (Wiley-VCH, Weinheim, 2003), pp. 355–395
19. M. Eigen, Naturwissenschaften **58**, 465 (1971)
20. O.E. Rossler, Z. Naturforsch. **26b**, 741 (1971)
21. E. Seneta, *Non-negative Matrices* (George Allen and Unwin, London, 1973)
22. J. Padgett, in *Networks and Markets*, ed. by J.E. Rauch, A. Casella (Russel Sage, New York, NY, 2001), pp. 211–257
23. M.D. Cohen, R.L. Riolo, R. Axelrod, Rationality Soc. **13**, 5 (2001)
24. J.M. Carlson, J. Doyle, Phys. Rev. E **60**, 1412 (1999)
25. H. Gutfreund, *Kinetics for the Life Sciences* (Cambridge University Press, Cambridge, 1995)

# Part II
# From Random to Complex Structures: The Concept of Self-Organization for Galaxies, Asters, and Spindles

# Chapter 6
# How Stochastic Dynamics Far from Equilibrium Can Create Nonrandom Patterns

Gunter M. Schütz

**Abstract** We describe several models for interacting particle systems far from thermal equilibrium and show that for both deterministic and stochastic dynamics ordered patterns decay or emerge. For these models we demonstrate in detail how random processes are capable of generating patterns with high complexity (defined by the Shannon information) in a short period of time.

## 6.1 Some Very Small Numbers

One of the big miracles of evolution is the emergence of highly complex, functional structures. To be astounded one does not need to consider the complexity of human beings or not even that of bacteria. It suffices to study the blueprint for the functionality of a cell, the DNA, to start wondering how this could have evolved in the course of time, just by the action of purposeless chemical processes. In fact, there are a large number of people who wonder so much that they cannot imagine that such an evolution could have taken place without the purposeful action of a supreme "engineer", usually identified with some notion of divine action. Evolution is a theme where even today Science and Religion meet.

While "meeting" of religion (of any persuasion) and science may turn into a very enlightening encounter for scientists and theologians alike, an unhealthy entanglement of the two has taken place in the last decade, known under the label "Intelligent Design" (ID). This term refers to a movement, prominent in the US, but existent worldwide, that claims to be able to provide hard scientific evidence for purposeful intervention of some supreme being in the process of evolution. One of the tenets of their belief is the allegedly negligibly small probability that an information-carrying structure such as DNA could have evolved by natural processes that are within the scope of science as we know it.

G.M. Schütz (✉)
Institut für Festkörperforschung, Forschungszentrum Jülich GmbH, D-52425 Jülich, Germany
e-mail: g.schuetz@fz-juelich.de

H. Meyer-Ortmanns, S. Thurner (eds.), *Principles of Evolution*,
The Frontiers Collection, DOI 10.1007/978-3-642-18137-5_6,
© Springer-Verlag Berlin Heidelberg 2011

It will transpire from the discussion below that this belief originates in a profound misunderstanding of the relation between random processes and the significance of Shannon information. However, one does not need to be an ID protagonist to be surprised what processes that are entirely random (in the sense of being unpredictable by any law of nature) can achieve. Not only biological evolution, but also simpler pattern formation processes in nature intrigue because they are often driven by processes where randomness plays an important part. How can that be?

To sharpen the question, we consider a well-known problem, viz. the probability that a monkey who wildly hacks symbols into a computer would in this way accidentally type a complete poem by Shakespeare. Will it ever happen, if we give the monkey enough time (say, the age of the earth)? Maybe not, so let us be a bit more modest and hope only for the first two lines of the short piece

> True, I talk of dreams,
> Which are the children of an idle brain,
> Begot of nothing but vain fantasy.

from Romeo and Juliet, Act 1, Scene 4. The first two lines have 64 characters, including commas and spaces. If we assume, for argument's sake, that the computer keyboard has 64 keys, and we let the monkey type 64 characters, then the probability that he would recreate Shakespeare's beautiful lines (ignoring capitals) is

$$p = (1/64)^{64} = 2^{-384} \, .$$

To put this result into a language more appropriate for dealing with texts we use the Shannon definition of information

$$I = -\log_2 p \, . \tag{6.1}$$

Based on the hypothetical 64-character set of the keyboard, which corresponds to 6 bit encoding of an individual character, the information content of the Shakespeare text is then $6 \times 64 = 384$ bit, corresponding to the probability we just computed. Generally, the magnitude of the Shannon information of a structure may be taken as a measure of its complexity. A given word of, say, just four letters can easily be generated by the action of a monkey. It has low complexity. A long sentence with a high information content is much more complex. Monkeys are not capable of generating them.

Therefore, if we see a poem by Shakespeare on a computer screen and ask ourselves how it got there, we are led to conclude that it got there by a sequence of deterministic events: From a data file (or other data storage device) to the screen, by some other deterministic process to the storage device, and so on. There is no problem with understanding deterministic transmission of complex information via different one-to-one encoding schemes. Ultimately, however, if we follow the flow of such deterministic events backwards, we are faced with the question: Where did the information originally come from? From deterministic processes in Shakespeare's brain? If this is the right answer, we can in principle follow these processes further back in time until we reach the Big Bang! This formal reasoning, which does not look into the details, is evidently not a meaningful procedure of scientific inquiry into the origin of complex information, not least because at some stage quantum mechanical randomness will enter into the deterministic flow and then we are back to the problem of small probabilities.

The probability $p$ computed above is evidently a pretty small number, but to appreciate how small it really is, and that it is a real problem within the framework of the formal arguments presented so far, let us consider the following. According to various sources in the internet, the total number of baryons in the known universe is somewhere around $10^{80}$ (this may be wrong, but for the following arguments it does not matter if we are off by a few or even many orders of magnitude). On the other hand, the age of the earth is approximately $10^{17}$ s. Let us now assume that instead of a monkey, each baryon in the universe somehow triggers the creation of a complete random string of 64 characters on our computer keyboard whenever it somehow changes its quantum state. Let this happen every femtosecond ($10^{-15}$ s). Then the total number of attempts to recreate Shakespeare's two lines by this random process would be

$$10^{17} \times 10^{15} \times 10^{80} = 10^{112} \approx 2^{373} \ .$$

Very roughly speaking we can identify this number with all baryonic processes that have ever taken place on the femtosecond scale in the observable universe since the formation of the earth.

Now we can compute the probability that all baryons in the known universe that have been trying to generate this text with a frequency of $10^{15}$ per second since the formation of the earth some 10 billion years ago have been successful at least once. Since the probability of obtaining the Shakespeare text in one attempt is $2^{-384}$, the probability $p_s$ of successfully creating this text at least once after $2^{373}$ attempts is

$$p_s = 1 - \left(1 - 2^{-384}\right)^{2^{373}} \approx 5 \times 10^{-4} \ . \tag{6.2}$$

In other words, the probability of successfully recreating the lines by Shakespeare (and only the first two) through all baryonic processes – fantastically quick processes – that have ever taken place in the universe since the formation of the earth is only about $5 \times 10^{-4}$! This example should give us some idea how unlikely the random generation of even a short specified text of just 384 bits is.

It is trivial to translate this into the probability that our $10^{80}$ fantastically fast baryonic random text generators would generate a specific string of DNA. There are four letters in a DNA "text", so each letter represents a 2-bit piece of information. Hence the probability of generating a given, fairly short DNA sequence of only $3 \times 64 = 192$ letters is equal to that of generating the Shakespeare text with 64 characters, it has the same Shannon information of 384 bits. Therefore the probability that all baryonic process in the universe could have caused the creation a specific DNA string of that length during the existence of the earth in this random fashion is only about $5 \times 10^{-4}$. Of course, the interpretation of the total number of baryonic processes should not be taken too literally, since one can hardly claim that any sort of change of the quantum state of a baryon would happen every $10^{-15}$ s.

Nevertheless, this interpretation provides a rough idea of how vastly we are exaggerating the chance of generating a very short DNA text in the monkey fashion. Chemical processes that generate DNA texts occur definitely with much smaller frequencies and with many orders of magnitude less chemical agents than baryons in the universe. Therefore the number of all chemical processes that have taken place on the earth (or even in the observable universe) is smaller by many orders of magnitude than $2^{373}$. Then the number of potentially useful chemical processes, involving the right chemicals for DNA, is again smaller by many orders of magnitude, leading to a ridiculously small probability of success $p_s$.

A final, still simpler example: What is the probability of generating a given binary string, say, the alternating pattern $010101010\ldots0101$ of length $L = 384$? The Shannon information is again 384 bits and again we obtain the same probability estimates as for the Shakespeare text and the DNA string. The probability that a monkey would type the two lines of Shakespeare is the same as if the monkey (or anyone else) would toss a coin and obtain a sequence $010101\ldots$ of length 384. Correspondingly, the probability that all baryons in the universe would have triggered the generation of this pattern during the lifetime of the earth has only probability 0.0005. In terms of probability considerations for a random process as described above, all three problems, the Shakespeare text of 64 characters, the DNA strand of length 192, and the binary string of length 384, are all equivalent problems since they have the same Shannon information content.

If indeed that were the full story about pattern formation through random events then we would be forced to conclude that such simple sequences, let alone the much more complex real DNA sequence of a human, could never have been generated by chance. To believe that an event with a probability as small as $2^{-384}$ would actually have occurred is highly irrational.

However, this is of course not the story of how evolution has happened. The underlying assumption behind all the small numbers presented above is that all possible outcomes of the random process occur with the same probability and that consecutive attempts are not correlated. This assumption is true for coin tossing, and may also be approximately true for the monkey. Thermodynamically speaking this equiprobability assumption describes an infinite temperature scenario where thermal fluctuations lead to a situation in which the Boltzmann weights of all microstates are equal. However, this certainly does not apply to the far-from-equilibrium process of generating a DNA molecule.

How in the course of evolution the first complex self-replicating "being" arose and how DNA-based replication came about is an entirely open problem and not the topic of these notes. Here we want to show that ordered structures can arise far from equilibrum out of random processes, despite the small probabilities derived above. Moreover, we want to demonstrate that whether dynamics are deterministic or random has in general no bearing on the question of the origin of complex patterns. The basic character of the process, deterministic or random, does not allow for any generally valid prediction of whether ordered structures could emerge or would disappear under such dynamics. For simplicity we focus our attention on a binary string $010101\ldots01$ of length $L$.

## 6.2 Some Models for Nonequilibrium Dynamics

All that we discussed above has long been known, but we want to elaborate on these issues in the context of a nonequilibrium lattice model that has been the subject of intense study in the last two decades. This process, called the totally asymmetric simple exclusion process (TASEP), shows very rich dynamical behavior and is tractable by rigorous mathematical methods [1]. Hence our understanding of nonequilibrium phenomena that we can derive from this process and some of its variants does not have the status of plausible conjecture (with possibly subtle pitfalls), but can be phrased in rigorous mathematical theorems [2, 3]. This model exhibits generic nonequilibrium behavior encountered in real systems. Often it is called the Ising model of nonequilibrium physics. This model has many applications, ranging from describing the kinetics of protein synthesis [4] to modeling the formation of traffic jams on motorways [5]. In the remainder of this section we define this model and some of its variants. To avoid trivial ambiguities we shall consider only lattices with an even number of sites.

### 6.2.1 Model 1: The Totally Asymmetric Simple Exclusion Process

The TASEP in its standard form is a continuous-time Markov process that models the dynamics of an interacting particle system with short-range interactions, driven by some random noise. Here comes an informal, but rigorous definition of the one-dimensional version of this process: Consider a periodic lattice of $L$ lattice sites. Each lattice site $i$ (defined modulo $L$) may be either empty or occupied by at most one particle. Each particle attempts to jump to its right (clockwise) neighboring site after an exponentially distributed random time of mean $\tau = 1/w$. All jump attempts occur independently. If the target site is empty, the jump is executed, i.e., a particle on site $i$ moves to $i + 1$. If the target site is occupied, the jump attempt is rejected.

This is the complete definition of the process. The microstates of this system, which we shall often refer to as configurations, denoted by the vector $\eta = (\eta_1, \eta_2, \ldots, \eta_L)$, are completely characterized by the set of occupation numbers $\eta_i$, which may take values 0 or 1. Therefore the phase space of the system, defined by the set of microstates $\{0, 1\}^L$, has dimension $2^L$. For a specified particle number $N$, the number of microstates is

$$C(L, N) = \binom{L}{N}. \tag{6.3}$$

The dynamics described above may be encoded in transition rates $w_{\eta',\eta}$ for a transition from configuration $\eta$ to $\eta'$. All these transition rates are equal to either zero or $w$.

To get a feeling for the physics that this process models, we make a few observations:

(i) We have a pure jump dynamics. Hence this process describes a many-body system with conserved total particle number.

(ii) The exclusion principle, which forbids multiple occupancy of a lattice site, corresponds to a hard-core repulsion between these particles. Hence we have an interacting particle system with short-range interactions. Under equilibrium conditions (no hopping bias) the equilibrium distribution would be the usual Gibbs measure with a repulsive $\delta$-function interaction. On a lattice (where this interaction is represented by a Kronecker $\delta$-function with infinite interaction energy) this means that the equilibrium distribution $P^*(\eta)$ of a microstate $\eta$ in the grand canonical ensemble reduces to a factorized distribution (see below).

(iii) For $N = 1$ this single particle performs a directed random walk. The same is true for a single particle in the many-body case, as long as this particle does not encounter another particle ahead. It is known that on large scales a directed random walk converges to Brownian motion with a constant drift. So on large scales an isolated particle in the TASEP models Brownian motion under the influence of a constant driving force. Notice, however, that because of periodic boundary conditions this constant driving force cannot be derived from a linear potential. Hence there is no Hamiltonian for this system. The particle will move forever with constant mean velocity around the ring. The latter will be true also for all particles in the many-body case. Therefore a current will be flowing, which keeps the system out of equilibrium at all times.

Notice that the notion of equilibrium is defined in the sense of invariance under time reversal, which in this stochastic setting means the condition of detailed balance

$$P^*(\eta)w_{\eta',\eta} = P^*(\eta')w_{\eta,\eta'} \, . \tag{6.4}$$

The presence of the steady-state current indicates the breaking of time-reversal invariance, as time reversal would change the direction of the current. Hence any current-carrying system is out of equilibrium.[1]

The TASEP has finite phase space and it is easy to prove that for any fixed particle number the dynamics is ergodic. Hence there exists a unique stationary distribution that is the analogue of the Gibbs distribution in equilibrium systems. The general problem with nonequilibrium systems is the absence of a Hamiltonian that would provide the energy $E(\eta)$ of a microstate. Therefore there are no Boltzmann weights $\exp(-\beta E)$ for microstates and consequently no knowledge in general about the form of the stationary distribution. It is this fact, always true, that explains why

---

[1] To avoid confusion we note that in systems with many internal degrees of freedom or large separation of intrinsic time scales some degrees of freedom may be in thermal equilibrium while others are not. For example, the water that flows out of a faucet has a current and hence is in a nonequilibrium state while it is flowing. Nevertheless this water may have a well-defined temperature, in equilibrium with its environment if the environment has the same temperature. The water that is then collected in the sink under the faucet would be in full equilibrium (after internal turbulence has relaxed).

there is no such thing as a generally valid thermodynamic theory of nonequilibrium systems, analogous to the usual equilibrium thermodynamics that is generally valid for all additive particle systems (i.e., systems with short-range interactions).

In the case of the TASEP, however, we are lucky. It is possible to obtain the exact stationary distribution by various approaches, which are not of interest here. It turns out that the stationary nonequilibrium distribution is the same product measure that describes the Gibbs distribution of the equilibrium version of this process (similar dynamics, but symmetric hopping rates in both directions on the lattice). Therefore, in a grand canonical ensemble with fugacity

$$z = \frac{\rho}{1 - \rho}$$

the stationary distribution of a configuration $\eta$ is given by

$$P^*_{\text{gcan}}(\eta) = \prod_{i=1}^{L} ((1 - \rho)(1 - \eta_i) + \rho\eta_i) = \rho^N (1 - \rho)^{L-N} , \qquad (6.5)$$

where $N$ is the number of particles in the configuration $\eta$ and $\rho$ is the particle density of this grand canonical ensemble with fugacity $z$. Notice that the stationary probability (6.5) depends on the configuration $\eta$ only through the total particle number. Therefore, in the canonical ensemble with fixed particle number $N$ the stationary distribution is uniform and given by the inverse of the number of configurations

$$P^*_{\text{can}}(\eta) = \frac{1}{C(L, N)} = \frac{N!(L - N)!}{L!} . \qquad (6.6)$$

### 6.2.2 Model 2: The TASEP with Random Sequential Update

Even though most quantities of interest can be computed for the TASEP by exact analytical methods, it is still important to perform numerical simulations of this process. However, the continuous-time TASEP defined above cannot – strictly speaking – be simulated on a computer for two reasons: (i) Time on a computer is by necessity discrete since any number is represented by a finite set of binary numbers. (ii) There is no such thing as randomness in a computer (even if with certain applications and operating systems one sometimes has the feeling that there is).

To approximate the dynamics of the TASEP on a computer one considers the so-called random sequential update, which works as follows: One stores a configuration $\eta(t)$, which represents the state of the system at time time $t$. Here $t$ is a discrete integer. In one update time step one performs the following operations: (i) One picks uniformly a "random number" $i$ from the set $1, \ldots, L$. Then one checks whether $\eta_i(t) = 1$ and $\eta_{i+1}(t) = 0$. If this is the case, the configuration $\eta(t)$ is updated corresponding to a particle jump from $i$ to $i + 1$. Otherwise nothing happens. The time variable $t$ remains unchanged after this elementary jump attempt. (ii) This operation is repeated $L$ times. (iii) The time variable $t$ is incremented by one unit. Then the steps (i)–(iii) are repeated. This algorithm produces one particular realization (history) of the process.

Notice not only the discreteness of time, but also the absence of any randomness. This is because a computer cannot generate genuine random numbers. Each "random number" is computed according to some deterministic algorithm, starting from a seed that one fixes and which is part of the computer code that generates these dynamics. Therefore, if one starts from the same seed and the same initial configuration, the computer will each time generate the same history.

We remark that if, on the other hand, one could pick a site $i$ genuinely at random, then the discrete-time nature of this process would not matter very much. The stationary distribution would be same uniform distribution (6.7) as for the continuous-time TASEP. For large $L$ even the dynamics would very closely follow that of the continuous-time TASEP.

### 6.2.3 Model 3: TASEP with Sublattice Parallel Update

Here we describe another discrete-time update that is purely deterministic. Instead of choosing a site $L$ times at random and performing a jump when it is possible, we define the dynamics in a single time step as follows: (i) We divide the lattice into neighboring pairs $(2, 3)$, $(4, 5)$ ... $(L, 1)$. In each pair we execute a jump to the right if it is possible, i.e., if and only if $\eta_{2k}(t) = 1$ and $\eta_{2k+1}(t) = 0$. Any other configuration of neighboring pairs remains unchanged. Time is not incremented. (ii) We divide the lattice into pairs $(1, 2)$, $(3, 4)$, ..., $(L - 1, L)$ and perform jumps to the right in all pairs where it is possible. (iii) Time is incremented by one unit. Then steps (i)–(iii) are repeated for the next full time step.

We note that this process is a limiting case of a similar discrete-time stochastic dynamics in which jumps are executed with a probability $p$ whenever possible. The case $p = 1$ is then the deterministic limit defined above. The stochastic case can be approximated on a computer by choosing for each pair of sites a (pseudo)random number $\xi_i$ uniformly in $[0, 1]$. If $\xi_i \leq p$ the jump is executed (if allowed). When $p$ is small (of order $1/L$) and $L$ is large, then $L$ discrete steps of the sublattice parallel update are not much different from one time step in random sequential update.

### 6.2.4 Model 4: TASEP with Next-Nearest-Neighbor Interaction

We return to genuinely stochastic dynamics. We define jump rules as in the TASEP defined above, but with the extra constraint that a jump attempt from site $i$ to site $i + 1$ is forbidden if also the next-nearest-neighbor site $i + 2$ is occupied. This models a repulsive interaction that is still short ranged but extends over more than just one lattice site. One can picture this as follows: Particles on nearest neighbor sites have a very high energy due to a repulsive potential. Therefore, pairs of particles tend to separate.

We remark that the difference from the usual TASEP is minute in the sense that both models describe driven random walks, with only a very small change in the range of the repulsive interaction.

### *6.2.5 Emergence of Order and Relaxation to Disorder*

We have introduced four variants of asymmetric exclusion processes that have in common driven dynamics that maintains a nonequilibrium steady state and short-ranged interactions. Two of the processes are genuinely stochastic, the other two deterministic. The stochastic cases (1) and (4) could be experimentally realized, for example by taking little balls as particles and a set of boxes as the lattice. The jump events could be triggered by a radioactive signal, which we know is truly random, uncorrelated, and yields an exponential jump time distribution. We wish to stress that under such an experimental realization, the TASEP and its variant (4) would not just be a mathematical model but would describe a genuine natural process, subject to the usual laws of nature that govern all other physical phenomena. Of course, also the deterministic processes can be realized as natural physical systems, obeying the laws of physics.

We wish to address the question whether the general features of these processes (nonequilibrium, short-ranged interaction, stochastic versus deterministic) allow us to make any prediction about the corresponding qualitative nature of the dynamics. In particular, we wish to investigate whether these processes are capable of generating ordered patterns with an information content equal to that of the Shakespeare text in a reasonable length of time. Conversely, given such a pattern as the initial configuration, we shall also investigate whether it persists or whether a disordered state with a random pattern arises in the course of time.

To this end, we consider for each model a half-filled lattice ($N = L/2$) of $L = 384$ sites. As the initial configuration we study two cases: (a) the ordered state $010101\ldots01$ and (b) a random initial configuration that can be generated by coin tossing of $L$ coins, conditioned on yielding heads and tails each exactly $L/2 = 192$ times. Notice that this coin tossing could in principle generate the ordered alternating initial state, but the probability that this happens is exactly the tiny number computed in the introduction. Hence it will not happen, even if we try for many ages of the universe. These two initial configurations have a natural interpretation in more general terms. The ordered state is imposed by design (the intention of the author of this chapter). The disordered state may be interpreted as resulting from some highly noisy dynamics (coin tossing), which after a strong change in the environment of the system (stopping coin tossing) turns into more gently hopping dynamics. So case (a) may be attributed to deliberate, purposeful intervention, while case (b) corresponds to some change in the environment of the particle system. We discuss each model separately.

*Model 1*

(a) *Ordered Initial State*: The state will change with each jump. The relaxation time
to reach stationarity is known to be on the order of $L^{3/2}\tau \approx 7,500\tau$. Hence
after this relatively short time the system will have "forgotten" its ordered initial state and take some random configuration that is drawn from the stationary
distribution (6.7). The stationary probability of the ordered alternating configuration equals

$$P^*(0101\dots01) = \frac{192!^2}{384!} \approx 5.52^{-384} , \qquad (6.7)$$

which is essentially as small (except for the factor of 5.5) as the probability of
obtaining this configuration through coin tossing. Therefore it will not happen,
even if we let the process run for many times the age of the universe. In other
words, the ordered pattern disappears quickly under these random dynamics
and will never reemerge again.

(b) *Disordered Initial State*: Since the system is ergodic, the same as for the
ordered initial state will be true for any initial configuration. The ordered pattern
$01010\dots01$ will never be generated in the TASEP with 384 sites.

*Model 2*

This model is deterministic and we cannot straightforwardly speak about ergodicity.
In fact, since the pseudorandom number generator of the computer is deterministic,
each random number will reappear in a deterministic cyclic fashion. Therefore also
all configurations of the particle system will reappear in a deterministic cyclic fashion, including the ordered state $010101\dots$. This is the analogue of the Poincaré
cycle in Newtonian dynamics. The length of this cycle depends on the quality of
the pseudorandom number generator: the longer the cycle, the better the generator. Notice, however, that each time a pseudorandom number reappears, it will do
so with a different configuration of the particle system. Hence the length of the
"Poincaré" cycle of this process may be estimated to be the cycle time of the pseudorandom number generator times the number of microstates (which is the inverse
of (6.7)). Hence this cycle length is a more-than-astronomic number that plays no
role for any practical consideration.

For any reasonable amount of time the system will behave essentially like the
stochastic TASEP. With a good pseudorandom number generator it would not be
possible to tell from the data alone whether the underlying dynamics is genuinely
random or not. Therefore, just from simulation measurements, the scenario concerning the evolution of ordered or disordered initial states would be indistinguishable.
The ordered state would quickly disappear and never be observed again, even though
the underlying dynamics are deterministic.

*Model 3*

(a) *Ordered Initial State*: It follows directly from the definition of the dynamics that the ordered initial state will not decay: In the first half-step of one full step the configuration $0101\ldots01$ is shifted to $1010\ldots10$ and then becomes after one complete time step again $0101\ldots01$. This deterministic dynamics preserves the ordered initial pattern, much like a computer preserves the information of a Shakespeare poem on its way from the data storage device to the computer screen.

(b) *Disordered Initial State*: This is more interesting. One might imagine that this deterministic dynamics would also preserve the disorder (as the computer would preserve the disorder of a random sequence from storage to screen). This is, however, not the case. It is not difficult to see that after a certain number of time steps (of order $L$) *every* initial state will evolve into the ordered pattern $0101\ldots01$. This follows from the fact that the ordered state is the unique stationary state. Hence this is an example where deterministic dynamics generate order out of disorder. The deterministic mapping of the TASEP with sublattice parallel update is not one-to-one.

*Model 4*

(a) *Ordered Initial State*: We can almost repeat the discussion of the previous model, even though the dynamics is genuinely random. It follows directly from the definition of the dynamics that the ordered initial state $0101\ldots01$ will not decay. It is a "frozen" state that is stable under the stochastic dynamics. The same is true for the ordered configuration $1010\ldots10$.

(b) *Disordered Initial State*: Now one might imagine that the random dynamics would at least preserve the disorder of the initial configuration. Also this guess is wrong. After a not very large number of steps the system will reach one of the ordered states $0101\ldots01$ or $1010\ldots10$. Which one depends on the random history and (probabilistically) on the initial configuration. In any case, however, these genuinely random dynamics generate an ordered state out of disorder after a short period of time (of the order of the system size), just like the deterministic dynamics of model 3 does.

## 6.3 Some Conclusions

Model 1, the TASEP, behaves as naive intuition would suggest. Order disappears under random dynamics, essentially forever, if sufficiently complex. Likewise, no ordered structure emerges from randomness. Model 2 has the same features, but offers a surprise in the sense that the dynamics is deterministic. The surprise comes from the formal argument that the deterministic dynamics of the computer are reversible and hence any state that arises during the evolution can be traced back to the initial state. This implies that there is no loss or gain of complexity during the evolution. Notice, however, that this general argument, while formally correct, has no meaningful application to this practical situation, since it does not take into account the length of the Poincaré cycle for this process. For any finite observation time, the process behaves as if it were random, i.e., given only observation data, we would not be able to distinguish whether these data come from model 1 or model 2, if a good pseudorandom number generator is used and the data set is finite. So one could speculate about conservation of the information contained in the initial state, but would not be able to decide on this matter based on empirical data (which in physics is the only thing we have to base decisions on).

Model 3 behaves as naive intuition would predict for the ordered initial state. The deterministic dynamics preserves the order. This happens in a different way than in model 2 in the sense that this preservation is actually observable in finite time, not only after collecting either (practically) infinitely many data or waiting for several ages of the universe. However, this model surprises naive intuition in the sense that is creates order out of an entirely disordered state, again in finite time (and, in fact, in a fairly short time on the order of a few hundred units of a single time step). Model 4 behaves contrary to naive expectation for both initial scenarios. Even though the dynamics is entirely stochastic, order is not only preserved, but even generated out of disordered initial states after a (short) finite length of time.

Again we find that from an empirical viewpoint (and this is the only one we have for sufficiently complex real systems) one cannot generally distinguish whether the underlying dynamics that generate observed data are random or deterministic. More importantly we conclude that whether dynamics are deterministic or stochastic is not a criterion for whether ordered patterns disappear or emerge. We see that random processes can generate order with the same Shannon information as Shakespeare lines. This observation can be easily extended to larger $L$ (and hence longer texts), since the times required to generate these strings grows only with a power law in $L$. Hence very long texts with very high Shannon information can be created through random processes in a short time. The basic issue where unguided intuition goes wrong is the interpretation of the Shannon information in terms of probabilities. The Shannon information is based on the probability generated by coin tossing, while in general random processes may generate the same patterns with much higher probability.

To understand this, it is instructive to approximate the exponential waiting time distribution by a geometric waiting time distribution as follows: Instead of waiting for a radioactive decay to trigger a jump attempt in the exclusion process, we could throw coins or dice and trigger a jump event as soon as heads (or a six in the case of dice) shows. Also under these dynamics the general observations made above for models 1 and 4 would apply. However, in the case of model 4, that means that we have constructed a random coin-tossing process (or dice-throwing) that leads to a pattern with the same Shannon information as a Shakespeare poem after only a few thousand (or perhaps million) tosses, as opposed to the zillions computed above. This is possible because the exclusion dynamics of model 4 direct these random events into the 010101... pattern. It is the combination of uncorrelated, equiprobable events (which alone cannot generate complex patterns) and simple rules ("laws of nature") that deceives unguided intuition. The rules that govern the possible motions of the system as a whole shape unstructured randomness into complex patterns.

Let us return to the beginning of this chapter: At present we are witnessing in the scientific and religious public a severe but pointless struggle between one segment of society that believes that explaining the presence of complexity requires the action of some intelligent designer, and another segment, represented by apostles of new atheism, who believe, with similar religious fervor, that scientific progress precludes an intellectually tenable faith in God. The intensity of this controversy, well characterized in a recent book by Michael Ruse [6], demands an evaluation of our discussion against this background. The fundamental insights in the interplay of randomness and rules that we can derive from considering simple generic models have shown that no supreme engineer is required to clarify basic principles of the emergence of ordered patterns from unpredictable and unguided random processes. Evolution is not the right place to find God.

On the other hand, our considerations should not obscure the fact that it remains legitimate to speak of pattern formation phenomena and more generally of evolution as a "miracle": With mathematical models (which we employ in all fields of physics) we can only *describe* what nature does, not explain why the natural processes act in the way they do (and not differently), and we cannot say what keeps natural "laws" (a strange metaphor for empirical regularities) going and going. Metaphysical questions like these leave room for awe and for religion. Declaring such questions for irrelevant is no less religious than religion itself.

This fundamental openness, that scientific inquiry leads us to by its very nature, marks one of the realms where religion and science can have a fruitful encounter. However, anyone getting into this is well-advised to listen to what no one less than the great biblical prophet Isaiah said more than 2,000 years ago. In a long passage he speaks about God the Creator and comes to the conclusion "Truly you are a hidden God" (Isaiah 45:15). Sounds almost as if he knew how science works.

## References


1. G.M. Schütz, in *Phase Transitions and Critical Phenomena*, Vol. 19, ed. by C. Domb, J. Lebowitz (Academic, London, 2001)
2. T.M. Liggett, *Interacting Particle Systems* (Springer, Berlin, Heidelberg, 1985)
3. T.M. Liggett, *Stochastic Interacting Systems: Contact, Voter and Exclusion Processes* (Springer, Berlin, Heidelberg, 1999)
4. J.T. MacDonald, J.H. Gibbs, A.C. Pipkin, Biopolymers **6**, 1 (1968)
5. D. Chowdhury, L. Santen, A. Schadschneider, Phys. Rep. **329**, 199 (2000)
6. M. Ruse, *The Evolution–Creation Struggle* (Harvard University Press, Cambridge, MA, 2005)

# Chapter 7
# Structure Formation in the Universe

**Matthias Bartelmann**

**Abstract** Two simple symmetry assumptions combined with general relativity lead to the class of Friedmann cosmological models on which the standard model for the structure and the evolution of the Universe is built. Within this model, dark matter dominates structures on the scales of galaxies and larger, and dark energy has dominated the expansion of the Universe since about half its present age. This chapter summarizes how cosmic structures could have developed under these circumstances and what they are characterized by. As to the origin of cosmic structures, the scenario of an early inflationary phase suggests that they arose from vacuum fluctuations of a primordial quantum field.

## 7.1 The Framework

### 7.1.1 Concepts

The standard model of cosmology [1], which has emerged over the past decade or so, rests on Einstein's theory of general relativity. Gravity is the only one of the four fundamental forces that is relevant for the evolution of the Universe as a whole. The strong and the weak interactions are confined to interactions between elementary particles. The electromagnetic interaction, although in principle long-ranged, is restricted in practice by shielding of unlike charges. Gravity alone dominates on the largest scales, and general relativity is the accepted theory of gravity.

General relativity introduces the space-time metric as a dynamical field whose evolution is controlled by Einstein's field equations. These equations couple the geometry of space-time with its matter-energy content. Since the flow of the matter depends on geometry itself, these equations are necessarily nonlinear. Unlike in electrodynamics, which is a linear field theory, no general scheme can be given

M. Bartelmann (✉)
Institut für Theoretische Astrophysik, Zentrum für Astronomie, Universität Heidelberg,
D-69120 Heidelberg, Germany
e-mail: mbartelmann@ita.uni-heidelberg.de

H. Meyer-Ortmanns, S. Thurner (eds.), *Principles of Evolution*,
The Frontiers Collection, DOI 10.1007/978-3-642-18137-5_7,
© Springer-Verlag Berlin Heidelberg 2011

for solving Einstein's equations. Solutions are typically constructed starting from symmetry assumptions simplifying the admissible form of the metric.

Two such symmetry assumptions were made by the Russian mathematician Alexander Friedmann soon after Einstein had published the final form of his field equations [2]. Purely for mathematical simplicity, he constructed solutions that were homogeneous and isotropic. Every (freely falling) observer in a universe described by such a solution would observe the same physical properties of the Universe independent of the direction of observation, and this would hold for all observers alike. Under this apparently vast simplification, Friedmann derived the class of cosmological models named after him.

Clearly, these models can depend on time only, not on space any more because of homogeneity. Their spatial sections are either flat or Euclidean, positively or negatively curved, and expand or shrink with time as given by a dimensionless scale factor $a(t)$. Any two freely falling particles would increase their separation in proportion to $a(t)$. Einstein's field equations then reduce to two ordinary differential equations for the scale factor, which can be written in the form

$$\left(\frac{\dot{a}}{a}\right)^2 = \frac{8\pi G}{3}\rho + \frac{\Lambda c^2}{3} - \frac{kc^2}{a^2} \,,$$
$$0 = \mathrm{d}(\rho c^2 a^3) + P\mathrm{d}(a^3) \,. \tag{7.1}$$

The first equation, often called Friedmann's equation, describes how the scale factor $a(t)$ changes with time in a universe that has energy density $\rho c^2$, cosmological constant $\Lambda$, and spatial curvature $k$. If space is flat, $k = 0$. The density $\rho$ is the sum of all densities contributing, i.e., all forms of matter and radiation. The nature of the cosmological constant is unclear. Einstein introduced it to allow static universes in which $\dot{a} = 0$. From the point of view of a classical field theory, it has to appear unless empirically ruled out. The consistent interpretation of the cosmological observations requires it to be there.

The second equation is the first law of thermodynamics. The first term on the right-hand side is the change in internal energy in a volume expanding or shrinking as given by the scale factor, while the second term is the work done by the pressure $P$.

These equations need to be complemented by an equation of state relating the pressure $P$ to the density $\rho$. Different types of matter or energy are characterized by their different equations of state. Ordinary, cold matter has $P = 0$, ultrarelativistic matter such as radiation has $P = \rho c^2/3$. If the cosmological constant is seen as a contribution to energy, it has $P = -\rho c^2$.

### 7.1.2 Isotropy on Average

The Friedmann models are thus a simple class of cosmological models derived from general relativity based on two far-reaching symmetry assumptions. They are characterized by a few parameters, namely the densities of all constituents of matter and radiation, the cosmological constant, the equation of state parameter(s), and the initial condition set by the cosmic expansion rate $\dot{a}/a$ at any given time, usually today. Can they have anything to do with our physical reality? Several questions arise:

- Can their symmetry assumptions be experimentally justified?
- Can cosmological findings be accommodated in their framework?
- How must their parameters be chosen?
- Which conclusions follow from them?

At first sight, it appears hopeless to justify the isotropy of the Universe surrounding us. Clearly, the sky looks different in different directions. Even on large scales, there are pronounced structures in the distribution of galaxies. They can be as large as some ten megaparsecs[1] and thus more than a thousand times larger than individual galaxies. The only possible justification for the assumption of isotropy is that the observable Universe is large enough to contain very many even of such structures. Its scale is characterized by the Hubble radius, which is more than 4,000 Mpc. It is thus possible to average over cosmic structures on scales that are much larger than the structures themselves and yet much smaller than the scale of the observable Universe. In this averaged sense, the matter distribution in the Universe does indeed approach isotropy. The most convincing evidence of isotropy, however, is provided by the temperature fluctuations in the Cosmic Microwave Background (CMB). They reach about ten parts per million compared to the average CMB temperature. Given the rich variety of structures in the present Universe, the CMB is phantastically isotropic.

Once we can accept the assumption of isotropy, the assumption of homogeneity does not seem problematic any more. Since the Copernican Revolution, we have become used to the concept that our position in the Universe is by no means preferred to any other. If we observe isotropy, any observer should see the Universe as isotropic as well. A space, however, that is isotropic about all of its points, is also homogeneous.

### 7.1.3 The Cosmic Expansion

It was found in the 1920s that the Universe expands. The American astronomer Vesto Slipher realized by spectral analysis that absorption lines in galaxy spectra typically appear redshifted, signaling that the galaxies are moving away from us [3]. Using the then largest telescope in the world, Edwin Hubble showed that this recession velocity grows with the distance to the galaxies, which is exactly the behavior expected from an expanding Friedmann model [4]. Every observer in such a universe should see nearby objects moving away with a velocity linearly

---

[1] A megaparsec (Mpc) is an astronomical unit of length. 1 Mpc $= 10^6$ pc $= 3.1 \times 10^{24}$ cm.

proportional to their distance. The facts that Hubble found this proportionality only by leaving out dubious data points and that his proportionality constant came out much larger than we know it today should not concern us here. The systematic recession of the galaxies shows that the Universe is not static, and the scaling of the local recession velocities with distance agrees with the expansion behavior expected in a Friedmann model.

If the Universe is expanding today, it is natural to assume that it kept growing with time and should thus be considered ever smaller the further we go back in time. This is not necessarily true. In certain regions of parameter space, Friedmann models exist that are expanding today but stalled or shrunk in the past. Nonetheless, a few simple observations show that if we live in a Friedmann universe at all, it cannot be of this type and must have grown with finite expansion velocity throughout its entire history. Going back in cosmic evolution, we must thus find a time before which the entire observable Universe must have been as small as quantum scales. At the latest when this point is reached, we lose all confidence in our physical description of the Universe, because then general relativity should be replaced by a quantum theory of gravity, which we do not yet have. We thus assume that prior to some finite time in the past, the Universe originated from a very hot and dense state, which we call the Big Bang. For our purposes, the main result of this discussion is that, if the Universe can be described by Friedmann models, it must have a finite age, and a Big Bang was inevitable.

### 7.1.4 Origin of the Light Elements and the Cosmic Microwave Background

These are already remarkable conclusions that lead to testable predictions. If the Universe was smaller in the past, it must have been hotter. We have seen that the first law of thermodynamics also holds in a Friedmann universe, and it implies that the temperature of matter and radiation increases as the available volume shrinks. Thus, there must have been a time in the past when the entire observable Universe was as hot as the interiors of stars are now, where hydrogen and other light elements are fused to form more massive nuclei [5].

That the entire Universe must have acted as a fusion reactor is also supported by the observation that approximately 25% of the ordinary, baryonic[2] matter consists of helium-4 rather than hydrogen. Although stars produce most of their energy by hydrogen fusion, their helium-4 production would by no means suffice to explain the amount of helium-4 observed in the Universe.

The amount of helium-4 in the Universe led Gamow, Alpher, and Herman to the prediction in the 1940s that the Universe must have gone through an early phase hot enough to fuse hydrogen to helium, and that there should be thermal radiation

---

[2] Particles participating in the strong interaction are called hadrons. Baryons are hadrons composed of three quarks, such as protons and neutrons.

left over from this time [6, 7]. Realizing that the fusion of helium-4 must have gone through the bottleneck of deuterium fusion first, and that this fusion step must have proceeded just right for part but not all of the hydrogen to be converted to helium-4, they concluded that this thermal relic radiation should by now have cooled to between 1 and 5 K, and thus have shifted to the regime of microwaves. Not only the existence of the CMB but also its temperature could be predicted from the observed amount of helium.

From the expansion rate of the Universe, its present temperature, and the cross sections for nuclear fusion, it is straightforward to calculate that the fusion of hydrogen to deuterium, helium, and other light elements must have happened when the Universe was approximately 150 s old and stopped soon thereafter when it became too cold.

Initially, this thermal radiation could not freely propagate because it kept scattering off the charged particles around, the electrons and the nuclei. As the Universe expanded and cooled, electrons and nuclei could finally combine to form neutral atoms. The number of charged particles and thus of scatterers dropped drastically, setting the thermal radiation free. Again, it is quite straightforward within the class of Friedmann models to calculate when this recombination process occurred and how long it lasted. The Universe was a little less than 400,000 years old when it became transparent within approximately 40,000 years.

The CMB was serendipitously discovered in 1965 by a measurement at a frequency of 4,080 MHz, or 7.4 cm wavelength [8]. This signal was immediately interpreted as being due to the CMB [9], but only the spectrometer on board the COBE satellite could firmly establish in 1992 that this radiation does indeed have the Planck, or blackbody, spectrum. In fact, the electromagnetic spectrum of the CMB is still the best-measured Planck curve ever. It reveals that the CMB has by now cooled down to 2.726 K [10].


### 7.1.5 Structures in the Cosmic Microwave Background

With its Differential Microwave Radiometer, the COBE satellite also detected the long-sought temperature fluctuations in the CMB [11] (see Fig. 7.1). Although we have started from the assumption of a perfectly isotropic universe, pronounced structures evidently exist. The most natural assumption is that they grew by gravitational instability from tiny seed fluctuations produced by some process in the very early Universe. But then, the imprint of these seed fluctuations should be visible as temperature fluctuations in the CMB. The expected amplitude of these fluctuations is easily calculated. It was predicted at a level of one part in a thousand, i.e., in the millikelvin regime [12, 13].

However, it was not found there even long after the experiments had reached sufficient sensitivity. A possible explanation, which turned out to be extremely fruitful, was proposed by Peebles [14]. He speculated that the imprint of later cosmic structures on the CMB could be much below the hitherto expected level if cosmic structures were not predominantly composed of ordinary, baryonic matter, but of

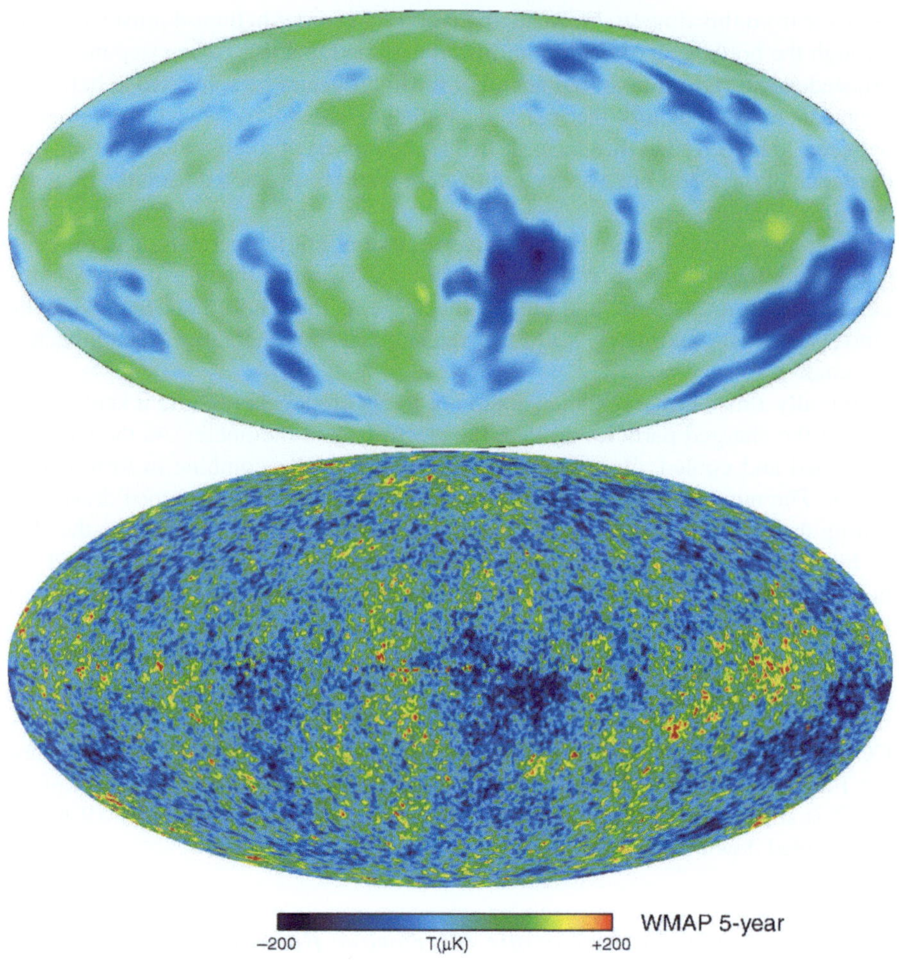



**Fig. 7.1** These projected full-sky maps are the icons of modern cosmology. They show the temperature fluctuations in the CMB as observed by the COBE (*top*) and WMAP satellites (*bottom*). COBE had an angular resolution of $7°$, WMAP has one of $15'$. The statistics of these fluctuations, as quantified by their power spectra, contain an enormous amount of precise cosmological information. (Courtesy of the WMAP team)

an unknown form of dark matter not participating in the electromagnetic interaction. Then, cosmic structure formation could have proceeded well before the release of the CMB at the recombination time without having left more than microkelvin fluctuations in the CMB temperature. The detection of fluctuations at this level by COBE in 1992 was a relief and a sensation for cosmology.

Now the CMB temperature fluctuations are much better known. While COBE had an angular resolution of $7°$, two satellites are currently being operated with much improved sensitivity and with a resolution down to a few arc minutes. One of them, the WMAP satellite, was launched in the summer of 2001, the other, Planck, started observing in the summer of 2009. While WMAP has already refined the cosmological model to an enormous degree (see Fig. 7.1), Planck is expected to drive the precision of the CMB observations considerably further.

Observations of temperature and polarization fluctuations in the CMB are so important for cosmology because they reveal a huge amount of information. Three primary physical effects have contributed to producing these fluctuations. On the largest scales, fluctuations of the gravitational potential were imprinted on the energy density of the radiation. On intermediate scales, gravity and pressure drove oscillations in the mixture of baryons, photons, and dark matter, which resemble sound waves and are thus called acoustic oscillations. On scales of a few arc minutes, the fluctuations are exponentially damped mainly by photon diffusion. These three effects are well understood and allow a firm prediction of how fluctuations in the CMB should be structured.

They can be quantified by a power spectrum, which is a function showing the (squared) fluctuation amplitude found in the CMB temperature as a function of angular scale (Fig. 7.2). According to the three physical effects shaping the CMB fluctuations, the power spectrum has three distinct parts: a featureless part corresponding to the potential fluctuations at large scales, the exponentially damped tail

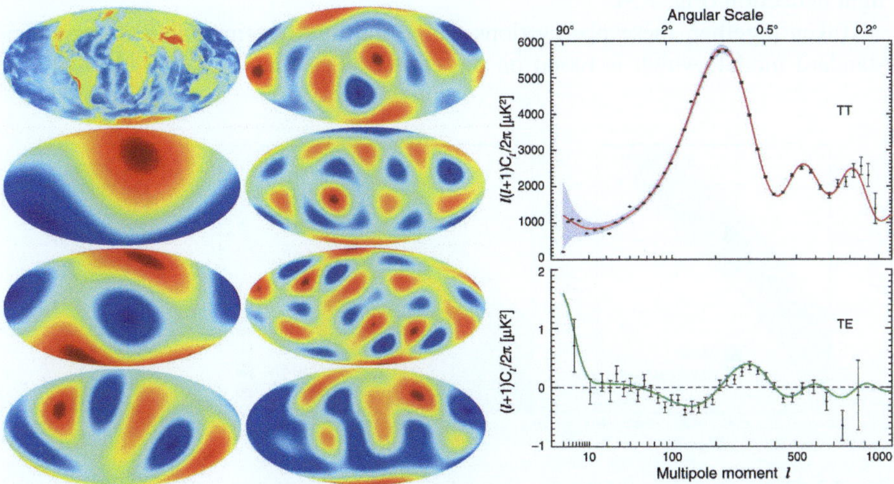


**Fig. 7.2** *Left*: The height of the Earth's surface above sea level (*top left*), decomposed into spherical harmonic functions of decreasing scale. From *top* to *bottom*, the *left* and *right columns* show the modes with $l = 1, 2, 3$ and $l = 4, 5, 6$, respectively. The map at the *bottom right* shows the structures with $l \leq 7$. *Right*: Power spectra of the CMB temperature fluctuations (*top*) and of the cross-correlation between temperature and linear polarization (*bottom*). The *continuous lines* are the theoretical fits to the data points measured by the WMAP satellite. They illustrate the wealth of information in the CMB and show an astounding agreement between the measurements and the theoretical expectation. (*Left panel*: data taken from the ETOPO-5 project; *right panel*: courtesy of the WMAP team)

towards small scales, and a sequence of pronounced maxima and minima caused by the acoustic oscillations.

It is a most remarkable success of the Friedmann cosmological models that they provide a framework in which the CMB power spectrum can be quantitatively predicted and related to their parameters, i.e., the densities of its constituents, the cosmological constant, and the spatial curvature. By fitting the predicted to the measured power spectrum, many of these parameters can be accurately determined. This process leaves narrow islands in parameter space to those Friedmann models which are compatible with the data and thus constrains them very tightly.

### 7.1.6 Cosmic Consistency

We owe much of our knowledge of the Universe to the CMB observations. Yet they leave some of the parameters of a Friedmann model poorly constrained because their effect can be compensated by other parameters. Such parameter degeneracies can be broken if the CMB observations are combined with other types of cosmological data. Most notable are the inference of the late cosmic expansion rate through a special class of stellar explosions, the so-called type-Ia supernovae; the constraint on the cosmic matter density from the preferred scale imprinted on the galaxy distribution; and the measurement of the amplitude of matter fluctuations by gravitational light deflection (Fig. 7.3).

Taken together, these observations define the highly remarkable cosmological standard model, which is based on the class of Friedmann models. Virtually all

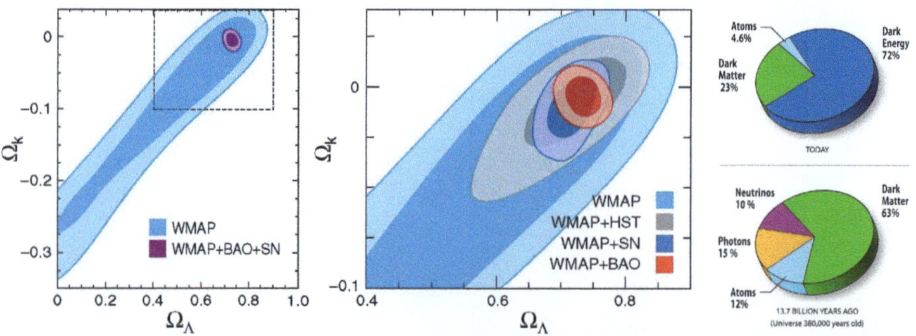


**Fig. 7.3** The two panels to the *left* show a cut through the multidimensional cosmological parameter space together with the contours of the likelihood of different cosmological data sets. The two parameters taken as an example here are the cosmological constant on the abscissa and the spatial curvature on the ordinate. Areas coulored *light blue* are allowed by the CMB data taken by WMAP; tighter contours take information from type-Ia supernovae (SN), the spatial galaxy distribution (BAO), and measurements of the Hubble constant (HST) into account. The diagrams illustrate the consistency of these different and independent data sets and the precision of the combined parameter constraints. The panel to the *right* shows the composition of the Universe today (*top*) and when the CMB was released (*bottom*). (*Left panels*: from [15]; *right panel*: courtesy of the WMAP team)

cosmological observations turn out to fit smoothly into a single Friedmann model. This is a highly astounding fact, firstly because of the simplicity of the symmetry assumptions on which these models are founded. They do seem to allow a consistent physical model for our Universe. Secondly, the data on which the cosmological standard model are based probe the physical state of the Universe at times between a few minutes after the Big Bang until now, when the Universe is almost 14 billion years old. One standard model, i.e., a Friedmann model specified by a single set of parameters, is evidently capable of accommodating essentially all types of cosmological information taken at several instances over nearly 14 billion years.

## 7.2 Structure Formation in the Universe

### 7.2.1 Concepts and Assumptions

In the framework of the cosmological standard model, we find ourselves living in a Universe whose matter content is dominated by some form of dark matter, which is presumably composed of weakly interacting, massive elementary particles. They cannot participate in the electromagnetic interaction because otherwise they would leave a pronounced and unobserved imprint on the CMB. Ordinary, baryonic matter as we know it only forms a small contribution to the matter budget of the Universe. Even more puzzling is that today the energy content of the Universe seems to be dominated not by matter but by the cosmological constant or something that behaves similarly. Since we do not know what this form of energy could be contributed by, we call it dark energy.

Specifically, the Universe today contains 23% dark matter, 4.6% baryonic matter, and 72% dark energy (Fig. 7.3). Going from now into the past, the dark energy changes its density very little or not at all. The density of nonrelativistic matter increases with $a^{-3}$ just because of volume compression, while the relativistic matter density grows as $a^{-4}$. Thus, the relative contributions of the various constituents of the cosmic fluid vary with time. When the CMB was released, dark matter contributed 63%, baryonic matter 12%, dark energy was unimportant, the CMB photons themselves provided 15% and neutrinos 10% of the energy density [15–17].

We are thus faced with the question of how the weak structures seen in the temperature of the CMB may have developed into the pronounced cosmic structures we see in the Universe today on a broad variety of scales, such as galaxies, galaxy clusters, and the even larger, filamentary structures traced by the galaxy distribution.

It turns out that the essential physical process of structure formation can be described by Newtonian hydrodynamics. This may seem surprising. Firstly, hydrodynamics is an approximation of the collective motion of particles in a system whose mean free path is very much smaller than all other scales appearing in the system. If the majority of matter in the Universe is composed of weakly interacting, dark-matter particles, their mean free path is expected to be enormous. Yet hydrodynamics is a valid approximation even in this context because any volume of interesting size contains so many particles that they can be described as a continuous fluid moving under the influence of the gravitational potential. Dark matter is approximated as pressureless because of the weak interaction of its particles.

For similar reasons, it suffices to express the gravitational interaction by Newtonian gravity. Compared to general relativity, this means in particular that the curvature of space-time is neglected and that the gravitational force is approximated as propagating instantaneously rather than at a finite speed. Both approximations are usually legitimate in the late Universe because the structures considered are typically much smaller than the curvature scale of the spatial sections and the propagation times for the gravitational force are much smaller than the evolutionary time scales of the structures.

Under these approximations, the equations governing the evolution of the dark-matter density are the continuity equation formulating mass conservation, the Euler equation expressing momentum conservation, and the Poisson equation, which is the gravitational field equation of Newtonian physics. These equations can be combined to obtain a single, second-order, ordinary differential equation for the density contrast $\delta$,

$$\ddot{\delta} + 2H\dot{\delta} - 4\pi G\bar{\rho}\delta = 4\pi G\bar{\rho}\delta^2 + \frac{1}{a^2}\nabla\delta \cdot \nabla\Phi + \frac{1}{a^2}\partial_i\partial_j\left[(1+\delta)u_i u_j\right] . \quad (7.2)$$

where $H$ is the Hubble function and $G$ the gravitational constant. The density contrast is the relative fluctuation of the density $\rho$ around the mean density $\bar{\rho}$, hence $\delta = (\rho - \bar{\rho})/\bar{\rho}$. The gravitational potential $\Phi$ provides the gravitational force $-\nabla\Phi$, and the matter moves with the velocity $u$ with respect to the mean Hubble expansion of the Universe.

### 7.2.2 Linear Structure Growth

Evidently, all terms on the right-hand side of (7.2) are nonlinear in the deviations from the mean cosmological background, defined by $\rho = \bar{\rho}$, thus $\delta = 0$, and $u = 0$. As long as the density fluctuations remain small, $\delta \ll 1$, the right-hand side can thus be set to zero. The resulting equation,

$$\ddot{\delta} + 2H\dot{\delta} - 4\pi G\bar{\rho}\delta = 0 , \quad (7.3)$$

then describes linear structure evolution. This equation has two linearly independent solutions, one decaying and one growing with time. In the context of structure formation, only the growing solution is of interest. The equation itself is independent of the size of the structure considered, thus its solution quantifies linear structure growth on all scales. It is called the growth factor, commonly written $D_+$. In a crude but sufficient approximation, $D_+$ grows like the scale factor, $D_+ \propto a$. As long as structures evolve linearly, their amplitude grows approximately like the Universe itself.

This immediately allows an interesting insight. We have seen that the fluctuation level in the CMB temperature is of order $10^{-5}$. These temperature fluctuations reflect fluctuations in the radiation density that are four times as large, but still of order $10^{-5}$. It is natural to assume that fluctuations in the radiation density traced fluctuations in the matter density. When the CMB was released, the Universe was about a thousand times smaller than today. Since then, structures can thus only have grown by a factor of $\sim 10^3$. Today, they should reach a relative amplitude near $10^{-2}$. Pronounced cosmic structures such as galaxies, galaxy clusters, and even larger-scale structures should therefore not exist.

How can the low temperature fluctuation level in the CMB be reconciled with the existence of highly nonlinear structures today? As mentioned above, Peebles [14] was the first to suggest that this could be achieved if cosmic structures were dominated not by ordinary matter, but by a form of hypothetic dark matter that cannot interact with light. Given that, structures in the matter distribution could have started growing way before the CMB was released, without leaving direct imprints in the CMB temperature fluctuations. Thus, the slow linear structure growth and the low amplitude of the CMB temperature fluctuations suggest that cosmic structures are composed mainly of dark matter rather than ordinary matter.

### 7.2.3 Cold Dark Matter

Under this condition, structure formation can be understood as starting from initial conditions reflected by the CMB temperature fluctuations and proceeding via gravitational collapse. As we have seen, this scenario requires dark matter. Its only property that we know of now is that it must not participate in the electromagnetic interaction. It could be composed of suitably massive, weakly interacting elementary particles. If so, another constraint on the nature of dark matter comes from the observation that galaxies and smaller-scale structures must have formed very early in the cosmic history, after just 1 or two billion years. This would be impossible if the hypothetic dark-matter particles were moving with high velocity, because then deep potential wells would be necessary to keep them bound, and such potential wells could only have been provided by large objects. Thus, the early formation of relatively small-scale objects implies that if dark matter consists of weakly interacting particles, they must at least be slow compared to the speed of light. If they originated under thermal-equilibrium conditions in the (very) early Universe, they must then be massive. The most economic assumption under these requirements is that the dark-matter particles move with velocities that can for all relevant purposes be set to zero. For this reason, such dark matter has been called cold.

Interestingly, the statistics of matter fluctuations composed of cold dark matter (CDM) can be accurately predicted based on a single further assumption, which has to do with the so-called horizon scale of the Universe. Since at any time the Universe has a finite age, there is a maximum distance that light can have traveled, which is called a horizon. Thus, at any given time, the horizon encompasses all particles that may have communicated with each other and thus have arranged for similar physical conditions. The horizon grows with time, i.e., it encloses increasingly larger scales, even though the Universe is expanding. A cosmic structure of a given scale will thus be larger than the horizon at very early times and smaller at sufficiently later times. There will thus be a time when the horizon has grown to the scale of the perturbation, at which point the perturbation is said to enter the horizon.

Again, it seems natural to assume that the mass contained in density fluctuations should be independent of the times when these fluctuations enter the horizon. Otherwise, it would either grow or shrink with time. If it grew, there would be time in the future when the observable Universe would collapse; if it shrank, this should have happened already in the past. Thus, the amplitudes of density fluctuations should be related to their scales in such a way that the mass entering the horizon in the form of these fluctuations was independent of time. This leads to the conclusion that the variance of the fluctuation amplitude, the so-called power spectrum, should decrease as $\lambda^{-1}$ with the scale of the perturbation, or increase with its Fourier wave number $k$ as $k = 2\pi\lambda^{-1}$ [18, 19].

Although the radiation density is today much smaller than any other density contribution, radiation becomes more and more important when going back in time. As the Universe shrinks, the density of ordinary (nonrelativistic) particles increases in indirect proportion to the volume: when the Universe had half its present size, the nonrelativistic matter density was eight times that today. Relativistic particles, however, such as photons, were redshifted and lost energy while the Universe was expanding. Going back in time, their density thus increases with one more power of the scale factor than the density of ordinary matter: when the scale factor had half its present value, the density of relativistic particles was 16 times that today. Thus, at some moment in the past, the radiation density was as high as that of ordinary matter, and there was a radiation-dominated epoch before.

Structures that entered the horizon during this radiation-dominated epoch could not continue growing because their collapse was slower than the expansion driven by the radiation. Structures large enough to have entered the horizon only after the end of the radiation-dominated epoch could continue growing without interruption. The smaller the structures are, the earlier they entered the horizon and the more they were suppressed. This allows the conclusion that, in the limit of $\lambda \to 0$ or $k \to \infty$, the power spectrum should asymptotically fall off as $k^{-3}$.

If CDM is assumed, the power spectrum thus has a well-predictable shape: It should rise $\propto k$ for small $k$ and fall as $k^{-3}$ for large $k$. The maximum in between is set by the size of the horizon at the time when the radiation-dominated epoch ended. Since the radiation density today is known from the temperature of the CMB, measuring the peak scale of the power spectrum returns information on the matter density. Only within the last few years have galaxy surveys grown beyond the peak scale (Fig. 7.4). They now provide additional constraints on the matter density, which are consistent with all the others.

Measurements of the power spectrum are now possible on a vast range of scales. On the smallest scales, it is constrained by absorption lines originating in cool

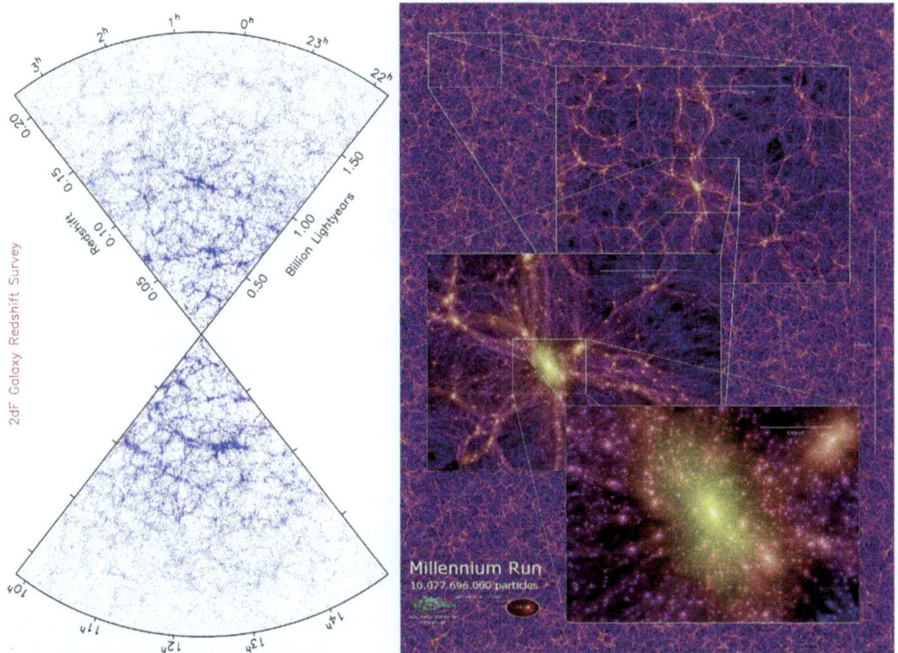


**Fig. 7.4** *Left*: The galaxy distribution in the local Universe as measured by the 2-degree Field Galaxy Redshift Survey (2dFGRS). *Right*: Image illustrating the dark-matter distribution at the present time produced by the Millennium Simulation of cosmic structure formation [20]. The measured galaxy distribution exhibits filamentary structures reproduced in the simulation. On smaller scales, dark matter clumps into so-called halos with a standard, but fundamentally unexplained radial density profile and abundant substructure. (*Left panel*: courtesy of the 2dFGRS team; *right panel*: courtesy of Volker Springel and the Virgo Consortium)

hydrogen gas clouds. Galaxies, galaxy clusters, gravitational lensing, and the CMB fix the power spectrum in a range of scales covering several orders of magnitude. The remarkable result is that the measured power spectrum so far agrees very well with the theoretical expectation in CDM, even though that was based on two very simple assumptions only. The CDM paradigm implies that structure formation in the Universe began with the smallest and proceeded to the largest structures. This is called the bottom-up scenario.

Owing to the central limit theorem, it is most plausible that the primordial fluctuations in the matter density were Gaussian, i.e., their probability distribution can be described by a Gaussian normal curve. The statistics of a Gaussian random field then predict that isotropic, spherical collapse is impossible, but sheet-like or filamentary structures must have formed first, which could later fragment into individual, smaller objects [21, 22]. It is thus a firm prediction of structure formation in a Gaussian random field of density fluctuations that sheets and filaments form, just like those that are observed in the large-scale galaxy surveys (Fig. 7.4).

### *7.2.4 Nonlinear Structure Growth*

Linear structure growth is applicable until the density contrast $\delta$ reaches approximately unity. The nonlinear evolution is complicated, but some aspects of it can remarkably be predicted analytically. Again starting from the assumption that the density fluctuations originally form a Gaussian random field, and drawing analogies between the linear and nonlinear collapse of a homogeneous sphere or ellipsoid, it is possible to predict not only the distribution of cosmological objects with mass, the so-called mass function, but also their correlations, i.e., the deviation of their spatial distribution from random. Considerable progress has also been made in analytic techniques working in the regime of mild nonlinearity based on higher-order perturbation theory. Detailed predictions and comparisons with observations of nonlinear structures, however, generally require numerical simulations.

Among a huge body of results obtained with increasingly sophisticated and extensive numerical simulations, one of the most important and puzzling is that gravitationally bound, nonlinear structures develop a universal density profile that asymptotically falls off as $r^{-3}$ far away from their centers and flattens towards $r^{-1}$ in their cores [23]. This is particularly remarkable because it is well known that self-gravitating systems have negative heat capacity because of the virial theorem, which implies that they do not have an equilibrium state. What causes the phenomenon that they approach a universal density profile and thus a long-lived transient state, and what defines their properties during this state, is fundamentally unknown.

Only massive nonlinear structures outside their core regions remain dominated by dark matter. The baryonic gas, which can dissipate energy and angular momentum in contrast to the dark matter, tends to shape structures small and dense enough for their cooling and hydrodynamic time scales to be sufficiently short compared to

**Fig. 7.5** On sufficiently small scales, baryons dominate over dark matter, as in this spiral galaxy, Messier 101. By their electromagnetic interaction, they can cool and thus dissipate energy and angular momentum and give rise to the rich phenomenology of hydrodynamic phenomena. (Hubble Space Telescope Archive, News Release STScI-2006-10)

the cosmic time. This is the case in galaxies (see Fig. 7.5 for a beautiful example), where complicated baryonic physics leads to star formation, and in some cores of galaxy clusters. Interestingly, there is an upper limit to the mass of an object in which stars can form. It is set by the atomic and molecular physics mainly of hydrogen and helium. In larger objects, the gas is too hot to cool efficiently. In galaxy clusters, for example, it remains hot and is visible in the X-ray regime.

### 7.2.5 *The Origin of Structures*

So far, we have described the evolution of comic structures from initial conditions defined mainly by the CMB. The question of where initial structures may have originated from has been left unanswered. Modern cosmology has a breathtaking suggestion for an explanation that is closely related to the concept of cosmological inflation.

It is essentially impossible to understand the appearance of the observable Universe without assuming that there was a phase during its very early evolution in which it was exponentially expanding. Such a phase is called cosmological inflation. Simple observational evidence such as the near-isotropy of the CMB requires a mechanism that may have established causal contact between any two points in the visible Universe that could otherwise never have communicated their physical conditions. The existence of a horizon, essentially caused by the finite age of the Universe, defines causally separated sections of the Universe which could never have reached thermal equilibrium and thus a common temperature unless inflation provided a physical mechanism for securing causal contact within the entire observable Universe.

At the same time, inflation provides a natural scenario for the origin of structures [24, 25]. Inevitable quantum fluctuations in the very early Universe would have been exponentially stretched by inflation, thus stabilized and magnified to cosmological scales. This inflationary scenario has testable predictions that are so far consistent with observations. First of all, in its simplest forms, it predicts that density fluctuations should be Gaussian, against which there is no convincing evidence. Second, it predicts a power spectrum of the density fluctuations that is on large scales almost, but not quite, proportional to the wave number $k$, but rather to $k^n$ with $n \approx 0.95$. The temperature fluctuations in the CMB show exactly that.

Thus, we need inflation for a consistent cosmological model, and inflation provides a physical mechanism not for the evolution but for the origin of cosmic structures. There is no direct evidence yet for inflation to have happened. However, the indirect evidence supporting it leaves us with the idea that cosmic structures seem to have originated from vacuum fluctuations in the very early Universe.

## References


1. M. Bartelmann, Rev. Mod. Phys. **82**, 331 (2010)
2. A. Friedman, Z. Phys. **10**, 377 (1922). doi 10.1007/BF01332580
3. V.M. Slipher, Pop. Astron. **29**, 128 (1921)

4. E. Hubble, M.L. Humason, Astrophys. J. **74**, 43 (1931). doi 10.1086/143323
5. G. Gamow, Phys. Rev. **70**, 572 (1946). doi 10.1103/PhysRev.70.572.2
6. G. Gamow, Phys. Rev. **74**, 505 (1948). doi 10.1103/PhysRev.74.505.2
7. R.A. Alpher, R.C. Herman, Phys. Rev. **75**, 1089 (1949). doi 10.1103/PhysRev.75.1089
8. A.A. Penzias, R.W. Wilson, Astrophys. J. **142**, 419 (1965)
9. R.H. Dicke, P.J.E. Peebles, P.G. Roll, D.T. Wilkinson, Astrophys. J. **142**, 414 (1965). doi 10.1086/148306
10. D.J. Fixsen, E.S. Cheng, J.M. Gales, J.C. Mather, R.A. Shafer, E.L. Wright, Astrophys. J. **473**, 576 (1996). doi 10.1086/178173
11. G.F. Smoot, C.L. Bennett, A. Kogut, E.L. Wright, J. Aymon, N.W. Boggess, E.S. Cheng, G. de Amici, S. Gulkis, M.G. Hauser, G. Hinshaw, P.D. Jackson, M. Janssen, E. Kaita, T. Kelsall, P. Keegstra, C. Lineweaver, K. Loewenstein, P. Lubin, J. Mather, S.S. Meyer, S.H. Moseley, T. Murdock, L. Rokke, R.F. Silverberg, L. Tenorio, R. Weiss, D.T. Wilkinson, Astrophys. J. Lett. **396**, L1 (1992). doi 10.1086/186504
12. P.J.E. Peebles, J.T. Yu, Astrophys. J. **162**, 815 (1970)
13. R.A. Sunyaev, Y.B. Zeldovich, Astrophys. Space Sci. **7**, 3 (1970). doi 10.1007/BF00653471
14. P.J.E. Peebles, Astrophys. J. Lett. **263**, L1 (1982). doi 10.1086/183911
15. E. Komatsu, J. Dunkley, M.R. Nolta, C.L. Bennett, B. Gold, G. Hinshaw, N. Jarosik, D. Larson, M. Limon, L. Page, D.N. Spergel, M. Halpern, R.S. Hill, A. Kogut, S.S. Meyer, G.S. Tucker, J.L. Weiland, E. Wollack, E.L. Wright, Astrophys. J. Supp. **180**, 330 (2009)
16. J. Dunkley, E. Komatsu, M.R. Nolta, D.N. Spergel, D. Larson, G. Hinshaw, L. Page, C.L. Bennett, B. Gold, N. Jarosik, J.L. Weiland, M. Halpern, R.S. Hill, A. Kogut, M. Limon, S.S. Meyer, G.S. Tucker, E. Wollack, E.L. Wright, Astrophys. J. Supp. **180**, 306 (2009)
17. D.N. Spergel, R. Bean, O. Doré, M.R. Nolta, C.L. Bennett, J. Dunkley, G. Hinshaw, N. Jarosik, E. Komatsu, L. Page, H.V. Peiris, L. Verde, M. Halpern, R.S. Hill, A. Kogut, M. Limon, S.S. Meyer, N. Odegard, G.S. Tucker, J.L. Weiland, E. Wollack, E.L. Wright, Astrophys. J. Suppl. **170**, 377 (2007). doi 10.1086/513700
18. E.R. Harrison, Phys. Rev. D **1**, 2726 (1970)
19. Y.B. Zeldovich, Mon. Not. R. Astron. Soc. **160**, 1P (1972)
20. V. Springel, S.D.M. White, A. Jenkins, C.S. Frenk, N. Yoshida, L. Gao, J. Navarro, R. Thacker, D. Croton, J. Helly, J.A. Peacock, S. Cole, P. Thomas, H. Couchman, A. Evrard, J. Colberg, F. Pearce, Nature **435**, 629 (2005). doi 10.1038/nature03597
21. M. Davis, G. Efstathiou, C.S. Frenk, S.D.M. White, Astrophys. J. **292**, 371 (1985). doi 10.1086/163168
22. S.D.M. White, C.S. Frenk, M. Davis, G. Efstathiou, Astrophys. J. **313**, 505 (1987). doi 10.1086/164990
23. J.F. Navarro, C.S. Frenk, S.D.M. White, Astrophys. J. **490**, 493 (1997). doi 10.1086/304888
24. A.H. Guth, S.Y. Pi, Phys. Rev. Lett. **49**, 1110 (1982)
25. V.F. Mukhanov, G.V. Chibisov, JETP Lett. **33**, 532 (1981)

# Chapter 8
# The Need for Quantum Cosmology

**Claus Kiefer**

**Abstract** In this contribution I argue that cosmology is incomplete without the implementation of quantum theory. The reasons are twofold. First, the beginning (and possibly the end) of the cosmic evolution cannot be described by general relativity. Second, the extreme sensitivity of quantum systems to their environment demands that the Universe as a whole must be described by quantum theory. I give an introduction to quantum gravity and explain how this is applied to cosmology. I discuss the role of boundary conditions and the semiclassical limit. Finally I explain how the arrow of time and structure formation can be obtained from quantum cosmology.

## 8.1 Introduction

Quantum cosmology is the application of quantum theory to the Universe as a whole. It may seem surprising at first glance that this is needed. Is it not sufficient to consider the standard picture of cosmology describing the Universe as expanding from a dense hot phase in the past to its present state with galaxies and clusters of galaxies (see Chap. 7)? The answer is negative for two reasons. First, general relativity is incomplete in that it predicts the occurrence of singularities in a wide range of situations. This concerns the origin of the Universe ("Big Bang") but also potentially its final fate; modern models using dark energy as an explanation for the current acceleration of the Universe can predict singularities in the future. Therefore, a more general theory is needed in order to encompass these situations. The general belief is that this theory is a quantum theory of gravity, for it was, after all, quantum mechanics that rescued the atom from the singularities of classical electrodynamics.

The second reason derives from a general feature of quantum theory. Except in microscopic cases, most quantum systems are not isolated. They interact with their natural environment, as a result of which a globally entangled state ensues, which includes the variables of the system and the environment. For macroscopic systems,

C. Kiefer (✉)
Institute for Theoretical Physics, University of Cologne, D-50937 Köln, Germany
e-mail: kiefer@thp.uni-koeln.de

H. Meyer-Ortmanns, S. Thurner (eds.), *Principles of Evolution*,
The Frontiers Collection, DOI 10.1007/978-3-642-18137-5_8,
© Springer-Verlag Berlin Heidelberg 2011

this entanglement leads to the emergence of classical properties for the system – a process called *decoherence* [1]. Since the environment of a system is again coupled to its environment, the only truly closed quantum system is the Universe as a whole. One arrives in this way at the notion of a "wave function of the Universe". Since gravity is the dominating interaction at cosmic scales, quantum cosmology must be based on a theory of quantum gravity [2].

Quantum cosmology is mathematically as well as conceptually demanding. Here is a list of the main questions that such a theory is supposed to answer:

- How does one properly impose boundary conditions in quantum cosmology?
- Is the classical singularity really being avoided?
- Will there be a genuine quantum phase in the future?
- How does the appearance of our classical universe follow from quantum cosmology?
- Can the arrow of time be understood from quantum cosmology?
- How does the origin of structure proceed?
- Is there a high probability for an inflationary phase (a phase where the Universe is accelerating at a very early stage)? Can inflation itself be understood from quantum cosmology?
- Can quantum cosmological results be justified from full quantum gravity?
- Which consequences can be drawn from quantum cosmology for the measurement problem in quantum theory and for the field of quantum information?
- Can quantum cosmology be experimentally tested?

In the following, I shall start by giving motivations for constructing a quantum theory of gravity. The main obstacles on the path to its construction are mentioned and one framework – quantum geometrodynamics – is presented. This is then applied to cosmology, where particular emphasis is put on the new concept of time as well as the central issue of boundary conditions. The two final sections are devoted to the semiclassical limit, that is, the bridge of quantum cosmology to classical cosmology, and the recovery of the arrow of time in the Universe as well as structure formation. In my presentation I shall rely on my earlier presentations in [2–5], where more details and references to original work can be found.

## 8.2 Quantum Gravity

In the first section I have argued that the Universe as a whole must be described by quantum theory. This is the field of quantum cosmology. Since it is the gravitational interaction that dominates at cosmic scales, we face the problem of *quantum gravity*, that is, the problem of requiring a consistent quantum theory of gravity. Such a theory is not yet available, although various approaches exist [2]. Independent of these particular approaches, one can put forward various arguments in support of the quantization of gravity.

- *Singularity theorems of general relativity*: Under very general conditions, the occurrence of a singularity, and therefore the breakdown of the theory, is unavoidable. A more fundamental theory is therefore needed to overcome these shortcomings, and the general expectation is that this fundamental theory is a quantum theory of gravity.
- *Initial conditions in cosmology*: This is related to the singularity theorems, since they predict the existence of a "big bang" where the known laws of physics break down. To fully understand the evolution of our Universe, its initial state must be amenable to a physical description.
- *Unification*: Apart from general relativity, all known fundamental theories are *quantum* theories. It would thus seem awkward if gravity, which couples to all other fields, should remain the only classical entity in a fundamental description.
- *Gravity as a regulator*: Many models indicate that the consistent inclusion of gravity in a quantum framework automatically eliminates the divergences that plague ordinary quantum field theory.
- *Problem of time*: In ordinary quantum theory, the presence of an external time parameter $t$ is crucial for the interpretation of the theory: "Measurements" take place at a certain time, matrix elements are evaluated at fixed times, and the norm of the wave function is conserved *in* time. In general relativity, on the other hand, time as part of space-time is a dynamical quantity. Both concepts of time must therefore be modified at a fundamental level. This will be discussed in some detail below.

Concerning currently discussed approaches to quantum gravity, one can mainly distinguish between the direct quantization of Einstein's theory of general relativity and string theory (or M-theory). The latter is more ambitious in the sense that it aims at a unification of all interactions within a single quantum framework. Quantum general relativity, on the other hand, attempts to construct a consistent, nonperturbative, quantum theory of the gravitational field on its own. This is done through the application of standard quantization rules to the general theory of relativity.

The fundamental length scales that are connected with these theories are the Planck length, $l_P = \sqrt{G\hbar/c^3}$, or the string length, $l_s$. It is generally assumed that the string length is somewhat larger than the Planck length. Although not fully established in quantitative detail, quantum general relativity should follow from superstring theory for scales $l \gg l_s > l_P$. Can one, in spite of this uncertainty about the fundamental theory, say something reliable about quantum gravity without knowing the exact theory? In [6] I have made the point that this is indeed possible. The situation is analogous to the role of the quantum-mechanical Schrödinger equation. Although this equation is not fundamental (it is nonrelativistic, it is not field-theoretic), important insights can be drawn from it. For example, in the case of the hydrogen atom, one has to impose boundary conditions for the wave function at the origin $r \to 0$, that is, at the center of the atom. This is certainly not a region where one would expect nonrelativistic quantum mechanics to be exactly valid, but its consequences, in particular the resulting spectrum, are empirically correct to an excellent approximation.

Erwin Schrödinger found his equation by "guessing" a wave equation from which the Hamilton–Jacobi equation of classical mechanics can be recovered in the limit of small wavelengths, analogously to the limit of geometric optics from wave optics. The same approach can be applied to general relativity. One can start from the Hamilton–Jacobi version of Einstein's equations and "guess" a wave equation from which they can be recovered in the classical limit. The only assumption that is required is the universal validity of quantum theory, that is, its linear structure. For this step it is not yet necessary to impose a Hilbert-space structure (a linear space with a scalar product). Such a structure is employed in quantum mechanics because of the probability interpretation, for which one needs a scalar product and its conservation in time (unitarity). The status of this interpretation in quantum gravity remains open.

The result of this approach is quantum geometrodynamics. Its central equation is the Wheeler–DeWitt equation, first discussed by Bryce DeWitt and John Wheeler in the 1960s. In a short notation, it is of the form

$$\hat{H}\Psi = 0 \,, \tag{8.1}$$

where $\hat{H}$ denotes the full Hamiltonian for both the gravitational field (here described by the three-metric) and all nongravitational fields. For the detailed structure of this equation I refer, for example, to the classic paper by DeWitt [7] or the general review in [2]. Two properties are especially important for our purpose here. First, this equation does not contain any classical time parameter $t$. The reason is that space-time as such has disappeared in the same way as particle trajectories have disappeared in quantum mechanics; here, only space (the three-geometry) remains. Second, inspection of $\hat{H}$ exhibits the local hyperbolic structure of the Hamiltonian, that is, the Wheeler–DeWitt equation possesses locally the structure of a Klein–Gordon equation (that is, a wave equation). In the vicinity of Friedmann universes, this hyperbolic structure is not only locally present, but also globally. One can thus define a new time variable that exists only intrinsically and that can be constructed from the three-metric (and nongravitational fields) itself. It is this absence of external time that could render the probability interpretation and the ensuing Hilbert-space structure obsolete in quantum gravity, for no conservation of probability may be needed.[1]

Independent of the exact theory, one can thus sensibly assume that quantum geometrodynamics should be a reasonable approximation at appropriate scales. This will be the framework on which the following discussion is based.

---

[1] The situation is different for an isolated quantum gravitational system such as a black hole; there, the semiclassical time of the rest of the Universe enters the description [8].

## 8.3 Quantum Cosmology

Cosmology can only be dealt with if one makes simplifying assumptions. Since the Universe looks approximately homogeneous and isotropic on large scales, one can impose this assumption on the metric of space-time. As a result, one obtains the Friedmann–Lemaître models usually employed. For historic reasons, such models are also called "minisuperspace models".

As we are aiming at a quantum theory, instead of an effective description of the cosmological fluid in terms of its energy density and pressure, we use a fundamental Lagrangian, namely that of a scalar field. The scalar field $\phi$ thus serves as a surrogate for the matter content of the universe. Our fundamental equation is then given by (see e.g. [2] or the Appendix of [4] for a derivation)

$$
\hat{\mathcal{H}}\Psi = \left( \frac{2\pi G\hbar^2}{3} \frac{\partial^2}{\partial \alpha^2} - \frac{\hbar^2}{2} \frac{\partial^2}{\partial \phi^2} \right.
$$
$$
\left. + e^{6\alpha}\left( V(\phi) + \frac{\Lambda}{8\pi G} \right) - 3e^{4\alpha}\frac{k}{8\pi G} \right) \Psi(\alpha, \phi) = 0 , \qquad (8.2)
$$

with cosmological constant $\Lambda$ and curvature index $k = \pm 1, 0$. The variable $\alpha = \ln a$, where $a$ stands for the scale factor, is introduced to obtain a convenient form of the equation.

The general structure of the Wheeler–DeWitt equation concerning the concept of time produces a peculiar notion of determinism at the level of quantum cosmology. Despite the absence of an *external* time parameter, the equation is of hyperbolic form thus suggesting that one use the 3-volume $v$ or $\alpha = \frac{1}{3}\ln v$ as an *intrinsic* time parameter. The term "intrinsic time parameter" denotes an evolution parameter of the equation, generally unrelated to any physical notion of time (which at the quantum level is anyway lost, as mentioned above). Exchanging the classical differential equations in time for a timeless differential equation hyperbolic in $\alpha$ alters the determinism of the theory. This, of course, changes the way in which boundary conditions can be imposed. Wave packets do not evolve with respect to Friedmann time but with respect to intrinsic time. This turns our notion of determinism on its head.

This is illustrated by the following example. Simplify the universe model with the two degrees of freedom $a$ (scale factor) and $\phi$ (scalar field) underlying (8.2) by the assumption $\Lambda = 0$. Take, moreover, the scalar field to be massless and the universe to be closed, $k = 1$. This model has a classical solution evolving from big bang to big crunch. The trajectories in configuration space are depicted in Fig. 8.1, where the arrow along the trajectory signifies increasing Friedmann time.

Classically, one imposes initial conditions at $t = t_0$, corresponding to the left intersection of the trajectory with the $\phi$-axis. These initial conditions determine the evolution of $a$ and $\phi$ into the big-crunch singularity. Not so in quantum cosmology. Here, initial conditions have to be imposed at $a = 0$. If the wave packet is to follow the classical trajectory, one has to impose two wave packets, one at each intersection

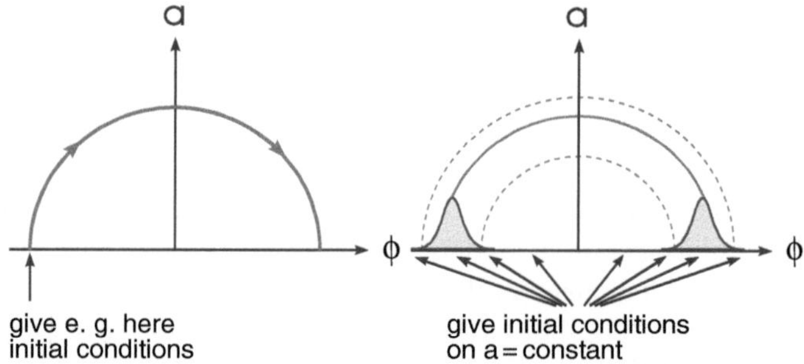



**Fig. 8.1** The classical and the quantum theory of gravity exhibit drastically different notions of determinism. The scalar field $\phi$ is shown on the *horizontal axis*, while the scale factor $a$ of the universe is shown on the *vertical axis*

point of the classical trajectory with the $a = 0$ line. Wave packets are evolved from both the classical big-bang and big-crunch singularities in the direction of increasing $a$; big bang and big crunch are intrinsically indistinguishable.

Given this new concept of time in quantum cosmology, one of the most important and nontrivial issues is to understand how boundary conditions are appropriately imposed.

## 8.4 Boundary Conditions

Implementing boundary conditions in quantum cosmology differs from the situation in both general relativity and ordinary quantum mechanics. In the following we shall briefly review two of the most widely discussed boundary conditions: the "no-boundary proposal" and the "tunneling proposal" [2].

### 8.4.1 No-Boundary Proposal

Also called the "Hartle–Hawking proposal" [9], the no-boundary proposal is basically of a topological nature. It is based on the Euclidean path integral representation for the wave function,

$$\Psi[h_{ab}] = \int \mathcal{D}g_{\mu\nu}(x)\, e^{-S[g_{\mu\nu}(x)]/\hbar} , \tag{8.3}$$

in which $S$ is the classical action of general relativity and $\mathcal{D}g_{\mu\nu}(x)$ stands for the integration measure – a sum over all four-geometries. (In general, one also sums over matter fields.) "Euclidean" means that the time variable is assumed to be imaginary ("imaginary time").

Since the full path integral cannot be evaluated exactly, one usually resorts to a saddle-point approximation in which only the dominating classical solutions are taken into account to evaluate $S$. The proposal, then, consists of two parts. First, it is assumed that the Euclidean form of the path integral is fundamental, and that the Lorentzian structure of the world only emerges in situations where the saddle point is complex. Second, it is assumed that one integrates over metrics with one boundary only (the boundary corresponding to the present universe), so that no "initial" boundary is present; this is the origin of the term "no-boundary proposal". The absence of an initial boundary is implemented through appropriate regularity conditions. In the simplest situation, one finds the dominating geometry depicted in Fig. 8.2, which is often called the "Hartle–Hawking instanton", but which was already introduced by Vilenkin [10]: the dominating contribution at small radii is (half of the) Euclidean four-sphere $S^4$, whereas for bigger radii it is (part of) de Sitter space, which is the analytic continuation of $S^4$. Both geometries are matched at a three-geometry with vanishing extrinsic curvature. The Lorentzian nature of our universe would thus only be an "emergent" phenomenon: standard time $t$ emerges only during the "transition" from the Euclidean regime (with its imaginary time) to the Lorentzian regime.

From the no-boundary proposal one can find for the above model with the massive scalar field (and vanishing $\Lambda$) the following wave function in the Lorentzian regime:

$$\psi_{\mathrm{NB}} \propto \left(a^2 V(\phi) - 1\right)^{-1/4} \exp\left(\frac{1}{3V(\phi)}\right) \cos\left(\frac{(a^2 V(\phi) - 1)^{3/2}}{3V(\phi)} - \frac{\pi}{4}\right). \quad (8.4)$$

In more general situations, one has to look for integration contours in the space of complex metrics that render the integral convergent. In concrete models, one can then find a class of wave functions that is a subclass of the solutions to the Wheeler–DeWitt equation. In this sense, the boundary condition picks out particular solutions. Unfortunately, the original hope that only one definite solution remains cannot be fulfilled.

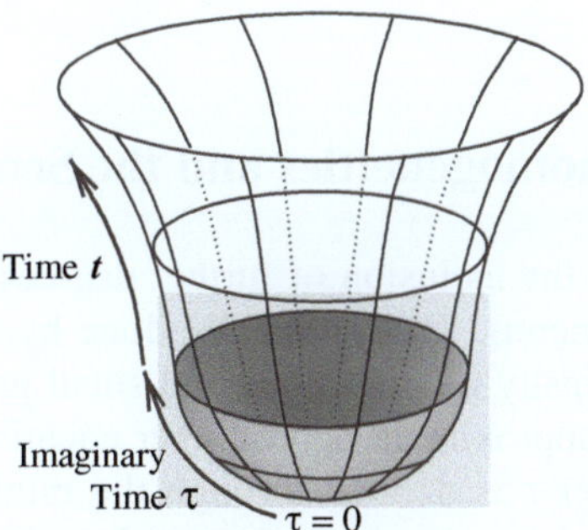


**Fig. 8.2** "Hartle–Hawking instanton": the dominating contribution to the Euclidean path integral is assumed to be half of a four-sphere attached to a part of de Sitter space. Obviously, this is a singularity-free four-geometry. This instanton demonstrates clearly the no-boundary proposal in that there is no boundary at $\tau = 0$

### 8.4.2 Tunneling Proposal

The tunneling proposal emerged from the work by Alexander Vilenkin and others, see [10–12] and references therein. It is most easily formulated in minisuperspace. In analogy with, for example, the process of $\alpha$-decay in quantum mechanics, it is proposed that the wave function consists solely of *outgoing* modes. More generally, it states that it consists solely of outgoing modes at singular boundaries of superspace (except the boundaries corresponding to vanishing three-geometry). In the minisuperspace example above, this is the region of infinite $a$ or $\phi$. What does "outgoing" mean? The answer is clear in quantum mechanics, since there one has a reference phase $\propto \exp(-i\omega t)$. An outgoing plane wave would then have a wave function $\propto \exp(ikx)$. But since there is no external time $t$ in quantum cosmology, one has to *define* what "outgoing" actually means. Independent of this reservation, the tunneling proposal picks out particular solutions from the Wheeler–DeWitt equation. The interesting fact is that these solutions usually differ from the solutions picked out by the no-boundary proposal: whereas the latter yields real solutions, the solutions from the tunneling proposal are complex; the real exponential prefactor differs in the sign of the exponent. Explicitly, one gets in the above model the following wave function:

$$\psi_{\mathrm{T}} \propto (a^2 V(\phi) - 1)^{-1/4} \exp\left(-\frac{1}{3V(\phi)}\right) \exp\left(-\frac{i}{3V(\phi)}(a^2 V(\phi) - 1)^{3/2}\right). \tag{8.5}$$

Comparing this with (8.4), one recognizes that the tunneling proposal leads to a wave function different from the no-boundary condition. Consequences of this difference arise, for example, if one asks for the probability of an inflationary phase to occur in the early universe: whereas the tunneling proposal seems to favor the occurrence of such a phase, the no-boundary proposal seems to disfavor it. No final word on this issue has, however, been spoken. It is interesting that the tunneling proposal allows the possibility that the Standard-Model Higgs field can play the role of the inflaton if a nonminimal coupling of the Higgs field to gravity is invoked [12].

## 8.5 Inclusion of Inhomogeneities and the Semiclassical Picture

Realistic models require the inclusion of further degrees of freedom; after all, our Universe is not homogeneous. This is usually done by adding a large number of multipoles describing density perturbations and small gravitational waves [2, 13]. One can then derive an approximate Schrödinger equation for these multipoles, in which the time parameter $t$ is defined through the minisuperspace variables (for example, $a$ and $\phi$). The derivation is performed by a Born–Oppenheimer type of approximation scheme. The result is that the total state (a solution of the Wheeler–DeWitt equation) is of the form

$$\Psi \approx \exp(iS_0[h_{ab}]/\hbar)\,\psi[h_{ab}, \{x_n\}]\,, \tag{8.6}$$

where $h_{ab}$ is here the three-metric, $S_0$ is a function of the three-metric only, and $\{x_n\}$ stands for the inhomogeneities ("multipoles"). In short, one has that

- $S_0$ obeys the Hamilton–Jacobi equation for the gravitational field and thereby defines a classical space-time that is a solution to Einstein's equations (this order is formally similar to the recovery of geometrical optics from wave optics via the eikonal equation).
- $\psi$ obeys an approximate (functional) Schrödinger equation,

$$i\hbar \underbrace{\nabla S_0 \nabla \psi}_{\equiv \frac{\partial \psi}{\partial t}} \approx H_{\mathrm{m}} \psi , \tag{8.7}$$

where $H_{\mathrm{m}}$ denotes the Hamiltonian for the multipole degrees of freedom. The $\nabla$-operator on the left-hand side of (8.7) is a shorthand notation for derivatives with respect to the minisuperspace variables (here: $a$ and $\phi$). Semiclassical time $t$ is thus defined in this limit from dynamical variables, and is *not* prescribed from the outside.

- The next order of the Born–Oppenheimer scheme yields quantum gravitational correction terms proportional to $G$ [2, 14]. The presence of such terms may in principle lead to observable effects, for example, in the anisotropy spectrum of the cosmic microwave background radiation.

The Born–Oppenheimer expansion scheme distinguishes a state of the form (8.6) from its complex conjugate. In fact, in a generic situation where the total state is real, being for example a superposition of (8.6) with its complex conjugate, both states will decohere from each other, that is, they will become dynamically independent [1]. This is a type of symmetry breaking, in analogy to the occurrence of parity violating states in chiral molecules. It is through this mechanism that the i in the Schrödinger equation emerges. Quite generally one can show how a classical geometry emerges from quantum gravity in the sense of decoherence [1]: irrelevant degrees of freedom (such as density perturbations or small gravitational waves) interact with the relevant ones (such as the scale factor or the relevant part of the density perturbations), which leads to quantum entanglement. Integrating out the irrelevant variables (which are contained in the above multipoles $\{x_n\}$) produces a density matrix for the relevant variables, in which nondiagonal (interference) terms become small. One can show that the universe assumes classical properties at the onset of inflation [1, 2].

The recovery of the Schrödinger equation (8.7) raises an interesting issue. It is well known that the notion of Hilbert space is connected with the conservation of probability (unitarity) and thus with the presence of an external time (with respect to which the probability is conserved). The question then arises whether the concept of a Hilbert space is still required in the *full* theory where no external time is present. It could be that this concept makes sense only at the semiclassical level where (8.7) holds.

Of course, the last word on quantum cosmology has not been spoken as long as we have no consensus on the interpretation of the wave function. What makes this issue so troublesome is the missing link of a wave function of the Universe to measurement. As remarked above, in standard quantum theory the Hilbert-space structure is needed for the probability interpretation. Expectation values are interpreted as possible outcomes of measurements with probability depending on the state the measured system is in. This interpretation entails the normalizability requirement for the wave function. Moreover, probabilities have to be conserved in time.

The problem is that we have no measurement crutch in quantum cosmology. This is a problem that persists also in the full theory and is a consequence of background independence. Only in a background of space and time can we make observations. An expectation value formulated in a theory deprived of that background is deprived of its interpretation (and justification) through measurement.

A background-independent quantum theory may thus be freed from a physical Hilbert space structure. It should keep linearity, since the superposition principle is not linked to observation, but it should dismiss the inner product as it is not clear how to endow it with a meaning in a timeless context. A Hilbert-space structure may, however, be needed at an effective level for quantum gravitational systems embedded in a semiclassical universe; a typical situation is a quantum black hole [8]. Owing to the linear structure of quantum gravity, the total quantum state is a superposition of many macroscopic branches even in the semiclassical situation, each branch containing a corresponding version of the observer (the various versions of the observer usually do not know of each other due to decoherence). This is often referred to as the "many-worlds (or Everett) interpretation of quantum theory", although only one *quantum* world (described by the full $\Psi$) exists [1].

We saw here that classical structures such as time arise only under certain conditions. It is in these regimes that we expect a physical Hilbert-space structure. Only here can we make connection with measurements.

## 8.6 Arrow of Time and Structure Formation

Although most fundamental laws are invariant under time reversal, there are several classes of phenomena in nature that exhibit an arrow of time [15]. It is generally expected that there is an underlying master arrow of time behind these phenomena, and that this master arrow can be found in cosmology. If there existed a special initial condition of low entropy, statistical arguments could be invoked to demonstrate that the entropy of the universe will increase with increasing size.

There are several subtle issues connected with this problem. First, a general expression for the entropy of the gravitational field is not yet known; the only exception is the black-hole entropy, which is given by the expression

$$S_{\mathrm{BH}} = \frac{k_{\mathrm{B}} c^3 A}{4 G \hbar} = k_{\mathrm{B}} \frac{A}{4 l_{\mathrm{P}}^2} \, , \tag{8.8}$$

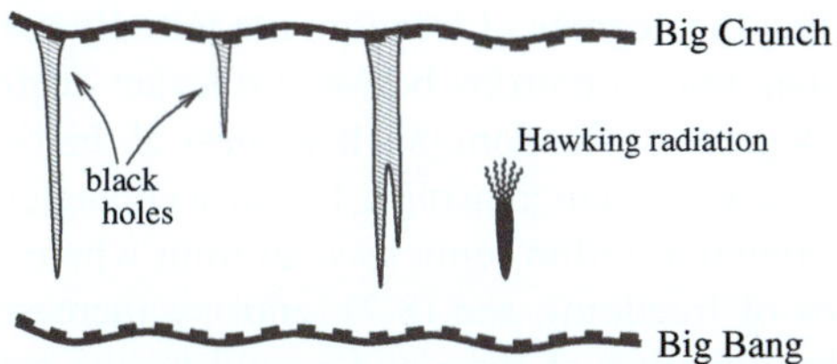


**Fig. 8.3** The classical situation for a recollapsing universe: the big crunch is fundamentally different from the big bang because the big bang is very smooth (low entropy) whereas the big crunch is very inhomogeneous (high entropy). Adapted from [15]

where $A$ is the surface area of the event horizon, $l_P$ is again the Planck length and $k_B$ denotes Boltzmann's constant. According to this formula, the most likely state for our universe would result if all matter assembled into a gigantic black hole; this would maximize (8.8). More generally, Roger Penrose has suggested using the Weyl tensor as a measure of gravitational entropy [15]. The cosmological situation is depicted in Fig. 8.3, which expresses the very special nature of the big bang (small Weyl tensor) and the generic nature of a big crunch (large Weyl tensor). Entropy would thus increase from big bang to big crunch.

Second, since these boundary conditions apply in the very early (or very late) universe, the problem has to be treated within quantum gravity. But as we have seen, there is no external time in quantum gravity – so what does the notion "arrow of time" mean?

We shall address this issue in quantum geometrodynamics, but the situation should not be very different in loop quantum cosmology or string cosmology. An important observation is that the Wheeler–DeWitt equation exhibits a fundamental asymmetry with respect to the "intrinsic time" defined by the sign of the kinetic term. Very schematically, one can write this equation as

$$H\,\Psi = \left(\frac{\partial^2}{\partial\alpha^2} + \sum_i \left[-\frac{\partial^2}{\partial x_i^2} + \underbrace{V_i(\alpha, x_i)}_{\to 0 \text{ for } \alpha \to -\infty}\right]\right)\Psi = 0\,, \qquad (8.9)$$

where again $\alpha = \ln a$, and the $\{x_i\}$ again denote inhomogeneous degrees of freedom describing perturbations of the Friedmann universe (see above); $V_i(\alpha, x_i)$ are the potentials of the inhomogeneities. The important property of the equation is that the potential becomes small for $\alpha \to -\infty$ (where the classical singularities would occur), but complicated for increasing $\alpha$; the Wheeler–DeWitt equation thus possesses an asymmetry with respect to "intrinsic time" $\alpha$. One can in particular impose the simple boundary condition

$$\Psi \xrightarrow{\alpha \to -\infty} \psi_0(\alpha) \prod_i \psi_i(x_i)\,, \qquad (8.10)$$

which would mean that the degrees of freedom are initially *not* entangled. Defining an entropy as the entanglement entropy between relevant degrees of freedom (such as $\alpha$) and irrelevant degrees of freedom (such as most of the $\{x_i\}$), this entropy vanishes initially but increases with increasing $\alpha$ because entanglement increases due to the presence of the potential. In the semiclassical limit where $t$ is constructed from $\alpha$ (and other degrees of freedom), see (8.7), entropy increases with increasing $t$. This then *defines* the direction of time and would be the origin of the observed irreversibility in the world. The expansion of the universe would then be a tautology. Due to the increasing entanglement, the universe rapidly assumes classical properties for the relevant degrees of freedom due to decoherence [1, 2]. Decoherence is here calculated by integrating out the $\{x_i\}$ in order to arrive at a reduced density matrix for $\alpha$.

This process has interesting consequences for a classically recollapsing universe [15, 16]. Since big bang and big crunch correspond to the same region in configuration space ($\alpha \rightarrow -\infty$), an initial condition for $\alpha \rightarrow -\infty$ would encompass both regions, see Fig. 8.1. This would mean that the above initial condition would always correlate increasing size of the universe with increasing entropy: the arrow of time would formally reverse at the classical turning point. Big bang and big crunch would be identical regions in configuration space. The resulting time-symmetric picture is depicted in Fig. 8.4, which has to be contrasted with Fig. 8.3. As it turns out, however, a reversal cannot be observed because the universe would enter a quantum phase [16]. Further consequences concern black holes in such a universe because no horizon and no singularity would ever form.

These considerations are certainly speculative. They demonstrate, however, that interesting consequences would result in quantum cosmology if the underlying equations were taken seriously. Quantum cosmology could yield a complete and consistent picture of the Universe.

Once the background (described by the scale factor and some other relevant variables) has assumed classical properties, the stage is set for the quantum-to-classical transition of the primordial fluctuations, which serve as the seeds for structure formation (galaxies and clusters of galaxies). This is thought to happen in

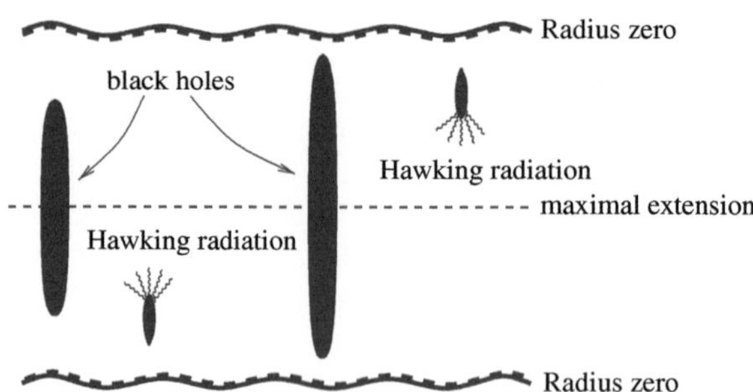


**Fig. 8.4** The quantum situation for a "recollapsing universe": big crunch and big bang correspond to the same region in configuration space. Adapted from [15]

the inflationary stage of the early Universe. The interaction with further irrelevant degrees of freedom (such as modes with short wavelengths) produces a classical behavior for the field amplitudes of these fluctuations [17]. These then manifest themselves in the form of classical stochastic fluctuations that leave their imprint in the anisotropy spectrum of the cosmic microwave background radiation. After the effective quantum-to-classical transition, the scenario proceeds as in the standard picture of cosmology described, for example, in Chap. 7.

## References


1. E. Joos, H.D. Zeh, C. Kiefer, D. Giulini, J. Kupsch, I.-O. Stamatescu, *Decoherence and the Appearance of a Classical World in Quantum Theory*, 2nd edn. (Springer, Berlin, Heidelberg, 2003)
2. C. Kiefer, *Quantum Gravity*, 2nd edn. (Oxford University Press, Oxford, 2007)
3. C. Kiefer, in *Towards Quantum Gravity*, ed. by J. Kowalski-Glikman (Springer, Berlin, Heidelberg, 2000), p. 158
4. C. Kiefer, B. Sandhöfer, *Quantum Cosmology*, arXiv:0804.0672v2 [gr-qc] (2008)
5. C. Kiefer, *Can the Arrow of Time Be Understood from Quantum Cosmology?* arXiv:0910.5836v1 [gr-qc] (2009)
6. C. Kiefer, Gen. Relativ. Gravit. **41**, 877 (2009); C. Kiefer, *Does Time Exist in Quantum Gravity?* arXiv:0909.3767v1 [gr-qc] (2009)
7. B.S. DeWitt, Phys. Rev. **160**, 1113 (1967)
8. C. Kiefer, J. Marto, P.V. Moniz, Ann. Phys. (Berlin) **18**, 722 (2009)
9. J.B. Hartle, S.W. Hawking, Phys. Rev. D **28**, 2960 (1983)
10. A. Vilenkin, Phys. Lett. B **117**, 25 (1982)
11. A. Vilenkin, in *The Future of Theoretical Physics and Cosmology*, ed. by G.W. Gibbons, E.P.S. Shellard, S.J. Rankin (Cambridge University Press, Cambridge, 2003), p. 649
12. A.O. Barvinsky, A.Yu. Kamenshchik, C. Kiefer, C. Steinwachs, Phys. Rev. D **81**, 043530 (2010)
13. J.J. Halliwell, S.W. Hawking, Phys. Rev. D **31**, 1777 (1985)
14. C. Kiefer, T.P. Singh, Phys. Rev. D **44**, 1067 (1991); A.O. Barvinsky, C. Kiefer, Nucl. Phys. B **526**, 509 (1998)
15. H.D. Zeh, *The Physical Basis of the Direction of Time*, 5th edn. (Springer, Berlin, Heidelberg, 2007)
16. C. Kiefer, H.D. Zeh, Phys. Rev. D **51**, 4145 (1995)
17. C. Kiefer, D. Polarski, A.A. Starobinsky, Int. J. Mod. Phys. D **7**, 455 (1998); C. Kiefer, D. Polarski, *Why Do Cosmological Perturbations Look Classical to Us?* arXiv:0810.0087v2 [astro-ph] (2008)

# Chapter 9
# Self-Organization in Cells

Leif Dehmelt and Philippe Bastiaens

**Abstract** Cells are dynamic, adaptable systems that operate far from thermodynamic equilibrium. Their function and structure is derived from complex biological mechanisms, which are based on several distinct organizational principles. On the one hand, master regulators, preformed templates or recipes can guide cellular structure and function. On the other hand, local interactions between fluctuating agents and growing work-in-progress can lead to de novo emergence of structures via self-organization. Here we discuss how these distinct principles are used in cellular organization. We highlight several examples of cellular self-organization, including intracellular gradient formation, growth based on stigmergy and force mediated feedbacks in spindle formation and contrast these to template-based mechanisms such as self-assembly. We conclude that an intimate interplay between distinct organizational principles, including template-based mechanisms and self-organization, forms the basis of cellular structure and function.

## 9.1 The Origin of Cellular Organization

In contrast to man-made structures, which are usually static and built for a specific purpose, cells are highly dynamic and able to adapt to varying external conditions. Upon external stimulation, many cell systems are capable of differentiating into specialized cell types to alter their behavior, function, or purpose. Such inherent plasticity of structure and function is characteristic of living organisms.

It can be helpful to rationalize the dynamic, adaptable building principles of cells in the context of their origin: evolution. Owing to natural selection, biological systems that are studied today are the result of a complex optimization process. Adaptability and dynamic properties that allowed continuous and/or stepwise progress are likely key factors for a successful evolvable species. However, other

L. Dehmelt (✉)
Department of Systemic Cell Biology, Max Planck Institute of Molecular Physiology,
D-44227 Dortmund, Germany; Department of Chemical Biology, Technische Universität,
D-44227 Dortmund, Germany
e-mail: leif.dehmelt@mpi-dortmund.mpg.de

H. Meyer-Ortmanns, S. Thurner (eds.), *Principles of Evolution*,
The Frontiers Collection, DOI 10.1007/978-3-642-18137-5_9,
© Springer-Verlag Berlin Heidelberg 2011

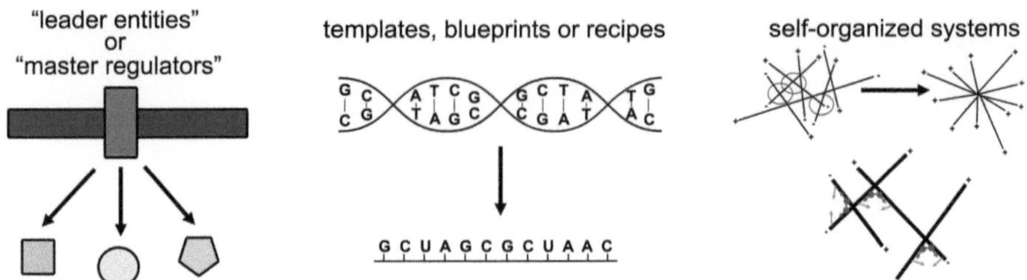


**Fig. 9.1** Organizational principles in cells. Simple signal transduction from a "leader entity" to subordinate entities is not considered to be related to self-organization. Likewise, organization based on information stored in templates, blueprints, or recipes, such as the base sequence in DNA, is also not related to self-organization. In contrast, self-organized processes display global pattern formation, based on dynamic interplay and teamwork of lower-order system components, such as in the depicted formation of ordered filament structures via sliding through molecular motors

features, such as robustness, will also play important roles. As discussed below, self-organization provides mechanisms by which dynamic biological structures can be generated that are both adaptable and robust. However, biological mechanisms often do not exclusively implement self-organization, but rather contain modules or mixtures that implement different organizational concepts (Fig. 9.1). In the following sections, we will explore distinct organizational principles through which cellular structure and function can be generated to solve biological problems.

## 9.2 Self-Organization and Other Organizational Principles in Cells

In a self-organized process, several entities can interact with each other, and team up to produce a behavior of the group as a whole. In such systems, the organizing entity is not imposed from outside, but rather internal to the system itself. In a simple definition, *self-organization* is a process in which a pattern at the global level emerges solely from numerous dynamic interactions among the lower-level components of the system. Moreover, the rules specifying interactions among the system's components are executed using local information, without reference to the global pattern [1]. The dynamic interactions between the lower-level components usually include both attraction and repulsion, as well as positive and negative feedbacks.

The emergence of a pattern is basically an increase in complexity or information in the self-organized system. According to the basic principles of thermodynamics, this is only possible in open, dissipative systems, which operate far from equilibrium. For most biological systems, this means that energy needs to be supplied, for example in the form of adenosine triphosphate (ATP), to produce and maintain the emerging pattern. In such systems, local interactions among the system components can change the internal organization of the system without being guided by an outside source. Moreover, the global behavior of the collective system often can display patterns that are not easily understood in a direct intuitive manner simply based on the knowledge of the underlying rules and behaviors of the interacting entities. Such unexpected, complex global patterns and behaviors are often referred to as "emergent properties".

Before a global pattern emerges, self-organized systems usually start with a phase of exploration, which is characterized by random fluctuations of the lower-level system components. These fluctuations allow a dynamic phase of trial-and-error, in which the system can make a transition from a state of high entropy and low information content towards the emergence of a global pattern with lower entropy and higher information content. In a self-organized system, this phase of trial-and-error is based on local interactions between the lower-level system components. After the emergence of the pattern, the phase of trial-and-error can continue to maintain the emerged pattern. Such patterns can therefore be reconstructed continuously and can thereby adapt themselves to changing external conditions.

Other organizational principles that are not considered to be self-organized include:

1. a "leader entity" or "master regulator", which can take command and dictate a specific behavior on subordinate entities, and
2. a template, blueprint, or a recipe, which is used to reproduce a specific cellular process based on hardcoded information.

These two alternative organizational principles are typical for man-made structures, such as buildings, which are constructed by a hierarchically organized roster of leaders and workers. Ultimately, the workers follow instructions based on a template or blueprint of the building via execution of pre-determined recipes. In such systems, the organization is imposed from an external source (leaders, blueprints), and not from within the system (composed of the workers and their actual work). Such externally imposed organizational principles are also used in many cellular processes: some master regulators, such as growth factors, their receptors, and immediate signal mediators, can be understood as "leaders" in the context of an isolated cellular process, as they relay signals to many target regulators. Such master regulators are often controlled by global feedbacks that regulate their overall activity.

A typical example of a cellular template is the genomic DNA. The information stored in the base sequence of the genomic DNA is used to generate a complementary sequence of messenger RNA, which is then decoded to generate a protein. In that way, the genomic DNA can be understood as a recipe for generating a protein, based on a fixed sequence of events – hardcoded as base triplets, which are translated by a fixed sequence of biomolecular steps. In principle, any component that contains interpretable spatial or temporal information can be regarded as a template for this information.

Self-organization should not be confused with simpler mechanisms of self-assembly, in which preexisting building blocks are combined into a stable non-dynamic structure, such as a crystal formed from interacting molecules, or the two- and three-dimensional structures formed by DNA origami [2]. Such self-assembled structures are rather formed by a template-based process. In contrast to self-organized dynamic and adaptable structures, which can be continuously rebuilt by dynamic trial-and-error processes including both repulsive and attractive interactions, self-assembled structures are usually produced by static attractive interactions, which approach thermodynamic equilibrium during the pattern-forming process. Furthermore, the information contained in complex self-assembled structures, such as the complex shapes of DNA origami, do not emerge from teamwork among lower-level components, but are instead derived from hardcoded blueprint or recipe types of information. In the example of DNA origami, this information is stored in the DNA sequences of the individual building blocks.

Typical examples of self-organized systems (Fig. 9.2) include (a) dynamic feedback systems, which lead to the formation of wave patterns in sand dunes [4], (b) building principles, which depend on feedback interactions and reference to the building in progress, such as building of termite colonies [5], (c) stable, dynamic structures, which arise from complex force interactions, such as convection in Bénard-cells [6]. In cells, self-organized systems that are similar in concept to these large-scale systems are found at much smaller length scales. Examples of such cellular systems that are related to the examples above include (a) dynamic activity gradients based on feedback systems, such as $Ca^{2+}$ signal waves [7], (b) directional transport and growth systems that display feedback regulation, such as neurite outgrowth [8, 9], and (c) dynamic structures that are formed by complex force feedback interactions, such as the mitotic spindle [10]. These types of systems are described in further detail in the following sections. See also Table 9.1.

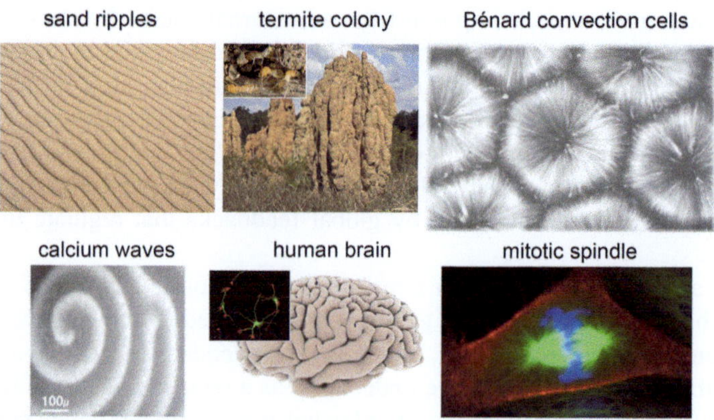


**Fig. 9.2** Examples of self-organized systems in cells and related macroscopic self-organized systems. (Image sources: sand ripples: © iStockphoto.com/pixonaut; termite colony: © iStockphoto.com/AAndromeda; termites (inset termite colony): © iStockphoto.com/jeridu; human brain: © iStockphoto.com/Henrik5000; Bénard convection cells: M.G. Velarde, Universidad Complutense Madrid; calcium waves: Reprinted from [3], with permission from Elsevier; primary hippocampal neurons (inset human brain): Dehmelt and Halpain, unpublished; mitotic spindle in Ptk1 cells: Dehmelt)

**Table 9.1** Comparison of organizational principles in cellular regulation

| Leaders or master regulators | Blueprints, templates, recipes | Self-organized systems |
| --- | --- | --- |
| A single biomolecule transduces a signal to many receiving effectors | The information for building a structure or performing a task is stored in a biomolecule | A global structure or dynamic behavior emerges from dynamic interactions between lower-level system components |
| Cells contain several examples of proteins that act at important signal hubs in cells, and many of these proteins can be seen as master regulators. Such hubs often also receive inputs from the cellular signal machinery, which can qualify some master regulators as "well-informed" leaders | In cells, blueprints can be present as sequences of biopolymers. Such sequences can also be understood as recipes to perform sequential tasks – for example during transcription or translation | The rules specifying interactions between the system components are based on local information only, without reference to the global pattern |
| In a cellular context, such master regulators have a limited repertoire of possible decisions on external conditions. These capabilities are hardcoded in the regulator via their protein sequence and the resulting biochemical function | Templates can provide information for building cellular structures that span several micrometers in size. For example, chromatin can provide a spatial cue for restricting the formation of the mitotic spindle. Such spatial restriction can be involved in the control of the overall size of self-organized structures | The dynamic interactions between the lower-level components usually include both attraction and repulsion and positive or negative feedbacks |
| | Templates can also be hardcoded into protein-complex subunits by predetermining interactions between complex components. Such complexes can arrange themselves by self-assembly processes, which are distinct from self-organization | Self-organized systems are open, dissipative systems. They require energy to be maintained. This energy is usually provided by a chemical energy source such as ATP |
| | | The emergence of global patterns via self-organization usually begins with a phase of exploration, which is characterized by fluctuations |
| | | Interactions between system components are not necessarily direct, and can also be communicated by a growing work-in-progress. Such mechanisms are related to stigmergy |

## 9.3 Emergence of Spatio-Temporal Gradients via Dynamic Feedback Systems

Many cellular functions need to be regulated in a spatio-temporal manner. Such regulation is often accomplished by gradients of protein activities [11]. These gradients are formed by a reaction–diffusion mechanism: a localized protein can activate a signal molecule, which then diffuses away from the site of activation. Subsequent deactivation of the signal molecule, for example by a uniformly distributed inhibitor, can lead to the formation of a stable gradient. In its basic form, such a reaction–diffusion mechanism does not display the hallmarks of self-organization, and the complexity of the emerging pattern is limited and closely resembles the pattern of the localized activator. However, if the components of reaction–diffusion processes are embedded in positive-feedback loops, more complex spatio-temporal patterns can emerge.

A classic example of such a reaction–diffusion-based self-organization process in cells is the formation of calcium waves [7, 12] (Fig. 9.3). Calcium is an important biological signal and changes in its intracellular concentration plays a role in many biological processes, such as cell contraction, propagation of nerve impulses and cell fate decisions [13]. The bulk of calcium in cells is not present in the cytosol, but instead concentrated in cellular compartments, such as the endoplasmic reticulum. In fact, the intracellular concentration of calcium is much lower

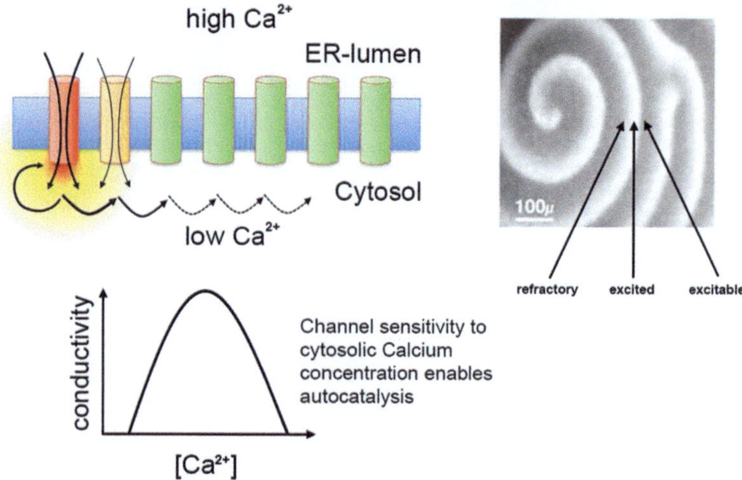


**Fig. 9.3** Calcium wave propagation in cells. The conductivity of calcium channels in the endoplasmic reticulum (ER) is stimulated by intermediate cytosolic calcium concentration. Therefore, calcium conductance is autocatalytic at low to medium calcium concentrations and can be excited by local calcium increases. At higher concentrations, calcium conductance is again reduced and the the calcium channel can enter a refractory period. These features allow formation of calcium waves, which travel through the cell. (Image of calcium waves reprinted from [3], with permission from Elsevier)

(ca. 100 nM) than either the extracellular level (1–3 mM) and the level in the endoplasmic reticulum (ca. 500 μM). This steep gradient is maintained by ATPases, which consume energy in the form of ATP to transport calcium into the endoplasmic reticulum [14].

A suitable extracellular stimulation, such as activation of G protein-coupled receptors (GPCRs) or receptor tyrosine kinases, can lead to release of a second messenger, the signal molecule IP3, which will activate a calcium channel on the endoplasmic reticulum, leading to release of calcium into the cytosol [13]. Interestingly, the opening probability of these calcium channels has a bell-shaped dependence on the cytosolic calcium levels themselves [15]. Thus, with increasing calcium release, they conduct even more calcium. This local property of the calcium channels represents a positive feedback mechanism, which can lead to self-amplification of an initially weak calcium signal. Owing to the bell-shaped calcium dependence, the amplification process is limited, and the channel conductance is again reduced at high calcium concentrations. After experiencing local high calcium concentrations, the channels enter a refractive state, in which they do not conduct calcium efficiently – even at previously optimal calcium concentrations.

In the geometric arrangement of calcium channels in the endoplasmic reticulum, which occupies a large portion of the cytosol, this self-amplifying and self-limiting mechanism can lead to the formation of waves of calcium release throughout the whole cytosol. Such waves can move through cells via a fire–diffuse–fire mechanism, which lends both directionality and robustness to the signal propagation [16]. Due to the high sensitivity in the positive feedback mechanisms of calcium release from the endoplasmic reticulum, the information transfer follows a wave propagation mechanism, which is faster than information transfer via diffusion on the scale of cells [17]. Typically, the speed of information transfer in wave propagation is constant, in contrast to diffusion, where it is inversely proportional to the square root of time.

The phenomenon of calcium wave propagation exhibits all the hallmarks of self-organized systems: It is a process in which a pattern at the global level, the calcium wave, emerges solely from numerous dynamic interactions among lower-level components, the calcium channels. These channels are operating on rules using only local information, the local calcium concentration, without reference to the global pattern, the calcium wave. This increase in complexity and information in the system is made possible via energy consumption through ATPases, which maintain steep gradients of calcium between the cytosol and endoplasmic reticulum.

The formation of calcium waves is a special case of a reaction-diffusion mechanism, which can lead to the emergence of higher order patterns – in this case to propagating wave fronts. Here, the reaction is the opening of the calcium channel, which produces a diffusing reactant – the calcium itself. Other examples of pattern formation by self-organization in reaction-diffusion processes include Turing patterns, which are formed in systems exhibiting short-range activation, long-range inhibition, and positive-feedback regulation [18]. Such systems can lead to the emergence of stable spatial patterns, which are thought to play a role in body segmentation processes during animal development.

Self-organization also plays an important role in cell polarization. Here, a cell's "front" and "back", with distinct morphologies and biological function, emerges during directional cell migration. A simple model for cell polarization, the so-called local excitation and global inhibition (LEGI) model, can account for one aspect of directed cell motility: the amplification of an external, shallow gradient into an intracellular gradient of a signal molecule (Fig. 9.4a) [19]. In this model, an external signal activates both an activator A and an inhibitor I for the substrate R. If the diffusion of the activator A is much slower than the diffusion of inhibitor I, then an intracellular gradient is formed inside the cells, which mimics the extracellular gradient. If the external gradient is changed in direction, the intracellular gradient follows that change in LEGI-type models. Such simple models of cell polarization therefore mainly mimic the behavior of the template gradient and thus the intracellular pattern does not exhibit a gain in complexity.

However, many cells also show a persistent polarization that is fairly insensitive to changes in the external signals, once the cell has become polarized. Such persistence after cell polarization can be observed in Turing-type models [20] (Fig. 9.4b). The main difference between the LEGI and Turing models is the addition of positive feedback regulation of the activator A on itself, which assures the maintenance of an initially formed intracellular gradient, even in the absence of external stimulation. New approaches, such as wave-pinning [21] can account for a mixture of persistence and plasticity, which is observed in many types of motile cells. In contrast

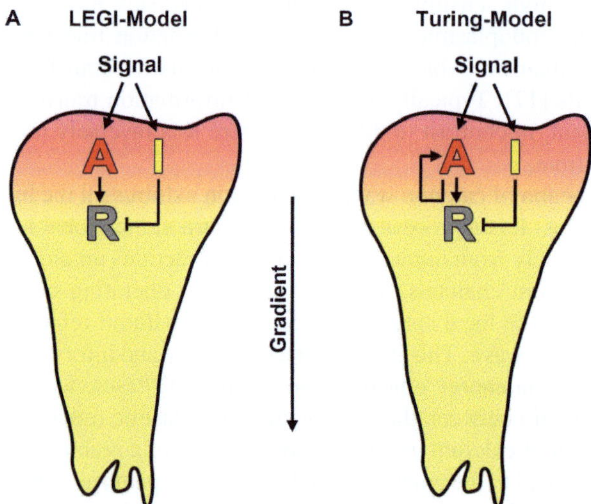


**Fig. 9.4** Comparison of models for cell polarization. An extracellular signal – either in the form of a gradient or a small localized elevation – activates both an intracellular activator and an intracellular inhibitor. The inhibitor diffuses much faster than the activator. In the simpler LEGI-type model (**a**), this extracellular stimulation is mimicked inside the cells. If the extracellular gradient is removed, the intracellular gradient disappears. In the more complex Turing-type model (**b**), self-amplification of the activator is introduced. Under permissive model parameters, a previously initiated gradient is maintained, even after the extracellular signal gradient is removed

to Turing-type models, in which a stable intracellular gradient is formed by growth of small local perturbations, an adaptable gradient is formed in the wave-pinning model by propagation of a wave, which slows down until it comes to a halt.

## 9.4 Stigmergy and Feedback Regulation in Directional Morphogenetic Growth or Transport Processes

The intracellular patterns and gradients formed by reaction–diffusion mechanisms discussed above play important roles in many biological processes, such as intracellular signal transduction, cell fate decisions, and cell polarization. The initial phase in the formation of such gradients involves an exploratory phase, in which individual lower-level components sample their environment by means of diffusion. On the molecular level, the diffusing components perform a random walk, in any direction except for physical barriers, such as intracellular organelles or the cell border. However, in cells, this random walk can be biased or steered into more directional movements via the cytoskeleton. This directionality can either derive from directional, asymmetric growth processes or from motor-based transport along directional cytoskeletal tracks, or from a combination of the two. The resulting growth processes often involve a specific form of self-organization, called stigmergy.

The term stigmergy was introduced by Grassé to describe mechanisms in termite nest building [5]. The basic principle underlying stigmergy is the organization of a building process based on the work in progress, which provides marks (stigmata) for further work (ergon). Classical examples of stigmergy include wall [22] or chamber [5] building by ants. In contrast to other self-organizing processes, the low-level system components do not team up with each other directly. Rather, they interact and communicate indirectly by altering the work in progress, which serves both as the goal to be built and as a working memory during its building process. Furthermore, the work in progress is often an integral part of a positive feedback loop, which can promote the emergence of complex growth patterns.

One example of a positive feedback mechanism that involves a growing "work-in-progress" has been proposed for the axonal growth cone turning towards a shallow gradient of the neurotransmitter GABA ($\gamma$-aminobutyric acid) [9] (Fig. 9.5a). In the proposed mechanism, such a gradient first induces an asymmetric calcium response on initially symmetrically distributed GABA receptors. The asymmetric calcium release then leads to asymmetric growth of microtubules, which in turn lead to asymmetric transport and redistribution of GABA receptors. The positive feedback loop is closed by a resulting increase in asymmetry of the calcium response. In this mechanism, the work in progress – the growth of microtubules – is a key factor in the spatial organization of a positive feedback mechanism and it provides a memory of the system on which the feedback can be propagated. At the same time, this work in progress also gives rise to the emerging biological effect: The steering of the growth cone towards a shallow gradient of the GABA neurotransmitter.

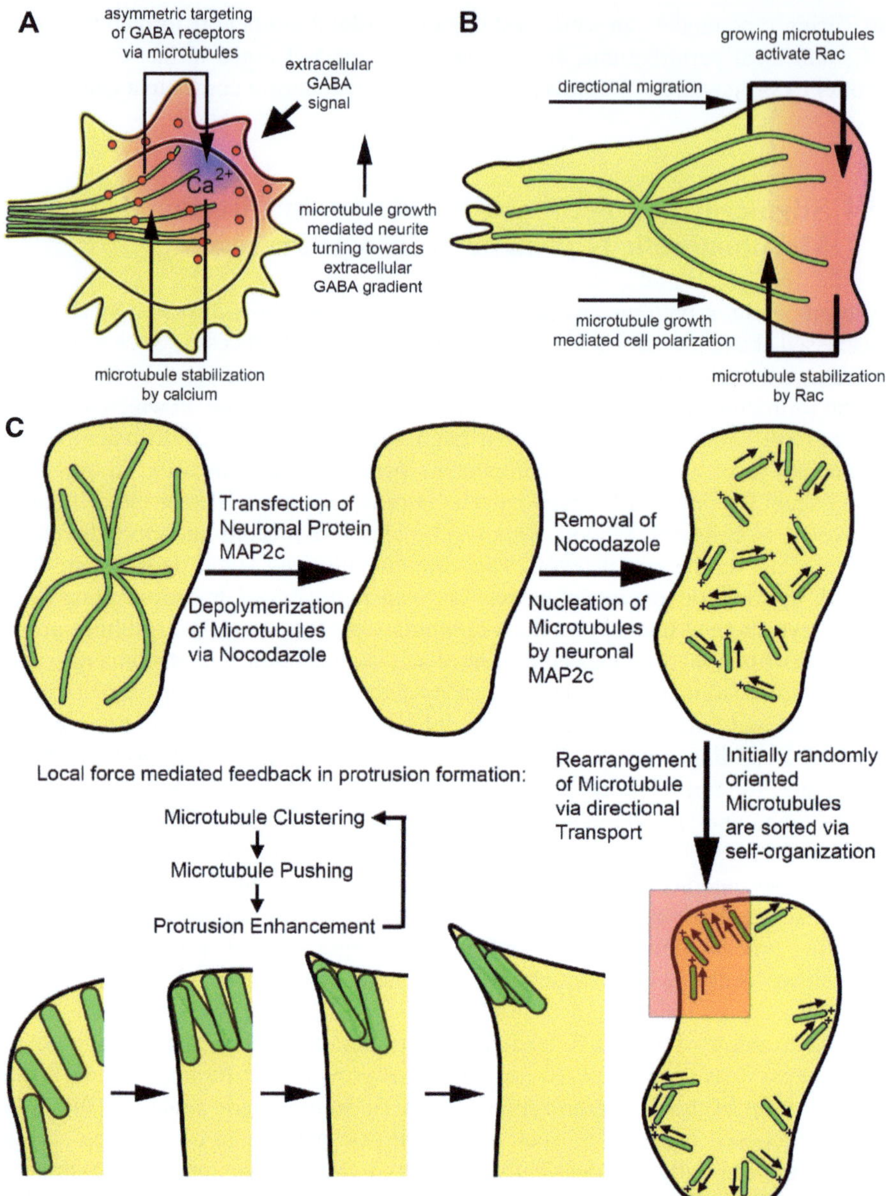


**Fig. 9.5** Cell polarization via growth processes related to stigmergy. **a** Positive feedback between microtubules and calcium release in growth cone steering. Local calcium release enhances microtubule stability and preferred microtubule growth towards regions of higher calcium concentration. Transport of additional calcium-conducting receptors via these microtubules amplifies the calcium release. **b** Positive feedback between Rac activity and microtubule growth in cell motility. Growing microtubules activate the small GTPase Rac, which in turn deactivates the microtubule growth inhibitor stathmin. Local inhibition of stathmin leads to increased microtubule stability

Similar positive feedback mechanisms have been proposed in directional cell migration [23] (Fig. 9.5b). In that model, activation of a signal transducer, the small GTPase Rac1, via microtubule polymerization can lead to the inhibition of a microtubule growth inhibitor called stathmin. This can in turn lead to increased microtubule growth and increased Rac activation. Again, the work in progress is the microtubule growth, which provides both a key node in the positive feedback and a memory of the current system state towards the emergence of directional cell motility. It is not entirely clear how growing microtubules can activate Rac, and both a directional transport mechanism via molecular motors, or targeting of signal molecules via specialized complexes concentrated at growing microtubule tips, the so-called "plus-tips", might play a role in activating Rac.

While the previous examples highlight the formation of self-organized signal gradients based on directional transport mechanisms or directional growth mechanisms, positive feedback from a work in progress can also arise by means of mechanical forces generated inside cells. One such example can be directly visualized in a model experiment for neurite outgrowth [8]. If COS7 cells are treated with the microtubule-disrupting drug nocodazole and simultaneously transfected with a neuronal-microtubule-stabilizing protein called MAP2C, small motile microtubule fragments are generated following drug washout (Fig. 9.5c). These microtubule fragments move directionally through cells by means of the microtubule motor dynein. As microtubules reach the cell periphery, they slide along the membrane until they get caught in small membrane protrusions. Once immobilized in such small protrusions, the microtubules can push the plasma membrane further, enhancing the protrusion. A larger protrusion is able to catch more pushing microtubules, thus further enhancing the protrusion. Here, the work in progress is the growing protrusion, which is amplified by a positive feedback involving the accumulation of microtubule fragments and their ability to enhance the cell protrusion. Similar mechanisms involving local accumulation and amplification of pushing microtubules might be involved in the initiation of neurites.

Cellular systems that involve stigmergy are likely embedded in multiple levels of feedback regulation. For example, the overall regulation of cell shape likely involves growth mechanisms involving positive feedback loops that can give rise to focused patterns such as cell polarization or neurite formation. On the other hand, negative feedback loops or limiting factors will regulate the extent of such focused growth processes [24] and will keep the overall cell structure in a steady state.

**Fig. 9.5** (continued) and increased, localized microtubule growth. **c** Transport-mediated polarity sorting and geometry-mediated protrusion amplification in a cellular microtubule reorganization assay. Microtubule fragments of random orientation regrow in cells after nocodazole washout in the presence of the neuronal microtubule stabilizer MAP2c. Directional transport of microtubule fragments orients microtubule fragments with their plus-ends towards the cell periphery. In a self-amplifying manner, small perturbations induced by pushing microtubules induce a convex disturbance of the cell periphery, which can collect more microtubule fragments to induce a larger protrusion

## 9.5 Emergence of Complex Structures via Self-Organization of Microtubules and Associated Motors

The previous examples of intracellular self-organized systems are composed of a relatively small number of component types, which interact to build comparably simple higher-order structures such as waves, gradients, or foci. However, many cellular structures, such as the lamellipodium in motile cells or the mitotic spindle, display much higher levels of complexity. Detailed mechanisms of how these structures emerge are still lacking, however, many of the underlying organizational principles have been uncovered. A key factor in the generation of these complex structures involves an intimate interplay between growing cytoskeletal filaments, their regulators, and mechanical forces generated by the growing filaments themselves, by associated motor proteins, or both. In addition to mechanisms based on self-organization, other means of organization are usually also used in building such complex structures. Deciphering which aspects of a generated structure emerge from self-organization and which from a template-based mechanism is not always easy owing to functional overlap and redundancies.

Early experiments to elucidate the organizational principles of spindle formation made use of mitotic *Xenopus* egg extracts, which contain many components necessary to build a mitotic spindle [25]. Interestingly, these extracts lack certain template structures, such as centrosomes, which usually form the spindle poles, and centromeric DNA, which forms anchor points for spindle microtubules – so called kinetochores – in the condensed chromatin at the spindle metaphase plate (Fig. 9.6). In one model, the spindle arises via a search-and-capture mechanism on the basis of two templates: the centrosomes, from which microtubules grow in all directions, and the kinetochores on the chromatin, which capture microtubule ends [26]. Indeed, if centrosomes and chromatin are added to *Xenopus* extracts, spindles are formed, as predicted by the search-and-capture model [27]. In a way, the search-and-capture model represents an exploratory phase – a period of trial-and-error – and the dynamic properties of microtubules, which undergo repeated growth and shrinkage phases, are capable of exploring a wide area of state space, which then converges to a defined emerging pattern – the mitotic spindle. As expected in a self-organized system, the global pattern is formed only by local interactions without reference to the global structure. Interestingly, the chromatin also serves an additional role in guiding microtubule exploration towards the emerging spindle structure by generating a gradient of microtubule stabilization [11]. This additional level of microtubule steering is necessary for efficient search-and-capture-based spindle formation [28]. While this process indeed displays the hallmarks of self-organization, it is nevertheless clear that several spatial templates exist in the search-and-capture model, which provide important spatial cues and restrictions in the exploratory phase. However, spindles are also formed in the absence of centrosomes in meiosis. Furthermore, in the absence of both centrosomes and kinetochores, the addition of beads that were coated with DNA lacking centromeric sequences, and therefore lacking kinetochores, is sufficient to induce the formation of a spindle [25].

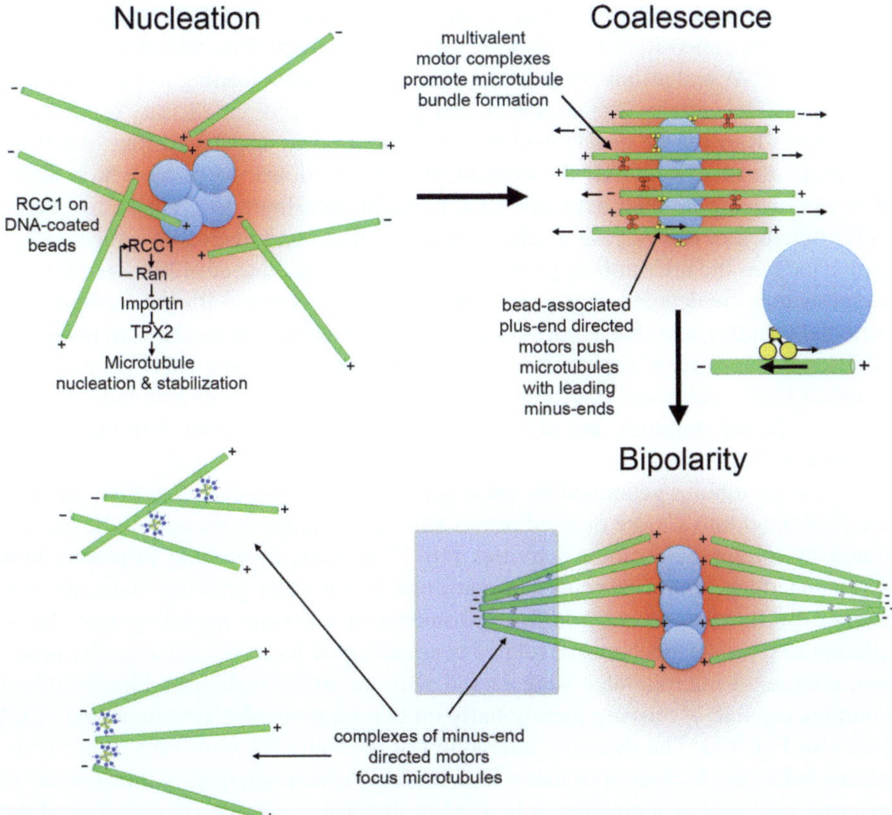



**Fig. 9.6** Template-based aspects of spindle assembly and self-organization of artificial spindle-like structures. Nucleation: In mitotic *Xenopus* egg extracts [25], DNA-coated beads bind the protein RCC1, which can locally activate the signal protein Ran. This signal protein can promote microtubule nucleation and stabilization indirectly by releasing the protein TPX2 from inhibition by importin. Via partially autocatalytic reaction–diffusion mechanisms, these proteins form an activity gradient, which leads to microtubule growth around DNA-coated beads. Coalescence: Subsequently, microtubules form a tight bundle around the DNA-coated beads, which is held together by multivalent microtubule motors. Bead-associated plus-end-directed motor activities are able to push microtubules in an orientation-dependent manner to sort microtubule polarity. Bipolarity: Finally, microtubules form foci at their poles by means of microtubule reorientation through forces generated by minus-end-directed motor complexes

The underlying mechanism of this reconstituted spindle formation is complex and involves several aspects of apparent self-organization. First, microtubules are nucleated in the vicinity of DNA-coated beads. Then microtubules form a bundled structure, which finally rearranges into a spindle with two focused poles arranged around the DNA-coated beads at the center of the structure [25]. The microtubule nucleation step is induced by a reaction–diffusion mechanism involving recruitment of protein activities to the DNA-coated beads [29]. The *Xenopus* extract contains a signaling molecule, called Ran, which can exist in an active, guanosine triphosphate

(GTP)-bound state and an inactive, guanosine diphosphate (GDP)-bound state. A Ran activator, called RCC1, is also present in the extract. This activator has a high affinity to DNA and it thus binds to the DNA-coated bead, leading to localized Ran activation in its vicinity. Active Ran is only present in the direct vicinity of the bead, as diffused Ran can be deactivated by a soluble, evenly distributed Ran deactivating protein called RanGAP. A potential positive feedback loop, by which active Ran can activate its own activator RCC1, might contribute to gradient formation [11]. Active Ran can release a microtubule stabilizer and nucleator, called TPX2, which induces the microtubule growth around DNA-coated beads [29]. Recent work suggests that the detailed shape of the Ran gradient is critical for the formation of artificial spindles, and that a more complex reaction–diffusion mechanism involving additional components appears to be critical for generating the exact shape of the gradient [30]. Additional microtubule regulators, such as the microtubule depolymerizing factor stathmin, are also regulated by signal gradients forming around chromatin [31].

The microtubules generated by these activities are disorganized at first, but then are reoriented into a more ordered array. This reorientation is thought to be orchestrated by microtubule motor activities [10]. The basic organizing principle, how force-generating motor proteins can rearrange dynamically growing filaments, was studied by in vitro reconstitution experiments in a simple model system: In an attempt to simulate mechanisms related to spindle pole focusing, dynamically growing, unordered microtubules were mixed with an artificial dimeric kinesin motor complex capable of moving directionally on two microtubules simultaneously [32] (see also Fig. 9.1). The dynamic interaction of microtubules with such motor complexes led to the formation of asters and vortices, which emerged in the absence of external guidance or templates. It is thought that asters emerge via focusing of the microtubule ends through forces generated by motors moving on two microtubules simultaneously towards the generated focus center (Fig. 9.6). Computer simulations of this process support this idea [33]. Studies on meiosis in oocytes suggest that similar mechanisms are used in cells if centrosomes are not present [34].

The emergence of such asters is a result of a dynamic force-generating mechanism that directly shifts the asymmetrical microtubules with respect to each other [32]. The system starts with an exploratory phase of random fluctuations, based on motor-based shifting of initially randomly oriented microtubules. The pattern of nonrandomly oriented and spatially organized microtubule asters emerges from the random fluctuations based on local rules, which affect motor-crosslinked microtubules differently depending on their crosslink angle. Through a series of trial-and-error events, which are characterized by large-scale microtubule movements, the regularly spaced microtubule asters emerge based on local interaction rules: If microtubules overlap, the directional motors lead to "attraction" of one type of microtubule end (in the study by Nedelec et al., the plus-ends), whereas the other type of end is repelled from each other (the minus-end). The attraction of plus-ends and repulsion of minus-ends leads to the growth of plus-end-centered foci, which leads to continuously more efficient plus-end focusing by a positive feedback mechanism similar to classical clustering. The attraction and repulsion is only effective while filaments overlap. Therefore, the system approaches a more stable structural configuration, a regular arrangement of foci, which is characterized by minimal microtubule overlap. The ordered microtubule arrangement is protected against randomization by Brownian fluctuations by ongoing dynamic force-generating interactions with crosslinking microtubule motors.

In the ensemble as a whole, asters emerge without the need for a template, purely based on the local interactions of microtubule motors and microtubules. The development of the aster itself can again be interpreted as a form of stigmergy. The lower-level components, the microtubule motors, do not team up by communicating or interacting directly with each other, but rather cooperate via the work in progress, the emerging aster, which also serves as a working memory of the system's current state.

The emergence of an individual aster through self-organized interactions between growing microtubule filaments recapitulates one aspect of the self-organization of the mitotic spindle. However, the bipolar organization of a mitotic spindle is more complex than a single aster. Computer simulations using a variety of different types of components and systematic analysis of parameter space were used to get a first idea of how a bipolar spindle could form from asters [35]. In these simulations, motor complexes were analyzed for their ability to influence a pair of microtubule asters. If only a single type of motor was included in a complex, these asters either converged via attraction or separated via repulsion. However, if two types of motors with opposite movement directionalities were combined in a complex, a stable but nevertheless dynamic interaction between two asters could be observed. By analyzing such simulations, it became clear that a balance of competing forces in the overlapping region is necessary for the emergence of this behavior. Moreover, the extent of interaction and overlap between the asters was regulated by feedback interactions, which were based on the extent of overlap between parallel and antiparallel microtubule overlaps in the resulting bipolar spindle structure.

While these simulations did not attempt to recapitulate a realistic set of components – the types of motor complexes used in these simulations do not have an obvious real-world counterpart – they were able to provide insight into organizational principles that are sufficient to generate a stable bipolar spindle-like structure.

## 9.6 Emergence of Dynamic Structures in Actin Filament Treadmill Systems

The emergence of actin-rich structures was also proposed to originate from self-organization mechanisms. One of these structures is the lamellipodium. This structure is composed of a branched network of treadmilling actin filaments, which polymerize at the peripheral border of the cell and depolymerize towards the center of the cell [36]. In general, treadmilling is thought to arise from the asymmetry of actin filaments, which tend to polymerize at their plus-end and depolymerize at their minus-ends. This asymmetry in polymerization kinetics is due to different kinetic association/dissociation rates for monomers at the filaments ends and coupled to the intrinsic hydrolysis of polymer-bound actin-ATP to actin-ADP. In cells, additional factors that stimulate regionally controlled nucleation, elongation, and depolymerization also play an important role [36].

A computational model for such treadmilling filaments, which contains only a few components – filaments and filament nucleators, which are transported directionally on the filament itself towards one filament end – shows aspects of self-organization [37]. Computational simulations of this simple system reveal stable solutions that are characterized by moving, crescent-shaped filament densities, which accumulate nucleators at their leading edge. This behavior is reminiscent of motile keratocyte fragments, which can migrate autonomously (Fig. 9.7) [38]. Such fragments display a similar shape and might be propelled by a related actin polymerization and treadmilling mechanism. In contrast to the example of the formation of the mitotic spindle, in which crosslinked motors produce forces that can directly shift filaments with respect to each other, a major component of the force generated by the lamelipodium is derived from the polymerization of the actin filaments themselves by an elastic Brownian ratchet mechanism [39]. Again, this system displays aspects of stigmergy, as the polymer array serves as a work in progress, to which additional subunits are added. It also serves as a source of positive feedback by mediating the transport of actin nucleators to the filament tip to initiate additional filaments.

Evidence that actin can indeed self-organize also in cells is provided by a simplified cellular assay, in which actin repolymerization is observed after washout of the actin-sequestering drug latrunculin [40]. In these experiments, waves of actin polymerization were observed, which propagate through the cell. Inhibition of the classical upstream activators of actin polymerization did not interfere with wave progression, suggesting that these waves were self-sustaining and that polymerization and depolymerization were somehow coupled within waves.

In lamellipodia, individual actin filaments form a dense, branching network, which is constantly rebuilt during treadmilling. Many of these branches arise from the activity of an actin nucleator called ARP2/3. This nucleator forms new fil-

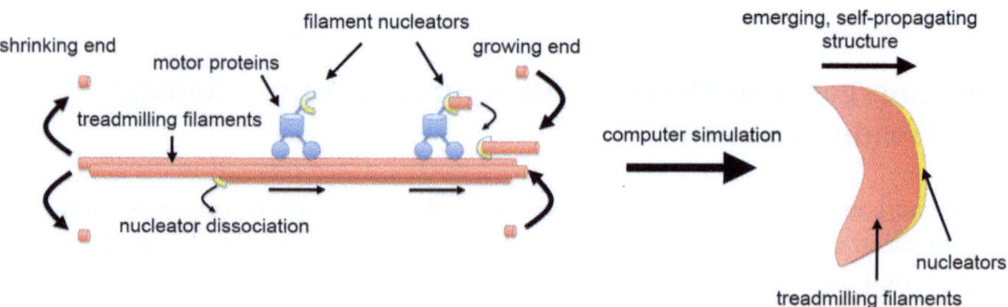


**Fig. 9.7** Self-organization of treadmilling filaments. Modeling of directional transport of a filament nucleator towards the growing end of a treadmilling filament is sufficient to generate solutions containing stable, treadmilling actin structures, which resemble the machinery of motile cells

aments from existing filaments in a defined, 70° branching angle. The formation of the higher-order structure of this branched actin network in the lamellipodium has been suggested to arise from a self-organization mechanism related to evolution [41, 42]. In this model, newly formed actin branches obtain orientation information from the previously formed filaments on which they are nucleated. This orientation information is modulated by fluctuations, and the resulting orientation is selected for its fitness to survive in the dynamic lamellipodium structure.

Thus, this model includes all the basic features of an evolutionary system: (a) inheritance in the form of orientation information transfer from mother filament to daughter filament, (b) mutation in the form of fluctuations in branching angle, and (c) selection on the basis of filament survival in the lamellipodium structure, which is dependent on the incident angle of the filaments towards the plasma membrane at the leading edge. Ultimately, such an evolutionary system can self-organize into a dynamic filament array, displaying structural features similar to the branched actin network seen in lamellipodia. The organizing principle of such a system is different from the previously discussed systems. Similar to systems organized via stigmergy, the resulting structure does not emerge from direct, dynamic repulsive and attractive interactions of the system components, but instead uses the work in progress, the filament network, as a means for indirect interactions. This work in progress also serves as a memory of the current system state. However, in contrast to the previous systems based on stigmergy, the work in progress does not function as a local amplifier in a positive feedback loop but in a more subtle optimization process, which tunes the most efficient geometry for the orientation of dynamic filaments similar to the natural selection process in evolution.

## 9.7 Limits of Self-Organization in Cells

As discussed in the previous examples, self-organization appears to play an important role in many biological processes. However, self-organization is certainly not the only organizing principle in cells, and most structures that display features of self-organization are also controlled by additional mechanisms. For example, the emergence of a mitotic spindle around a DNA-coated bead in a *Xenopus* extract as discussed above is not a pure process of self-organization, as a spatial template – the DNA-coated bead – is necessary to form the structure. In this case, however, the spatial template is very simple and does not contain all the information to build the spindle. It nevertheless contains some information: for example, the chromatin dimensions and most likely therefore also the resulting size of the gradient are related to the overall size of the emerging spindle [43]. In that respect, the organizing principle of the spindle around the DNA-based template bears a remarkable resemblance to the organization of wall building in ant colonies, in which either the colony size (mostly via mechanical stimuli) can act as a template for wall building [22], or else queen-specific pheromone release [44], which attracts workers and guides their behavior to deposit pebbles, can act as a template to guide the building of a chamber that can dynamically adapt to the queen's body size.

The opposite of self-organization, the action of a well-informed leader, in the form of a master regulator that receives input from various signaling pathways, is often found at the onset of many biological processes. For example, the cyclin system receives information about the current state of the cell, and whether it should proceed to mitosis [45]. For the mitotic machinery, the relevant cyclin can be considered a master regulator, and thus would play the role of a well-informed leader. However, as discussed above, while some limited types of templates do exist in the cell, such well-informed leaders usually do not themselves possess a recipe for performing a biological process or a blueprint for building a structure. The capabilities of master regulators are therefore very limited in terms of specifying a complex structure or behavior.

Some simple structures such as kinetochores or centrosomes might arise from nondynamic, crystallization-like aggregation or self-assembly mechanisms. The information for such self-assembly mechanisms is mainly stored in two ways: either as a blueprint in the DNA, indirectly encoding the interactions between the individual complex partners, or in pre-existing templates, which are inherited from the previous cell division. It is unclear for many cellular organelles to which extent they are built based on hardcoded blueprints, pre-existing templates, or self-organization mechanisms.

In a simple view, centrosomes might be self-assembled through biochemical interactions encoded ultimately in DNA blueprints. However, centrosomes usually form around two halves of an inherited pair of centrioles [46], which are therefore a pre-existing, inherited template. Are cells capable of de novo construction of a centriole pair? Or does the inherited half of the pair contain required epigenetic template information, which is not inherited via the genetic code? In the case of centrioles, this questions is easily answered, as centrioles can also assemble de novo during normal development [47] or after laser ablation [48]. In the case of the Golgi apparatus, it is less clear if this organelle can form de novo. In mammalian cells, the Golgi apparatus forms an elaborate structure that is dependent on specific matrix proteins, dynamic membrane fusion events, and motor-based interactions with microtubules [49]. If the microtubles are removed by pharmacological intervention or reorganized during mitosis, the Golgi apparatus fragments into a less structured, vesicular form. Its characteristic structure is rebuilt after microtubules resume their normal interphase configuration. Thus, the Golgi structure can be interpreted as an emerging pattern of a dynamic self-organization process. However, the Golgi apparatus never disappears completely and the vesicles of the fragmented organelle still retain specific properties in membrane and protein composition. Are Golgi fragments essential templates required for the formation of this organelle, or could the Golgi apparatus be formed de novo in the complete absence of Golgi-derived vesicles?

Similar questions could be raised for many other cellular compartments. Do mammalian cells possess all the biological features to generate these compartments de novo, or was this capability lost during evolution? These questions are not yet clearly answered in the case of the Golgi apparatus [49]. The genome of mammalian cells contains the blueprint for either making or importing all the Golgi components, however, as the Golgi apparatus is also an integral part of the cell's metabolic machinery, its complete loss might be too disrupting for the cells to rebuild it. Thus, Golgi-dependent metabolic or signaling pathways might be necessary to produce critical Golgi components. Even if all Golgi components were available to the cell, it is unclear whether they would self-assemble into a functional Golgi apparatus, or if Golgi-derived vesicles are necessary as a seed, or a template. In yeast, it has been suggested that structures identified by markers for the Golgi apparatus and its precursor, the transitional ER sites, can indeed form de novo. However, it was unclear whether this de novo formation required template structures, which were not visualized by these markers [50].

From the overall discussion in this chapter, it becomes clear that most higher-order structures in cells form through a mixture of many organizational mechanisms. In particular, larger-scale, adaptable, dynamic structures, such as the overall morphological shape of the cell, require not only template-based and blueprint-type information to build and maintain themselves and to adapt their shape. Such dynamic structures typically also use mechanisms related to self-organization, in which the individual system components dynamically interact with each other, or their work in progress. Many of the studies mentioned here incorporate a combination of detailed microscopic analysis, mathematical modeling, and in vitro reconstitution of biological systems. This combination of approaches is best suited to fully understand the underlying principles of the often unexpected and not intuitive emerging properties of self-organized systems in cells, as it incorporates a detailed description of the system (microscopic analysis), its theoretical foundation (mathematical modeling), and an experimental proof (in vitro reconstitution). Furthermore, reconstitution and simulation of subcellular systems allows a precise delineation of the mixture of organizational principles used to build and maintain their characteristic structures and dynamic behaviors.

**Acknowledgments** We thank Markus Grabenbauer and Ali Kinkhabwala for helpful discussions, and Tomáš Mazel and Perihan Nalbant for critical reading of the manuscript.

# References

1. S. Camazine, J.-L. Deneubough, N.R. Franks, J. Sneyd, G. Theraulaz, E. Bonabeau, *Self-Organization in Biological Systems* (Princeton University Press, Princeton, NJ, 2001)
2. A. Somoza, Angew. Chem. Int. Ed. **48**, 9406 (2009)
3. M. Falcke, J.L. Hudson, P. Camacho, J.D. Lechleiter, Biophys. J. **77**, 37 (1999)
4. H. Nishimori, N. Ouchi, Phys. Rev. Lett. **71**, 197 (1993)
5. P.P. Grasse, Insectes Sociaux **6**, 41 (1959)
6. M.G. Velarde, C. Normand, Sci. Am. **243**, 93 (1980)
7. J. Lechleiter, S. Girard, E. Peralta, D. Clapham, Science **252**, 123 (1991)
8. L. Dehmelt, P. Nalbant, W. Steffen, S. Halpain, Brain Cell Biol. **35**, 39 (2006)

9. C. Bouzigues, M. Morel, A. Triller, M. Dahan, Proc. Natl. Acad. Sci. USA **104**, 11251 (2007)
10. E. Karsenti, Nat. Rev. Mol. Cell Biol. **9**, 255 (2008)
11. P. Bastiaens, M. Caudron, P. Niethammer, E. Karsenti, Trends Cell Biol. **16**, 125 (2006)
12. J.D. Lechleiter, L.M. John, P. Camacho, Biophys. Chem. **72**, 123 (1998)
13. M.J. Berridge, M.D. Bootman, H.L. Roderick, Nat. Rev. Mol. Cell Biol. **4**, 517 (2003)
14. D.E. Clapham, Cell **131**, 1047 (2007)
15. I. Bezprozvanny, J. Watras, B.E. Ehrlich, Nature **351**, 751 (1991)
16. S.P. Dawson, J. Keizer, J.E. Pearson, Proc. Natl. Acad. Sci. USA **96**, 6060 (1999)
17. J. Keizer, G.D. Smith, S. Ponce-Dawson, J.E. Pearson, Biophys. J. **75**, 595 (1998)
18. A.M. Turing, Philos. Trans. R. Soc. Lond., Ser. B, Biol. Sci. **237**, 37 (1952)
19. B. Kutscher, P. Devreotes, P.A. Iglesias, Sci. STKE **2004**, pl3 (2004)
20. A. Gierer, H. Meinhardt, Kybernetik **12**, 30 (1972)
21. Y. Mori, A. Jilkine, L. Edelstein-Keshet, Biophys. J. **94**, 3684 (2008)
22. N.R. Franks, J. Deneubourg, Anim. Behav. **54**, 779 (1997)
23. T. Wittmann, C.M. Waterman-Storer, J. Cell Sci. **114**, 3795 (2001)
24. M. Pinot, F. Chesnel, J.Z. Kubiak, I. Arnal, F.J. Nedelec, Z. Gueroui, Curr. Biol. **19**, 954 (2009)
25. R. Heald, R. Tournebize, T. Blank, R. Sandaltzopoulos, P. Becker, A. Hyman, E. Karsenti, Nature **382**, 420(1996)
26. M. Kirschner, T. Mitchison, Cell **45**, 329 (1986)
27. M.J. Lohka, J.L. Maller, J. Cell Biol. **101**, 518 (1985)
28. R. Wollman, E.N. Cytrynbaum, J.T. Jones, T. Meyer, J.M. Scholey, A. Mogilner, Curr. Biol. **15**, 828 (2005)
29. M. Caudron, G. Bunt, P. Bastiaens, E. Karsenti, Science **309**, 1373 (2005)
30. C.A. Athale, A. Dinarina, M. Mora-Coral, C. Pugieux, F. Nedelec, E. Karsenti, Science **322**, 1243 (2008)
31. P. Niethammer, P. Bastiaens, E. Karsenti, Science **303**, 1862 (2004)
32. F.J. Nedelec, T. Surrey, A.C. Maggs, S. Leibler, Nature **389**, 305 (1997)
33. T. Surrey, F. Nedelec, S. Leibler, E. Karsenti, Science **292**, 1167 (2001)
34. M. Schuh, J. Ellenberg, Cell **130**, 484 (2007)
35. F. Nedelec, J. Cell Biol. **158**, 1005 (2002)
36. T.D. Pollard, G.G. Borisy, Cell **112**, 453 (2003)
37. K. Doubrovinski, K. Kruse, Phys. Rev. Lett. **99**, 228104 (2007)
38. A.B. Verkhovsky T.M. Svitkina, G.G. Borisy, Curr. Biol. **9**, 11 (1999)
39. A. Mogilner, G. Oster, Biophys. J. **71**, 3030 (1996)
40. T. Bretschneider, K. Anderson, M. Ecke, A. Müller-Taubenberger, B. Schroth-Diez, H.C. Ishikawa-Ankerhold, G. Gerisch, Biophys. J. **96**, 2888 (2009)
41. I.V. Maly, G.G. Borisy, Proc. Natl. Acad. Sci. USA **98**, 11324 (2001)
42. T.E. Schaus, E.W. Taylor, G.G. Borisy, Proc. Natl. Acad. Sci. USA **104**, 7086 (2007)
43. A. Dinarina, C. Pugieux, M.M. Corral, M. Loose, J. Spatz, E. Karsenti, F. Nedelec, Cell **138**, 502 (2009)
44. J.F. Watkins, T.W. Cole, Texas J. Sci. **18**, 254 (1966)
45. J.J. Tyson, B. Novak, Curr. Biol. **18**, R759 (2008)
46. J. Loncarek, A. Khodjakov, Mol. Cells **27**, 135 (2009)
47. D. Szollosi, P. Calarco, R.P. Donahue, J. Cell Sci. **11**, 521 (1972)
48. A. Khodjakov, C.L. Rieder, G. Sluder, G. Cassels, O. Sibon, C.L. Wang, J. Cell Biol. **158**, 1171 (2002)
49. M. Lowe, F.A. Barr, Nat. Rev. Mol. Cell Biol. **8**, 429 (2007)
50. B.J. Bevis, A.T. Hammond, C.A. Reinke, B.S. Glick, Nat. Cell Biol. **4**, 750 (2002)

# Part III
# Protocells In Silico and In Vitro

# Chapter 10
# Approach of Complex-Systems Biology
# to Reproduction and Evolution

**Kunihiko Kaneko**

**Abstract** Two basic issues in biology – the origin of life and evolution of phenotypes – are discussed on the basis of statistical physics and dynamical systems. In section "A Bridge Between Catalytic Reaction Networks and Reproducing Cells", we survey recent developments in the origin of reproducing cells from an ensemble of catalytic reactions. After surveying several models of catalytic reaction networks briefly, we provide possible answers to the following three questions: (1) How are nonequilibrium states sustained in catalytic reaction dynamics? (2) How is recursive production of a cell maintaining composition of a variety of chemicals possible? (3) How does a specific molecule species carry information for heredity? In section "Evolution", general relationships between plasticity, robustness, and evolvability are presented in terms of phenotypic fluctuations. First, proportionality between evolution speed, phenotypic plasticity, and isogenic phenotypic fluctuation is proposed by extending the fluctuation–response relationship in physics. We then derive a general proportionality relationship between the phenotypic fluctuations of epigenetic and genetic origin: the former is the variance of phenotype due to noise in the developmental process, and the latter due to genetic mutation. The relationship also suggests a link between robustness to noise and to mutation. These relationships are confirmed in models of gene expression dynamics, as well as in laboratory experiments, and then are explained by a theory based on an evolutionary stability hypothesis For both sections "A Bridge Between Catalytic Reaction Networks and Reproducing Cells" and "Evolution", consistency between two levels of hierarchy (i.e., molecular and cellular, or genetic and phenotypic levels) is stressed as a principle for complex-systems biology.

K. Kaneko (✉)
Department of Pure and Applied Sciences, University of Tokyo, Tokyo 153-8902, Japan
e-mail: kaneko@complex.c.u-tokyo.ac.jp

H. Meyer-Ortmanns, S. Thurner (eds.), *Principles of Evolution*,
The Frontiers Collection, DOI 10.1007/978-3-642-18137-5_10,
© Springer-Verlag Berlin Heidelberg 2011

## 10.1 A Bridge Between Catalytic Reaction Networks and Reproducing Cells

### *10.1.1 Catalytic Reaction Network for a Protocell*

To understand what life is, we need to determine the universal features that all life systems have to satisfy at a minimum, irrespective of detailed biological processes. In the study of complex-systems biology, we aim to extract universal features that all biological systems have to satisfy [1]. In particular, we have formulated general laws for reproduction, adaptation, development, and differentiation.

Present-day organisms, however, include detailed, elaborated processes that have been captured throughout the history of evolution. For our purpose of extracting universal logic, it is desirable to study a system that is as simple as possible. Accordingly, we have proposed an approach called *constructive biology*, in which we set up experimental and theoretical models that possess a certain basic property of life and try to understand the conditions required to possess such a property [2, 3].

One of the most important steps in constructive biology is the construction of reproducing cells. However, despite considerable efforts and developments toward the experimental construction of artificial cells that reproduce themselves, there remain several difficulties [4–8]. We need to bridge the gap between "simple catalytic reaction networks" and reproducing cells.

Let us begin with a simple argument for a biochemical process that a cell that grows must satisfy at a minimum. In a cell, there exist a large variety of chemicals that catalyze each other and form a complex network. These molecules are spatially arranged in a cell, and for some problems the spatial arrangement is very important, whereas for others simply the knowledge of the composition of the chemicals in a cell is sufficient to determine the state of a cell. As a starting point, we disregard the spatial structure within a cell and consider only the composition of the chemicals in a cell. If there are $k$ chemical species in a cell, the cell state is characterized by the number of molecules of each species as $N_1, N_2, \ldots, N_k$. These molecules change their number by reactions between each other. Because most reactions are catalyzed by some other molecules, the reaction dynamics consist of a catalytic reaction network [1, 9–13].

As our constructive biology is aimed at neither making a complicated, realistic model for a cell nor imitating specific cellular functions, we set up a minimal model with a reaction network. Now, there exist several levels for the modeling, depending on the question we wish to address [11].

1. Type I model: Assuming that reactions for some molecules are fast, they can be adiabatically eliminated. Now, most biochemical reactions are catalyzed by some other chemicals. These catalysts are also synthesized as a result of intracellular reactions. A simple model for this situation is the following reversible, two-body reaction among catalysts $X_i$ ($i = 1, 2, \ldots, k = $ total species)

$$X_i + X_j(+S) \longleftrightarrow X_m + X_j(+S') \,,$$

where $X_j$ is a catalyst and the ratio between backward and forward reactions is chosen so that the detailed balance condition is satisfied.

2. Type II model: Under the flow of some chemicals into a cell, most slow reactions progress almost unidirectionally, where fast reversible reactions are regarded to be already balanced. For example, a simple two-body catalytic reaction is considered:

$$X_i + X_j(+S) \rightarrow X_m + X_j(+S') \,,$$

where $X_j$ catalyzes the reaction. If the catalysis progresses through several steps, this process is replaced by

$$X_i + mX_j + (S) \rightarrow X_\ell + mX_j(+S') \,,$$

leading to higher-order catalysis. For a cell to grow, some resource chemicals must be supplied through a membrane. Through the above catalytic reaction network, the resource chemicals are transformed to others, and as a result, a cell grows. Indeed, this class of model is adopted to study the condition for cell growth and clarify universal statistics for such cells.

3. Type III model (network consisting of replicating units, e.g., hypercycle [9]): For a cell to grow effectively, there should exist some positive feedback process to amplify the number of molecule species. Such a positive feedback process leads to an autocatalytic process to synthesize each molecule species. During the reproduction of a cell, all species are somehow synthesized. Then it would be possible to consider a replication reaction. For example, there exists a reaction $S+X+Y \rightarrow X'+Y : S'+X' \rightarrow 2X$. Then, as a total, $S+S'+X+Y \rightarrow 2X+Y$. Assuming that the resources $S$ and $S'$ are constantly supplied, we can consider the replication reaction $X + Y \rightarrow 2X + Y$ catalyzed by $Y$. At this level, we can consider a unit of a replicator and consider a replication reaction network. This model was first discussed as the hypercycle by Eigen and Schuster [9]. By choosing each model level, we address the "basic" questions to be discussed below. Through such a study, one can obtain insight into how to bridge the gap between only a set of chemical reactions and a reproducing cell; this should also help to construct an artificial protocell. In addition, it is interesting to check whether the universal properties discovered in a simple protocell model are preserved in the present cell.

### 10.1.2 Long-Term Sustainment of Nonequilibrium State

A cell consists of a network of catalytic reactions. This reaction process progresses at a nonequilibrium condition, and indeed, it is occasionally considered that the (dissipative) structure that appears far from equilibrium is a prerequisite for a biological system [14] However, in a cell, such a nonequilibrium condition is sustained by itself. Then, how is the system prevented from falling into equilibrium while maintaining catalytic activity? We address this question here.

Most theoretical models assume that chemical reactions are set at a nonequilibrium condition, whereas a biological cell has to sustain the nonequilibrium condition by itself. Can such a nonequilibrium condition be sustained even once a transient dissipative structure is formed? This question is answered in the affirmative for a class of catalytic reaction network systems [15]. The core mechanism for this is a negative correlation between the abundances of resource chemicals and the catalysts. Assume that the concentration of resources exhibits a spatially inhomogeneous pattern, and that of the catalyst exhibits an anticorrelated pattern. In this case, in a region with abundant resource chemicals, the abundance of catalysts necessary for the reaction to consume the resource is suppressed, whereas in a region with less resource chemicals, it is obvious that no more consumption of resources occurs. Hence, the consumption of a resource is suppressed over the entire space, and therefore the relaxation to equilibrium is hindered and a structure at the nonequilibrium condition is sustained for longer.

Next, let us consider a catalytic reaction network in which each chemical species is assigned an energy, and the rate for each reaction is determined by the energy difference to satisfy the detailed condition. In a closed thermodynamic system consisting of chemical reactions, equilibrium is ultimately attained after a certain relaxation time. The relaxation process is exponential with the time scale given by the reaction kinetic coefficients, as long as we start from an initial state close to equilibrium. This is also true for any initial condition for linear reaction kinetics, that is, reactions without catalysts or catalytic reactions with fixed concentrations of catalysts. In contrast, the reaction kinetics whose catalysts are synthesized by themselves involve nonlinear terms, because the rate of such catalytic reaction is given by the product of the concentrations of the substrate and the catalyst. In such catalytic reaction networks, we have recently found a mechanism to slow down the relaxation to equilibrium, even in a well-mixed condition assuring spatially homogeneous concentrations.

We consider a type I model as in Sect. 10.1.1, and assign energy $E_i$ to each molecule [16]. The ratio of the forward and backward reactions is given by $\exp(-\beta(E_j - E_i))$ to satisfy the detailed balance condition, where $\beta$ is the inverse temperature $1/kT$. Now, let $\varepsilon$ be the variance of energy of each chemical. When the temperature of the system is sufficiently lower than $\varepsilon$ (i.e., $\beta\varepsilon > 1$), overall $\log t$ relaxation appears. The deviation from the equilibrium decreases with $\log t$, whereas several plateaus appear successively through the course of relaxation. We have studied a variety of reaction networks to confirm that these two characteristics are universal. How many and which type of plateaus appear depend on the network and initial conditions; however, the existence of several plateaus itself is universal.

Thus, we have revealed a general mechanism for the emergence of plateaus. The plateaus are not metastable states in the energy landscape; rather, they are a result of kinetic constraints due to a reaction bottleneck, originating in the formation of local-equilibrium clusters and suppression of equilibration by the negative correlation between an excess chemical and its catalyst. The existence of such negative correlation depends both on the initial concentrations of chemicals and on the network structures; however, even in randomly chosen networks, there exist several sets of chemicals that satisfy the negative correlation, as long as the number of species is not small (say larger than five). The prediction of each plateau is possible by detecting such negative correlation, whereas a systematic procedure to extract the same has to be developed in the future.

In biochemical reaction processes, the energy variance is rather large, and therefore the above slow-relaxation is observed even if the temperature is not so low. Hence, the slow speed of relaxation to equilibrium is a rather common feature of catalytic reaction networks.

Of course, for the origin of life, initially at least, some nonequilibrium condition has to be supplied externally. Indeed, it is natural that there exists some nonequilibrium condition in nature, as, for example, is provided by a thermal vent. Then, a nonequilibrium condition supplied exogenously is embedded into the internal dynamics so that the relaxation is hindered and the activity is maintained endogenously.

Furthermore, we may expect mutual reinforcement between the sustainment of nonequilibrium conditions, spatial structure with compartmentalization, and reproduction. By taking advantage of nonequilibrium reaction processes, a structure is organized in network and in space, as was also discussed in the case of a dissipative structure. Then, spatial compartmentalization is possible. With such a compartmentalized structure, reproduction in molecules is possible. Such a reproduction process naturally enhances the spatial inhomogeneity in chemical compositions. This inhomogeneity further suppresses the relaxation to equilibrium.

This hindrance of relaxation to equilibrium is important for the origin of life; in addition, it will be relevant to understanding slow processes in present cells. For example, a plant seed, even though it is almost closed with regard to energetic and maternal flow, is "alive" over a large time span without falling into an equilibrium state. In dormant states that are ubiquitous in bacteria, intracellular processes almost stop but activity restarts when they are put under an appropriate culture condition.

### 10.1.3 Consistency Between Cell Reproduction and Molecule Replication

The second question concerns the consistency of cell reproduction and molecule replication [10, 13, 17]. For a cell to continue reproduction, at least catalytic activity should be preserved. Furthermore, at least a set of chemicals has to be synthesized. At a primitive stage of the cell, this reproduction need not be precise, and in fact is probably rather loose. The composition of chemicals is not fixed, but it exhibits some degree of similarity between generations ("recursive production").

The reproduction of a cell involves numerous reactions for membrane synthesis, and metabolic and genetic processes, as noted by Ganti in the Chemton model [18]. All components have to be replicated for cell reproduction. At the very least, a membrane that partly separates a cell from the outside has to be synthesized, and this process must maintain some degree of synchronization with the replication of other intracellular chemicals. How is such recursive production maintained while preserving chemical diversity? In other words, the question of consistency between cell reproduction and molecule replication is raised.

To answer this question, Furusawa and I [19] studied a cell model consisting of catalytic reaction networks of the type II model in Sect. 10.1.1. By the intracellular catalytic reaction networks, resource chemicals that are transported from the outside are transformed into other chemical species (see Fig. 10.1 for a schematic representation). Chemicals are successively transformed by means of these catalytic reactions starting from nutrients. When the number of total (or specific) molecule species increases beyond some threshold, the cell is assumed to divide into two. We have studied a variety of models within this class, with different types of networks, reaction kinetics, and transport processes of resources.

We have discovered that when a certain condition is satisfied, the cell continues reproduction, approximately maintaining the compositions of chemicals at which the growth speed of the cell is optimized. The reproduction of a cell with diversity in chemicals is generally possible, even in this simple setup with mutual catalytic reactions. Therefore, we investigated the statistical characteristics of such cells.

First, we found universal statistics of the abundance of chemicals, for a cell that maintains reproduction and chemical compositions. We measured the rank-ordered number distributions of chemical species by plotting the number of molecules $n_i$ as a function of their rank determined by $n_i$. The distribution exhibits a power law with an exponent of $-1$. In our model, this power law of gene expression is maintained

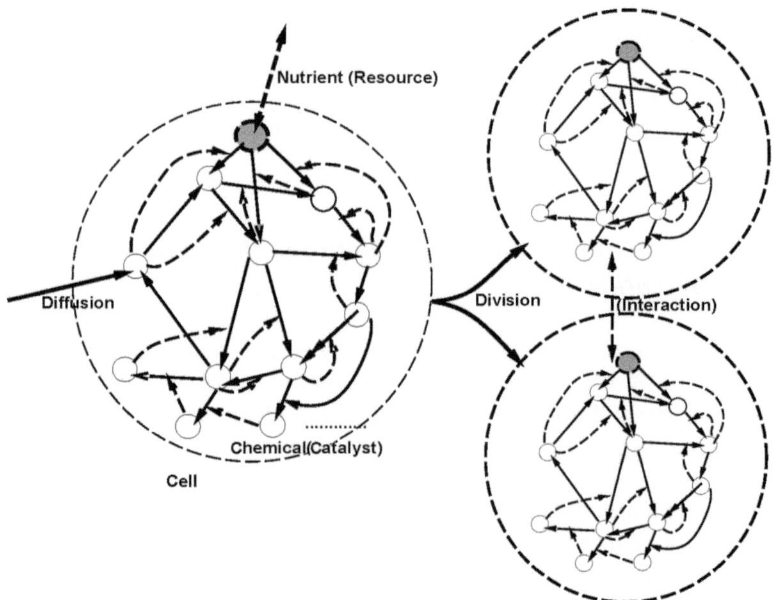


**Fig. 10.1** Schematic representation of a reproducing cell with internal catalytic chemical reaction

by a hierarchical organization of catalytic reactions. Major chemical species are synthesized and catalyzed by chemicals with slightly smaller abundances. The latter chemicals are synthesized by chemicals with much less abundance, and so forth. This hierarchy of catalytic reactions continues until it reaches the chemical species with minority in number. This power law is confirmed universally for a variety of cell models.

Furthermore, this power law was also confirmed by measuring the abundances of a large variety of mRNAs, over more than a hundred cell types, using microarray analysis [19–21]. Hence, the statistical law as a result of the recursive production of a protocell is also valid in present cells.

Second, we studied the cell-to-cell fluctuations of chemical compositions. Because the chemical reaction process is stochastic, the number of each type of molecule differs between cells. We then studied the distribution of each molecule number $n_i$, sampled over cells, and found that the number distribution is fitted reasonably well by the log-normal distribution, that is,

$$P(n_i) \approx \frac{1}{n_i} \exp\left( -\frac{(\log n_i - \overline{\log n_i})^2}{2\sigma} \right) , \tag{10.1}$$

where $\overline{\log n_i}$ indicates the average of $\log n_i$ over cells [22]. This implies that the distribution has a rather long tail on the side of larger numbers. This log-normal distribution holds for the abundances of all chemicals that are reproduced within a cell. In general, when successive catalytic reactions for recursive production exist in a biochemical reaction network, fluctuations are multiplied successively through the catalytic reaction cascade. Then, by taking the logarithms of concentrations, these successive multiplications are transformed into successive additions, and the problem is reduced to the addition of random noise. According to the central limit theorem, the distribution of $\log n_i$ is expected to approach the Gaussian distribution. Hence, the log-normal distribution of $n_i$ is derived. This log-normal distribution is also experimentally confirmed for present-day cells (e.g., for bacteria) [22, 23].

Note that the power law in abundances and log-normal distribution are a consequence of the reproduction of a cell. Both the laws studied here are universal and a result of "consistency between replication of molecules and reproduction of a cell." In fact, when recursive production does not occur, some deviation from these two laws is observed.

### 10.1.4 Minority Control: Origin of Genetic Information

The third question we address is rather naive. In present-day cells, we have molecules (DNA) that carry "genetic information" separated from a metabolic reaction: is such a separation a necessary course for a system with reproduction and evolvability? In a reproducing reaction network system consisting of a variety of molecule types, do some molecules carry the role for heredity, even if we do not assume DNA or RNA? If so, what properties must such molecules satisfy? To answer the question of the origin of genetic information, we have recently proposed the following hypothesis [24]: In a reproducing system consisting of mutually catalytic molecules, molecule species that in a minority play the role of heredity carriers, in the sense that they are preserved well and control the behavior of this protocell relatively strongly.

As a first step toward the investigation of the origin of genetic information, we study how some molecule species are preserved over cell generations and play an important role in controlling the growth of a cell. We consider a model of replicating molecules, that is, the type III model of Sect. 10.1.1, and encapsulate it into a protocell (see Fig. 10.2). For simplicity we consider two types of mutually catalyzing molecule species ($X$ and $Y$), each of which has catalytically active and inactive types. Here, most types are inactive, and only a specific pair of active types of $X$ and $Y$ mutually support the synthesis of the other. In other words, inactive types are synthesized by the active type of the other species, but they do not contribute to the others' synthesis. In this sense, they are parasites in the replication system. Indeed, most mutant molecules are such parasites, and the removal of such parasitic molecules is an important question in the origin of life [9]. Now, one of the species $Y$ is assumed to have a much slower synthesis speed than $X$. As long as active molecules exist, molecule numbers within the protocell increase, and we assume that at a certain number, it divides into two.

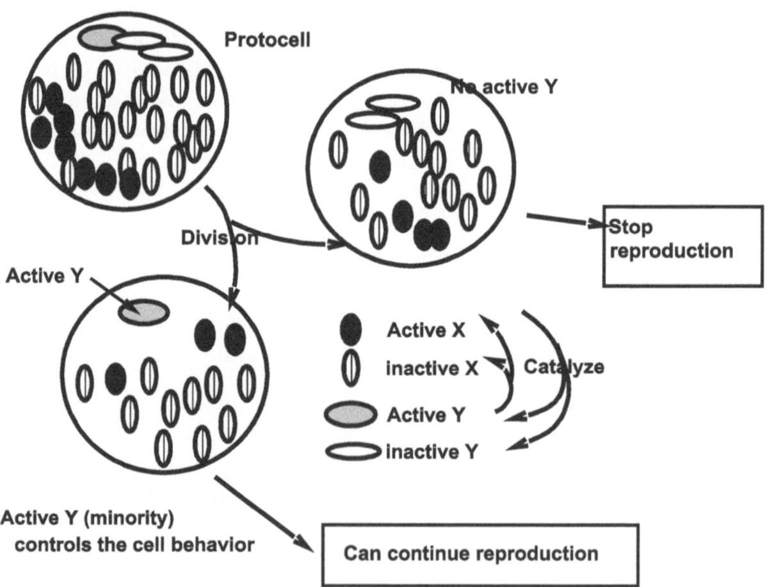


**Fig. 10.2** Schematic representation of a reproducing cell with mutually catalytic hypercycle system

Thus, through this growth–division process, the fraction of $Y$ species is much lower than $X$. As the inactive types are more common, the protocell is expected to reach a state in which there exist no active $Y$ molecules. Indeed, from the continuous rate equation, such a solution is expected. However, when there exist no active $Y$ molecules, $X$ molecules are no longer synthesized, and finally, active $X$ molecules become extinct, so that the reproduction of the protocell would stop. In contrast, through stochastic simulation of the present model, the protocell comes to and remains at a state in which only a few active $Y$ and almost no inactive $Y$ molecules exist. Such cells can continue the growth–division process. Probabilistically, such a state is very rare; however, when the number of molecules is small, it can appear due to fluctuation. Once it appears, it is selected, because such a state can continue to grow. Hence, a rare state with a few active $Y$ molecules and no inactive ones is preserved over many divisions of protocells (i.e., a rare initial condition is selected and frozen). Furthermore, these few active $Y$ molecules are shown to control (relatively strongly) the behavior of the protocell, because a slight change in such molecules strongly influences the replication of other molecules. The minority molecule species now acts as a heredity carrier because of the relatively discrete nature of its population, in comparison with the majority species, which behaves statistically in accordance with the law of large numbers. Hence, the kinetic origin of genetic information is demonstrated [1, 24].

Note that we assumed compartmentalization, that is, chemicals are encapsulated into a membrane that itself grows and divides as in the model of Sect. 10.1.3. The importance of compartmentalization to remove parasitic (inactive) molecules has been discussed for the last few decades [25–29]. Here, a minority of some molecules (whose number can go to zero frequently) is also essential for the removal of the parasitic molecules.

Thus far, we have shown the origin of "minimal" genetic information, that is, one-bit information to distinguish between the existence and nonexistence of an active $Y$ molecule. Minority control, however, provides a basis for genetic information with more bits. First, the minority-controlled state gives rise to a selection pressure for mechanisms that ensure the transmission of the minority molecule. Because the active ("minority") $Y$ molecule is transmitted faithfully, more chemicals will be synthesized with this minority molecule. Then, life-critical information is packaged into this minority molecule. Once the minority control mechanism is in place, the minority molecule becomes the ideal storage device for information to be transmitted across generations, thus giving rise to "genetic information" in the current sense.

An important consequence of this minority control is evolvability (see also [30]). Because only a few molecules of the $Y$ species exist in the minority-controlled state, a structural change to them strongly influences the overall catalytic activity of the protocell. On the other hand, a change to $X$ molecules has a weaker influence, on average, because the variation of the *average* catalytic activity caused by such a change is smaller, as can be deduced from the law of large numbers. Hence, the minority-controlled state is important for a protocell to realize evolvability [24, 31].

## 10.2 Evolution

### *10.2.1 Fluctuations and Robustness*

One consequence of the previous section is that fluctuations in the protein number in a cell are indeed rather large, as is highlighted by the log-normal distribution of the protein abundances. Recently, the distributions of protein abundances over isogenic individual cells have been measured using fluorescent proteins. The fluorescence level that gives an estimate of the protein concentration is measured either by flow cytometry or single-cell microscopy [32–35]. Note that in the model, the network and parameters are identical over cells, and in the experiment, isogenic bacteria are used. Nonetheless, there exist large phenotypic fluctuations, that is, the concentration of molecules exhibits a rather large variance over isogenic cells. Here, we discuss the relevance of such fluctuations to evolution, in relation to genotype–phenotype mapping.

Often, stochasticity in gene expression is thought to be an obstacle in tuning a system to achieve and maintain a state with higher functions. If the phenotype concerns the fitness, that is, crucially influences the survivability, a very large variability in it must be harmful. A phenotype that is concerned with fitness is expected to suppress such fluctuations. Indeed, the question most often asked is how some biological functions are robust to phenotypic noise.

In contrast, considering that relatively large phenotypic noise is preserved through the history of evolution, it should be important to also pursue the positive effects of phenotypic noise on biological functions. Indeed, the positive role of fluctuations on adaptation [36, 37] and development has been discussed earlier. In particular, in *isologous diversification theory*, the differentiation of cell states triggered by noise leads to noise-tolerant developmental processes [1, 38]. Then, does the noise in gene expression dynamics play some role in evolution? Does the degree of phenotypic fluctuations induced by noise change through evolution?

Robustness is strongly related to fluctuation. Robustness is defined as the insensitivity of a system to changes to it [39–43]. In any biological system, these changes have two distinct origins: genetic and epigenetic. The former concerns structural robustness of the phenotype, that is, rigidity of the phenotype with respect to genetic changes produced by mutations. On the other hand, the latter concerns robustness with respect to the stochasticity that can arise in an environment or during the developmental process, which includes fluctuations in initial states and stochasticity occurring during developmental dynamics or in the external environment.

These two types of robustness can be discussed in terms of phenotypic fluctuations. First, developmental robustness is represented by the degree of phenotype change as a result of a developmental process that generally involves noise, as mentioned previously. Accordingly, the phenotype as well as the fitness of isogenic individuals is distributed. Here, let us denote the variance of phenotypes of isogenic organisms as $V_{ip}$. When the phenotype is robust to noise, this phenotype is not changed so much by noise, and therefore the distribution is sharper. Hence, the (inverse of the) variance of isogenic phenotypic distribution, $V_{ip}$, gives an index for robustness with respect to noise in developmental dynamics [43, 44].

Genetic robustness is also measured in terms of fluctuations. Owing to mutation in genes, the phenotype (fitness) is distributed. Because even the phenotype of isogenic individuals is distributed, the variance of phenotype distribution of a heterogenic population includes both the contribution from phenotypic fluctuations in isogenic individuals and that due to genetic variation. To distinguish the two, first the average phenotype over isogenic individuals is measured and then the variance of this average phenotype over a heterogenic population is computed. Thus, this variance is due only to genetic heterogeneity. This variance is denoted by $V_g$. Then the robustness to mutation is estimated by this variance. If $V_g$ is smaller, genetic change influences the phenotype negligibly, implying larger genetic (or mutational) robustness.

## 10.2.2 Evolutionary Fluctuation–Response Relationship

One might still suspect that isogenic phenotype fluctuations $V_{ip}$ are not related to evolution, because phenotypic change without genetic change is not transferred to the next generation. However, the variance, a degree of fluctuation itself, can be determined by the gene, and accordingly it is inheritable [45]. Hence, there may exist a relationship between isogenic phenotypic fluctuation and evolution.

To verify this possibility, we carried out a selection experiment to increase the fluorescence in bacteria and investigated the possible relationship between the evolution speed and the isogenic phenotypic fluctuation [46]. First, by attaching a random sequence to the N terminus of a wild-type green fluorescent protein (GFP) gene, a protein with low fluorescence was obtained. The gene for this protein was introduced into *E. coli* as the initial generation for the evolution. By applying random mutagenesis only to the attached fragment in the gene, a mutant pool with a diversity of cells was prepared. Then, cells with the highest fluorescence intensity were selected for the next generation. With this procedure, the (average) fluorescence level of the selected cells increases in successive generations. The evolution speed at each generation was computed as the difference between the fluorescence levels of the two generations. The isogenic phenotypic variance was computed from the distribution of clone cells of the selected bacteria, measured with the help of flow cytometry for the fluorescence of cells. The data suggest that the evolution speed is proportional to, or at least positively correlated with, the variance of the isogenic phenotypic fluctuation.

To confirm this relationship between the evolution speed and isogenic phenotypic fluctuation quantitatively, we also numerically studied a model for reproducing cells consisting of the catalytic reaction networks mentioned in Sect. 10.1.3. Here, the reaction networks of mutant cells were slightly altered from the network of their mother cells. Among the mutants, those networks with a higher concentration of a given, specific chemical component were selected for the next generation. Again, the evolution speed was computed by the increase in the concentration at each generation and the fluctuation by the variance of the concentration over identical networks [44].

The origin of proportionality between the isogenic phenotypic fluctuation and genetic evolution has been discussed in light of the fluctuation–response relationship in statistical physics [47, 48]. Here, we refer to a measurable phenotype quantity (e.g., logarithm of the concentration of a protein) as a "variable" $x$ of the system, whereas we assume the existence of a "parameter" $a$ that controls the change in the variable $x$. In the present context, this parameter is an index of change in genes that governs the corresponding phenotype variable.

Consider the change in the parameter value $a \to a + \Delta a$. Then the proposed fluctuation–response relationship [1, 46] is given by

$$\frac{\langle x \rangle_{a+\Delta a} - \langle x \rangle_a}{\Delta a} \propto \langle (\delta x)^2 \rangle , \tag{10.2}$$

where $\langle x \rangle_a$ and $\langle (\delta x)^2 \rangle = \langle (x - \langle x \rangle)^2 \rangle$ are the average and variance of the variable $x$ for a given parameter value $a$, respectively. The above relationship is derived by assuming that the distribution $P(x; a)$ is approximately Gaussian and that the effect of the change in $a$ on the distribution is represented by a bilinear coupling between $x$ and $a$.

Note, however, that the argument based on the distribution $P(x; a)$ is not a first-principles derivation of the evolutionary fluctuation–response relationship. Rather, it is a phenomenological description. Here, the description by $P(x; a)$ itself is an assumption: for example, whether genotype change is represented by a scalar parameter $a$ is an assumption. A perturbative approach to obtaining the above relationship is based on the assumption that the change in $a$ as well as the response to $x$ is small. Nonetheless, it is interesting to note that both the laboratory experiment and in silico evolution support the relationship.

### 10.2.3 Relationship Between Fluctuations by Noise and by Mutation

The above-mentioned evolutionary fluctuation–response relationship gives rise to another question. There is an established relationship between the evolutionary speed and the phenotypic fluctuation, namely, the so-called fundamental theorem of natural selection proposed by Fisher [49], which states that the evolution speed is proportional to $V_g$, the variance of phenotypic fluctuation due to genetic variation. As mentioned, it is given by the variance of the *average* phenotype for each genotype over a heterogenic population. In contrast, the evolutionary fluctuation–response relationship proposed above concerns the phenotypic fluctuation of isogenic individuals, $V_{ip}$. Hence, the evolutionary fluctuation–response relationship and the relationship concerning $V_g$ from Fisher's theorem are not identical.

If $V_{ip}$ and $V_g$ are proportional through an evolutionary course, the two relationships are consistent. Such proportionality, however, is not self-evident, as $V_{ip}$ is related to variation resulting from developmental noise and $V_g$ is due to the variation resulting from mutation. The relationship between the two, if it exists, should be a result of a constraint on genotype–phenotype mapping achieved through evolution.

Hence, it is important to study the evolution of robustness and phenotypic fluctuations by using a model with the "developmental" dynamics of a phenotype. To introduce such a dynamical-systems model, one requires the following structure for development and evolution.

i. There is a population of organisms under a distribution of genotypes.
ii. A phenotype is determined by the genotype through "developmental" dynamics. This dynamics is not simple and involves many degrees of freedom. As already mentioned, it generally includes some noise, so that the phenotype (such as the abundance of some protein) from isogenic cells (organisms) fluctuates.
iii. The fitness for selection, that is, the number of offspring, is given by some phenotype.
iv. The genotypes change slightly by mutation, and with the selection process according to the fitness, the distribution of genotypes for the next generation may be altered.

We have simulated two models to satisfy the above postulates, by adopting a genetic algorithm for (i), (iii), and (iv). Here, it is important that the phenotype is determined only after (complex) "developmental" dynamics (ii), which are stochastic due to the noise therein.

In the first example, we again adopted the protocell model with a catalytic reaction network, as described in Sect. 10.1.3, by selecting cells that have a higher concentration of a specific chemical [44]. As another example, we studied gene expression dynamics that are governed by regulatory networks [43]. The developmental process (ii) is given by this gene expression dynamics under noise with amplitude $\sigma$. It shapes the final gene expression profile that determines the evolutionary fitness. The mutation at each generation alters the network slightly. Among the mutated networks, we select those with higher fitness values.

Results from several evolutionary simulations are summarized as follows (see Fig. 10.3):

1. There is a certain threshold noise level $\sigma_c$ beyond which the evolution of robustness progresses, so that both $V_g$ and $V_{ip}$ decrease. Here, most of the individuals take the highest fitness value. In contrast, for a lower level of noise $\sigma < \sigma_c$, mutants that have very low fitness values always remain. There exist many individuals taking the highest fitness value, whereas the fraction of individuals with much lower fitness values does not decrease.
2. Around the threshold noise level, $V_g$ approaches $V_{ip}$ and at $\sigma < \sigma_c$, $V_g \sim V_{ip}$ holds, whereas for $\sigma > \sigma_c$, $V_{ip} > V_g$ is satisfied. For robust evolution to progress, this inequality is satisfied.
3. When the noise is larger than this threshold, $V_g \propto V_{ip}$ holds (see Fig. 10.3) through the course of evolution after a few generations.

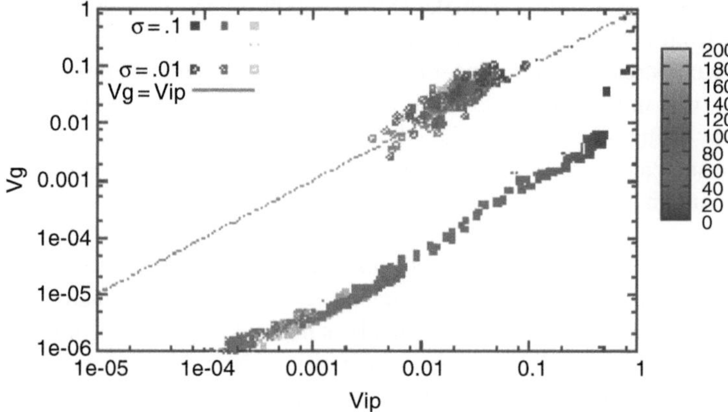

**Fig. 10.3** An example of the evolution process of $V_g$ and $V_{ip}$, computed using the gene regulation network model (see [43]). Each point is the variance at each generation up to 200 generations, with a gradual change in the density. *circles* give the result of a lower noise case, and *squares* that of a higher noise case, where the variances decrease in successive generations

Why does the system not maintain the highest fitness state under a small phenotypic noise level with $\sigma < \sigma_c$? Indeed, the dynamics of the top-fitness networks that evolved under such low noise levels have dynamics distinguishable from those that evolved under high noise levels. It was found that for networks evolved under $\sigma > \sigma_c$, a large portion of the initial conditions reach attractors that give the highest fitness values, whereas for those evolved under $\sigma < \sigma_c$, only a tiny fraction (that is, in the vicinity of all-off states) reached such attractors.

When the time course of gene expression dynamics to reach its final pattern (attractor) is represented as a motion falling along a potential valley, our results suggest that the potential landscape for development becomes smoother and simpler through evolution and loses its ruggedness after a few dozen generations, for $\sigma > \sigma_c$. In this case, the "developmental" dynamics gives a global, smooth attraction to the target (see Fig. 10.3). In fact, such a type of developmental dynamics with global attraction is known to be ubiquitous in protein-folding dynamics [50, 51], gene expression dynamics [52], and so on. On the other hand, the landscape evolved at $\sigma < \sigma_c$ is rugged. Except for the vicinity of the given initial conditions, the expression dynamics do not reach the target pattern.

Now consider mutation to a network to slightly alter the gene regulatory network. This introduces slight alterations in gene expression dynamics. In a smooth landscape with a global basin of attraction, such a perturbation in dynamics only slightly changes the final expression pattern (attractor). In contrast, under dynamics with a rugged developmental landscape, a slight change easily destroys the attraction to the target attractor. Then, low-fitness mutants appear. This explains why the network evolved at a low noise level is not robust to mutation. In other words, evolution to eliminate ruggedness in developmental potential is possible only for a sufficiently high noise level.

### 10.2.4 *Phenomenological Distribution Theory*

To understand the observed relationship between $V_{\mathrm{ip}}$ and $V_{\mathrm{g}}$, we have formulated a phenomenological theory (see Fig. 10.2) [44, 53]. Recall that the phenotype is originally given as a function of a gene; therefore we consider the conditional distribution function $P(x; a)$. However, the genotype distribution is influenced by the phenotype through the selection process, as the fitness for selection is a function of the phenotype. Now consider a gradual evolutionary process (see Fig. 10.4). Then, it should be possible to assume that some "quasi-static" distribution on the genotype and phenotype is obtained as a result of the feedback process from the phenotype to the genotype. Considering this point, we hypothesize that there exists a two-variable distribution $P(x, a)$ for both the phenotype $x$ and the genotype $a$.

By using this distribution, $V_{\mathrm{ip}}$, the variance of $x$ of the distribution for given $a$ can be written as $V_{\mathrm{ip}}(a) = \int (x - \overline{x(a)})^2 P(x, a) dx$, where $\overline{x(a)}$ is the average phenotype of a clonal population sharing the genotype $a$, namely $\overline{x(a)} = \int P(x, a) x dx$. $V_{\mathrm{g}}$ is defined as the variance of the average $\overline{x(a)}$ over genetically heterogeneous individuals and is given by $V_{\mathrm{g}} = \int (\overline{x(a)} - \langle \overline{x} \rangle)^2 p(a) da$, where $p(a)$ is the distribution of genotype $a$ and $\langle \overline{x} \rangle$ is the average of $\overline{x(a)}$ over all genotypes.

Assuming a Gaussian distribution again, we write the distribution $P(x, a)$ as

$$P(x, a) = \widehat{N} \exp\left( -\frac{(x - X_0)^2}{2\alpha(a)} + C(x - X_0)(a - a_0) - \frac{1}{2\mu}(a - a_0)^2 \right),$$

$$(10.3)$$

with $\widehat{N}$ as a normalization constant. The Gaussian distribution

$$\exp\left( -\frac{1}{2\mu}(a - a_0)^2 \right)$$

represents the distribution of genotypes around $a = a_0$, whose variance is (in a suitable unit) the mutation rate $\mu$. The coupling term $C(x - X_0)(a - a_0)$ represents the change in the phenotype $x$ with the change in genotype $a$. Recalling that the above distribution (10.3) can be rewritten as

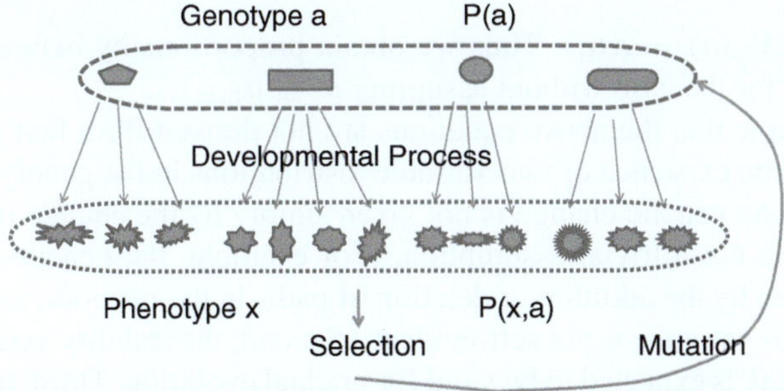


**Fig. 10.4** Schematic representation of evolutionary processes with genotype and phenotype distributions

$$P(x, a) = \widehat{N} \exp\left(-\frac{(x - X_0 - C(a - a_0)\alpha(a))^2}{2\alpha(a)} + \left(\frac{C^2\alpha(a)}{2} - \frac{1}{2\mu}\right)(a - a_0)^2\right),$$

(10.4)

the average phenotype value for the given genotype $a$ satisfies

$$\overline{x}_a \equiv \int x P(x, a)dx = X_0 + C(a - a_0)\alpha(a) .$$

(10.5)

Now we postulate *evolutionary stability*, that is, at each stage of the evolutionary course, the distribution has a single peak in $(x, a)$ space. This assumption is rather natural for evolution to progress robustly and gradually: the selection of a phenotype with larger $x$ to increase the phenotype value works if the distribution is concentrated. On the other hand, if the distribution is flattened, the selection cannot increase the average value of the phenotype $x$. Here, this stability condition leads to $\frac{\alpha C^2}{2} - \frac{1}{2\mu} \leq 0$, that is,

$$\mu \leq \frac{1}{\alpha C^2} \equiv \mu_{\max} .$$

(10.6)

This implies that the mutation rate has an upper bound $\mu_{\max}$ beyond which the distribution does not have a peak in the genotype–phenotype space. In the above form, the distribution becomes flat at $\mu = \mu_{\max}$, so that mutants with low fitness rate appear; this scenario was termed error catastrophe by Eigen [9].

We recall the definition of $V_{\mathrm{g}}$, $(C\alpha)^2\langle(\delta a)^2\rangle$. Hence, we obtain

$$V_{\mathrm{g}} = \frac{\mu C^2\alpha^2}{1 - \mu C^2\alpha} = \alpha\frac{\mu/\mu_{\max}}{1 - \mu/\mu_{\max}} .$$

(10.7)

If the mutation rate $\mu$ is sufficiently small to satisfy $\mu \ll \mu_{\max}$, we obtain

$$V_{\mathrm{g}} \sim \frac{\mu}{\mu_{\max}} V_{\mathrm{ip}} ,$$

(10.8)

recalling that $V_{\mathrm{ip}}(a) = \alpha(a)$. Thus we obtain proportionality between $V_{\mathrm{ip}}$ and $V_{\mathrm{g}}$ (see also [53] for the case without assuming $\mu \ll \mu_{\max}$).

Here, we note that the above equations are not derived from first principles. We first assumed the existence of two-variable distributions in the genotype and phenotype $P(x, a)$. As genetic change is not given simply by the change in a continuous parameter, it is not a trivial assumption. (For example, the genetic change in our models is given by the addition or deletion of paths in the network, and if expressed as a continuous variable, is not self-evident.) Second, the stability assumption assuring a single peak is expected to be valid for gradual evolution. Third, to adopt (10.3), the existence of error catastrophe (to produce mutants with very low fitness values at a large mutation rate) is implicitly assumed.

We note also that the proportionality is observed only after some generations in simulations, when evolution progresses steadily. This might correspond to the assumption for the quasi-static evolution of $P(x, a)$.


## *10.2.5 Discussion*

Evolvability may depend on the species. Some species, called living fossils, preserve their phenotype over many more generations than others. An origin of such differences in evolvability could be provided by the degree of rigidness in the developmental process. If the phenotype generated by a developmental process is rigid, then the phenotypic fluctuation as well as phenotypic plasticity against environmental change is smaller. The evolutionary fluctuation–response relationship, that is, $V_{ip} \propto$ evolution speed, can provide a quantitative expression for such an impression.

As mentioned, robustness is tightly correlated with the fluctuations. The developmental robustness to noise is characterized by the inverse of $V_{ip}$, whereas the mutational robustness is characterized by the inverse of $V_g$. We have found that the two variances decrease in proportion through the course of evolution under a fixed fitness condition, if there exists a certain level of noise in the development. Robustness evolves, as discussed by Schmalhausen's stabilizing selection [54] and Waddington's canalization [55]. On the other hand, the proportionality between the two variances implies correlation between the developmental and the mutational robustness. Note that this evolution of robustness is possible only under a sufficient level of noise during development. Robustness to noise during development induces robustness to mutation.

Waddington proposed genetic assimilation, in which phenotypic change due to environmental change is later embedded into genes [55] (see also [56–58]). The proportionality among phenotypic plasticity, $V_{ip}$, and $V_g$ is regarded as a quantitative expression of such genetic assimilation, at the level of fluctuations.

Note that the existence of the phenotype–genotype distribution $P(x, a)$ is based on the mutual relationship between the phenotype and the genotype. The genotype determines (probabilistically) the phenotype through developmental dynamics, whereas the phenotype restricts the genotype through a mutation–selection process. For a robust evolutionary process to progress, the consistency between the phenotype and the genotype levels is shaped, which is the origin of the distribution $P(phenotype, genotype)$ assumptions, and the general relationships proposed here.

In the discussion based on the evolutionary stability of $P(x, a)$, we need to consider such a "consistency" [59] principle. We hope that the quantitative formulation of Waddington's ideas on the basis of Einstein's thoughts will provide a coherent understanding of the evolution–development relationship.


**Acknowledgments** The work presented here is based on a collaboration with A. Awazu (Sect. 10.1.2), C. Furusawa (Sects. 10.1.3, 10.2), T. Yomo (Sect. 10.1.3, Sect. 10.2.2), and K. Sato (Sect. 10.2.2). I thank all of them for their collaboration and discussions.

## References


1. K. Kaneko, *Life: An Introduction to Complex Systems Biology* (Springer, Berlin, Heidelberg, 2006)
2. K. Kaneko, Complexity **3**, 53 (1998)
3. K. Kaneko, I. Tsuda, *Complex Systems: Chaos and Beyond – A Constructive Approach with Applications in Life Sciences* (Springer, Berlin, Heidelberg, 2000)
4. J.W. Szostak, D.P. Bartel, P.L. Luisi, Nature **409**, 387 (2001)
5. M. Hanczyc, S.M. Fujikawa, J.W. Szostak, Science **302**, 618 (2003)
6. W. Yu, K. Sato, M. Wakabayashi, T. Nakaishi, E.P. Ko-Mitamura, Y. Shima, I. Urabe, T. Yomo, J. Biosci. Bioeng. **92**, 590 (2001)
7. P.A. Bachmann, P.L. Luisi, J. Lang, Nature **357**, 57 (1992)
8. V. Noireaux, A. Libchaber, Proc. Natl. Acad. Sci. USA **101**, 17669 (2004)
9. M. Eigen, P. Schuster, *The Hypercycle* (Springer, Berlin, Heidelberg, 1979)
10. S.A. Kauffman, *The Origin of Order* (Oxford University Press, Oxford, 1993)
11. K. Kaneko, Adv. Chem. Phys. **130**, 543 (2005)
12. S. Jain, S. Krishna, Proc. Natl. Acad. Sci. USA **99**, 2055 (2002)
13. D. Segre, D. Ben-Eli, D. Lancet, Proc. Natl. Acad. Sci. USA **97**, 4112 (2000)
14. G. Nicolos, I. Prigogine, *Self-Organization in Nonequlibrium Systems* (Wiley, Chichester, 1977)
15. A. Awazu, K. Kaneko, Phys. Rev. Lett. **92**, 258302 (2004)
16. A. Awazu, K. Kaneko, Phys. Rev. E, **80**, 041931 (2009)
17. F. Dyson, *Origins of Life* (Cambridge University Press, Cambridge, 1985)
18. T. Ganti, Biosystems **7**, 189 (1975)
19. C. Furusawa, K. Kaneko, Phys. Rev. Lett. **90**, 088102 (2003)
20. V.A. Kuznetsov, G.D. Knott, R.F. Bonner, Genetics **161** 1321 (2002)
21. H.R. Ueda, S. Hayashi, S. Matsuyama, T. Yomo, S. Hashimoto, S.A. Kay, J.B. Hogenesch, M. Iino, Proc. Natl. Acad. Sci. USA **101**, 3765 (2004)
22. C. Furusawa, T. Suzuki, A. Kashiwagi, T. Yomo, K Kaneko, Biophysics **1**, 25 (2005)
23. S. Krishna, B. Banerjee, T.V. Ramakrishnan, G.V. Shivashankar, Proc. Natl. Acad. Sci. USA **102**, 4771 (2005)
24. K. Kaneko, T. Yomo, J. Theor. Biol. **214**, 563 (2002)
25. J. Maynard-Smith, Nature **280**, 445 (1979)
26. M. Eigen, *Steps Towards Life* (Oxford University Press, Oxford, 1992)
27. M. Boerlijst, P. Hogeweg, Physica D **48**, 17 (1991)
28. S. Altmeyer, J.S. McCaskill, Phys. Rev. Lett. **86**, 5819 (2001)
29. E. Szathmary, J. Theor. Biol. **157**, 383 (1992)
30. A.L. Koch, J. Mol. Evol. **20**, 71 (1984)
31. T. Matsuura, T. Yomo, M. Yamaguchi, N. Shibuya, E.P. Ko-Mitamura, Y. Shima, I. Urabe, Proc. Natl. Acad. Sci. USA **99**, 7514 (2002)
32. M.B. Elowitz, A.J. Levine, E.D. Siggia, P.S. Swain, Science **297**, 1183 (2002)
33. J. Hasty, J. Pradines, M. Dolnik, J.J. Collins, Proc. Natl. Acad. Sci. USA **97**, 2075 (2000)
34. A. Bar-Even, J. Paulsson, N. Maheshri, M. Carmi, E. O'Shea, Y. Pilpel, N. Barkai, Nat. Genet. **38**, 636 (2006)
35. M. Kærn, T.C. Elston, W.J. Blake, J.J. Collins, Nat. Rev. Genet. **6**, 451 (2005)
36. A. Kashiwagi, I. Urabe, K. Kaneko, T. Yomo, PLoS ONE **1**, e49 (2006); doi: 10.1371/journal.pone.0000049
37. C. Furusawa, K. Kaneko, PLoS Comput. Biol. **4**, e3 (2008)
38. K. Kaneko, T. Yomo, J. Theor. Biol. **199**, 243 (1999)
39. A. Wagner, Nat. Genet. **24**, 355 (2000)
40. G.P. Wagner, G. Booth, H. Bagheri-Chaichian, Evolution **51** 329 (1997)
41. N. Barkai, S. Leibler, Nature **387**, 913 (1997)

42. S. Ciliberti, O.C. Martin, A. Wagner, PLOS Comput. Biol. **3**, e15 (2007); doi: 10.1371/journal.pcbi.0030015
43. K. Kaneko, PLoS ONE **2**, e434 (2007); doi: 10.1371/journal.pone.0000434
44. K. Kaneko, C. Furusawa, J. Theor. Biol. **240**, 78 (2006)
45. Y. Ito, H. Toyota, K. Kaneko, T. Yomo, Mol. Syst. Biol. **5**, 264 (2009)
46. K. Sato, Y. Ito, T. Yomo, K. Kaneko, Proc. Natl. Acad. Sci. USA **100**, 14086 (2003)
47. A. Einstein, *Investigation on the Theory of of Brownian Movement* (Collection of papers ed. by R. Furth) (Dover, Mineola, NY, 1956); republication from 1926
48. R. Kubo, M. Toda, N. Hashitsume, *Statistical Physics II: Nonequilibrium Statistical Mechanics*, Springer Ser. Solid-State Sci., Vol. 31, 2nd edn. (Springer, Berlin, Heidelberg, 1991); first published 1985
49. R.A. Fisher, *The Genetical Theory of Natural Selection*, 2nd edn. (Oxford University Press, Oxford, 1958); first published 1930
50. H. Abe, N. Go, Biopolymers **20**, 1013 (1980)
51. J.N. Onuchic, P.G. Wolynes, Z. Luthey-Schulten, N.D. Socci, Proc. Natl. Acad. Sci. USA **92**, 3626 (1995)
52. F. Li, T. Long, Y. Lu, O. Qi, C. Tang, Proc. Natl. Acad. Sci. USA **101**, 4781 (2004)
53. K. Kaneko, J. Biol. Sci. **34**, 529 (2009)
54. I.I. Schmalhausen, *Factors of Evolution: The Theory of Stabilizing Selection* (University of Chicago Press, Chicago, 1949); reprinted 1986
55. C.H. Waddington, *The Strategy of the Genes* (Allen & Unwin, London, 1957)
56. L.W. Ancel, W. Fontana, J. Exp. Zool. **288**, 242 (2002)
57. M.W. Kirschner, J.C. Gerhart, *The Plausibility of Life* (Yale University Press, New Haven, CT, 2005)
58. M.J. West-Eberhard, *Developmental Plasticity and Evolution* (Oxford University Press, Oxford, 2003)
59. K. Kaneko, C. Furusawa, Theory Biosci. **127**, 195 (2008)

# Chapter 11
# Wet Artificial Life: The Construction of Artificial Living Systems

**Harold Fellermann**

**Abstract** The creation of artificial cell-like entities – chemical systems that are able to self-replicate and evolve – requires the integration of containers, metabolism, and information. In this chapter, we present possible candidates for these subsystems and the experimental achievements made toward their replication. The discussion focuses on several suggested designs to create artificial cells from nonliving material that are currently being pursued both experimentally and theoretically in several laboratories around the world. One particular approach toward wet artificial life is presented in detail. Finally, the evolutionary advantage of cellular aggregates over naked replicator systems and the evolutionary potential of the various approaches are discussed. The enormous progress toward man-made artificial cells nourishes the hope that wet artificial life might be achieved within the next several years.

## 11.1 Introduction

The possible creation of life has fascinated mankind throughout history. The questions "what is life?" and "where do we come from?" have driven intellectual curiosity for centuries. In fact, attempts to create artificial life-like entities can be followed back to the very roots of modern science [1]. Whereas historic attempts to create life were limited to just mimicking the phenomenological behavior of living organisms (such as growth or motion), the depth of understanding gained in the life sciences over recent decades now gives rise to endeavors to truly capture living organisms in their functional organization.

As we speak about the creation of artificial living systems, it seems mandatory to give a definition of what we mean by "life". Unfortunately, however, the concept has

H. Fellermann (✉)
Department of Physics and Chemistry, FLinT Center for Living Technology,
University of Southern Denmark, DK-5230 Odense M, Denmark
e-mail: harold@ifk.sdu.dk

H. Meyer-Ortmanns, S. Thurner (eds.), *Principles of Evolution*,
The Frontiers Collection, DOI 10.1007/978-3-642-18137-5_11,
© Springer-Verlag Berlin Heidelberg 2011

turned out to be notoriously difficult to define, and the search for a satisfying definition is still ongoing. Nevertheless, most of the suggested definitions agree that – in order to be called alive – an entity must first be able to self-replicate, that is, to produce a sufficiently similar copy of itself, and second, it must be subject to mutation and selection in order to undergo evolution.

The elementary unit of all modern life is the cell, which couples a container, a metabolism, and inheritable information into an integrated system able to replicate and evolve. The emergence of these first cells was thus a major transition in prebiotic evolution: if there ever was a period in prebiotic history that was dominated by naked molecular replicators – for example self-replicating RNA or DNA strands – then there must have been an evolutionary advantage in incorporating these replicators into a body – an advantage that was significant enough to outweigh the additional complexity needed to replicate not only the RNA/DNA but the whole container–metabolism–information system. The predominance of cellular life and the complete absence of cell-free molecular replicators today might even suggest that there was never a period of naked replicators but that life started immediately from the interactions of container, information, and metabolic molecules.

Therefore, the experimental approaches to creating artificial life discussed in this chapter concern the creation of cell-like entities, commonly referred to as "protocells". In general, there are two possible routes toward such building blocks of artificial life: on the one hand, top-down approaches start from the simplest forms of contemporary natural life and attempt to strip away everything that is not vital for the survival and replication of the original biological cells. The most popular of these top-down approaches is the work by Venter and Smith on the artificial bacterium *Mycoplasma laboratorium*, an organism with merely about 200 genes [2]. Bottom-up approaches, on the other hand, start with the building blocks of life – such as container and information molecules – and try to combine them in ways that support the emergence of cell-like entities from scratch. Bottom-up approaches differ in the way in which they use biochemical molecules and range from the most puristic endeavors that employ only nonbiological molecules to those that make use of ever more biological ingredients, eventually using extracts of natural cells.

The research community that is engaged in the creation of artificial life is truly interdisciplinary and relates to the origin-of-life and astrobiology community concerned with the history of "life as we know it", the computational artificial life community, which is concerned with more general organizing principles of "life as it might be", as well as the more modern research communities of synthetic biology ("life as we employ it") and systems chemistry ("life as we create it"). To actually achieve the creation of artificial life, the research draws from areas as diverse as physics, synthetic chemistry, physical chemistry, biochemistry, applied mathematics, computer science, and the science of complex systems.

In this chapter, we present in some detail the building blocks of artificial protocells (Sect. 11.2), namely chemical information, protocell containers, and metabolisms. In Sect. 11.3, we introduce several historic and contemporary approaches toward artificial cells, thereby focusing on approaches that can be identified as being bottom-up. In Sect. 11.4, we highlight one particular approach – the minimal protocell of Rasmussen and coworkers. Finally, in Sect. 11.5, we discuss the evolutionary potential of protocells compared to naked replicators and to modern biological cells. Unfortunately, the available space is not sufficient to properly attribute the enormous experimental efforts needed to create artificial cells, which underlie all the presented results. The reader is referred to the original literature.

## 11.2 Bits and Pieces

Before we present several bottom-up approaches toward wet artificial life, we set the stage by introducing the information, container, and metabolism subsystems that an artificial organism has to integrate.

### *11.2.1 Chemical Information*

Information refers to any configuration of a system that (i) can be communicated from one system to another and (ii) has an impact on the properties, dynamics, or behavior of the said system. In the context of prebiotic, biological, and artificial life, communication is realized via inheritance of system properties from an ancestor to its offspring, which is achieved via copying, that is, replication of the information carrier. Naturally, the communicated (inherited) information is subject to noise (mutations), which opens the door for evolution to select favorable from unfavorable instances.

In biological systems, information is primarily stored in DNA (deoxyribonucleic acid), where it affects the behavior of the cell via protein synthesis and subsequent metabolic regulation. DNA and its cousin RNA (ribonucleic acid) are linear heterogeneous polymers of pairwise complementary nucleotides that are able to align with each other by specific and energetically weak Watson–Crick binding. This structure immediately suggests the replication mechanism that is found in all contemporary cells [3]. Nevertheless, even in the most simple prokaryotes, biological DNA replication is controlled by enzymes at almost every step. A significantly less complex, nonenzymatic replication mechanism – possibly based on external factors – must have preceded contemporary replication [4]. Likewise, in a purely bottom-up approach to wet artificial life, it is desirable to employ a nonenzymatic replication mechanism.

Basically, the replication of biopolymers requires the processes of hybridization, ligation, and melting. Hybridization is the alignment of complementary nucleotides or oligomers along a template strand. Ligation is the forming of covalent bonds between the aligned nucleotides. Finally, melting separates the double strand into the original and its complementary copy. The melting temperature increases with strand length and fraction of C-G pairs. As the ligation reaction is energetically uphill, monomers or oligomers need to be activated in order to ligate spontaneously.

Nonenzymatic template-directed RNA polymerization of elementary activated nucleotides (monomers) shows little yield in aqueous solution. Template-directed

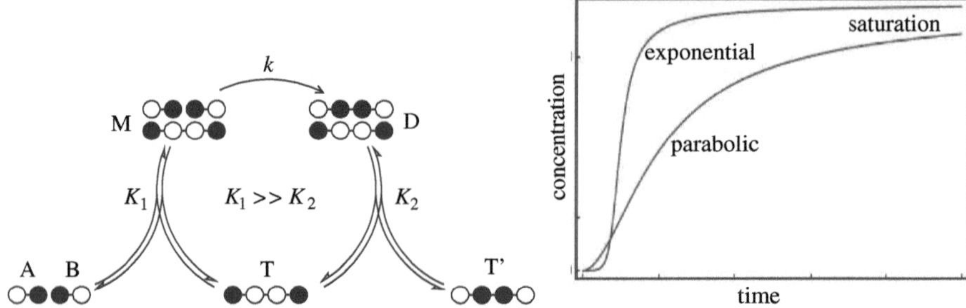



**Fig. 11.1** *Left*: Basic template-directed replication: an RNA or DNA strand T serves as a template. Complementary shorter strands A and B hybridize with the template and form the double strand M. The double strand configuration fixates A and B in a vicinity that promotes their chemical ligation. Once the resulting double strand complex D separates, a new copy of the template T is formed. ($K_1$ and $K_2$ are the dehybridization/hybridization equilibria for oligomers and template.) *Right*: Parabolic versus exponential growth

replication from shorter activated oligomers, on the other hand, can be achieved with high yields for both RNA and DNA [5, 6]. The shortest DNA strand that has been experimentally replicated in the absence of enzymes is a hexamer with complementary trimers [7]. The basic mechanism of this *minimal replicator* is schematically shown in Fig. 11.1.

Each cycle of the template-directed replication reaction doubles the number of templates present in solution. From this, one might expect that the replicator templates grow exponentially in time. However, von Kiedrowski and coworkers have demonstrated that the template species grows more slowly than exponential, and instead follows a parabolic growth law ($d[T]/dt \propto \sqrt{[T]}$) [8]. The reason for this counter-intuitive behavior is so-called *product inhibition*: as the hybridization tendency of complementary strands increases with the length of the strands, it is more likely for two template strands T to form a double strand configuration D than it is for one template T and two oligomers A and B to form the complex M. Thus, for a temperature regime in which the formation of M is possible, most of the template strands will be in the configuration D, where they are inaccessible for the reactants A and B. In other words, the very product of the reaction (T) inhibits its catalyst (also T), thereby reducing the reaction rate.

Interestingly, competition between replicator species that obey such a parabolic growth law are expected to support coexistence of all species, rather than the survival of only the fittest (fastest replicating) species [9]. This, in turn, prevents true Darwinian evolution, which is essentially based on survival of the fittest (and only the fittest). We can conclude that naked replicators based on complementary RNA or DNA strands do not suffice to support Darwinian evolution. Toward the end of this chapter, we will show how the incorporation of information polymers into replicating containers can recover exponential growth. Before that, however, we will discuss the general properties of protocellular containers and metabolisms and introduce current approaches to their realization.

## 11.2.2 Protocell Containers

By introducing containers, we advance from well-mixed homogeneous solutions of (possibly self-replicating) molecular species toward spatially organized chemical aggregates. By embodying the cell, containers introduce a separation of the system from its environment, which allows us to recognize the embodied cell as a specific, spatially confined entity.

These rather general statements immediately imply functionalities that protocell containers must provide: they must be spatially confined structures that are stable enough to maintain the integrity/identity of the cell over time, while being dynamic enough to allow for replication of the entire aggregate (container plus contents). Containers must impose a diffusion barrier to contain functional molecules of the protocell in its interior, while simultaneously being permeable to nutrient molecules and waste.

Premier molecules of biological containers (such as cells and organelles) are lipids. Lipids are a class of chemical components that comprise many different molecular species. Their unifying feature is their amphiphilic character: a lipid molecule is composed of two parts, one being soluble in water (hydrophilic), the other one being soluble in oil (lipophilic or hydrophobic). In aqueous solution, lipid molecules self-assemble into supramolecular aggregates (see Fig. 11.2) in which the hydrophobic part of the molecules is shielded from the aqueous environment.

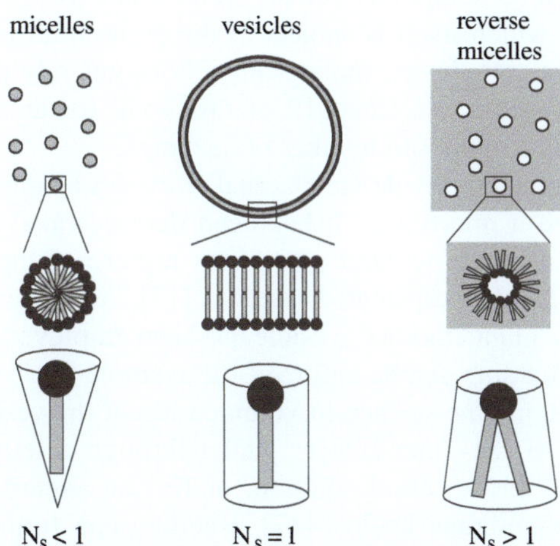


**Fig. 11.2** Typical amphiphile aggregates: micelles (*left*), vesicles (*center*), and reverse micelles (*right*). The *top row* depicts the respective phase diagrams, consisting of water (*white*), hydrophobic molecules, and hydrophobic sections of amphiphilic molecules (*light gray*), and hydrophilic sections of amphiphiles (*dark gray*). The *center row* shows the detailed molecular arrangement, and the *bottom row* shows the geometry of amphiphiles that support each structure formation. The molecular geometry can be roughly described by the molecular *packing parameter* $N_S = V/A_0 l$, which relates the molecular volume $V$, tail length $l$ and effective surface area $A_0$

The membranes of contemporary cells exhibit a complex composition of lipids and proteins, which is adapted to the specific environment and function of the cell. Contemporary protocell designs, in contrast, are based on much simpler surfactant compositions and hardly ever employ proteins [10–12]. In particular, phospholipids, fatty acids, and amphiphilic alcohols are well known to form supramolecular structures by spontaneous self-assembly in aqueous solution – micelles and vesicles are their most prominent examples. The phase diagram of these aggregates is subject to a variety of molecular and systemic parameters: whether an amphiphile solution forms micelles, vesicles, or other (less prominent or less defined) structures is influenced by the length and possible branching of the hydrocarbon chain, the characteristics of its head group, its $pK$ value, pH, temperature, and other systemic parameters. Ternary mixtures of oil, surfactant, and water exhibit an even richer phase behavior. Notably, fatty acid surfactants can stabilize otherwise unstable oil–water emulsions, giving rise to surfactant-coated oil droplets known as microemulsion compartments. Unfortunately, the subject of soft condensed matter systems is too broad to allow for a concise overview. The reader is referred to the standard literature. In the context of protocell research, micelles [13], reverse micelles [14], surfactant-coated oil droplets [15], and, prominently, vesicles [11, 16] have been suggested as container candidates for protocells.

Division of vesicles requires bending of the bilayer membrane to form a bud small enough for lipids of the adjacent bilayer sheets to rearrange. For phospholipid (and likely also fatty acid) membranes, the bending energy of the membrane imposes an energy barrier that is unlikely overcome by thermal motion. For this reason, the division of contemporary cells is orchestrated by the complex machinery of the cytoskeleton, which itself is subject to the proteomics of the cell cycle [3]. In order to avoid this complexity, most protocell designs rely on partially external means of vesicle division. See Chap. 12 of this book for an approach to vesicle division that employs a minimum number of enzymes.

Vesicles can be forced to divide via external work, for example by pressing vesicle solutions through a microfilter. It has been demonstrated that such *extrusion* can force vesicles to divide without significant rapture of the membrane, which would lead to leakage of encapsulated material [17]. Alternatively, budding off of small vesicles from a giant "mother" vesicle has been employed as a mechanism for vesicular division. Budding can be enforced for example by (i) mechanical energy, (ii) osmotically changing the surface-to-volume ratio of the vesicle, (iii) selectively increasing the area of the outer bilayer leaflet through a temperature difference between the internal and external solution, or (iv) an asymmetry in the density of inner and outer membrane leaflets [17]. Budding can further be supported by the boundaries of domain-forming lipid compositions [18]. Theoretical studies have suggested additional means of induced vesicle division based on adhesive nanoparticles [19] or heterogeneous osmotic pressure [20, 21].

For micelles and reverse micelles, an autonomous replication process has been suggested in the literature [13, 14, 22]. Bachmann et al. first reported self-replication of reverse micelles that hosts a catalyst for a metabolic reaction that constitutes an autonomous growth and division cycle: a hydrophobic ester is added to the organic solvent of the system and serves as a nutrient for the single metabolic reaction of cleaving the ester bond by hydrolysis. Reaction products are fatty acids and alcohols – which are essentially the surfactants of the ternary system. Ester cleavage is enhanced by a hydrophilic catalyst, which will reside in the aqueous interior of the reverse micelle. The setup guarantees that the metabolic turnover of nutrients occurs at the micellar interface.

The supposed replication process in these systems is best described for the original setup: surfactants that are newly produced by ester hydrolysis arrange at the lipid–water interface of the reverse micelles as a result of their amphiphilic properties. During the course of the reaction, this leads to a change in the surface-to-volume ratio of these aggregates, as the water is entrapped in the interior of the reverse micelles. It is thought that this induces an elongation of the structure up to a point where thermal fluctuation suffices to divide the aggregate in two.


### 11.2.3 Protocell Metabolisms

In the context of artificial life, a metabolism is any network of chemical reactions that allows a protocell to produce its building blocks from provided nutrients and energy. As any chemical reaction, metabolic processes must be designed energetically downhill (exergonic) or, if uphill, must be coupled to downhill reactions such that the net free energy change in the system is negative and products are formed with high yield.

One way to achieve this is to provide energy-rich precursors as nutrients that are broken down in the course of metabolic reactions (catabolism). An example is activated derivatives of nucleotides instead of plain nucleotides. This changes the equilibrium constant of a polymerization/ligation reaction toward the product side.

Another way is to exploit external free energy sources such as sunlight or electrochemical gradients to build up energy-rich chemicals (anabolism). The build-up of energy-rich components is significantly more difficult than their breakdown, which is why most protocell design pursued today relies on energy-rich precursors. Biological life is able to produce energy-rich chemicals such as adenosine triphosphate (ATP) in order to store energy for later use, which accounts for the autonomy of the living organism. None of the protocell designs we have today comes close to such functionality.

For a truly replicating protocell, the metabolism must not only produce all building blocks of the container and information subsystems, but also provide its own building blocks, that is, the reaction network must be autocatalytic. Autocatalytic closure – the appearance of catalytic cycles – has been shown to arise spontaneously in random catalytic networks once a threshold connectivity of the network is reached [23–25]. While these studies are particularly interesting within the context of origins-of-life research, they give little or no advice for the purposeful design of a protocellular metabolism from scratch. More closely related to experimental work is the observation that the citric acid cycle – the hub of biological reaction networks and also known as the Krebs cycle – can be made into an autocatalytic reaction network when run in its reductive (reverse) direction [26]. Experimental work on this system, for example its encapsulation into vesicles, is still to come.

## 11.3 Bottom-Up Approaches to Artificial Cells

Pioneering attempts to conceptually understand living organisms from the general perspective of systems theory date back to the early 1970s, when Maturana and Varela introduced the concept of *autopoiesis* [27, 28]. An autopoietic system is a spatially confined network of production processes of components that continuously regenerate and realize the network that produces them. This perception of life prescinds the self-generative power of living systems from its actual chemical materialization, and thus provides a framework in which it is possible to sketch out nonbiological living systems.

In order to construct a self-replicating protocell, container, metabolism, and information have to be arranged into a functionally coupled system with at least the following interactions: the information component has to influence the metabolic activity of the system in one way or another; the metabolism has to provide building blocks for all subsystems; the replication rates of the subsystems have to be synchronized in order for the entire aggregate to self-replicate.

In 1971, Gánti designed the first such integrated system, which he referred to as the *chemoton* (short for "chemical automaton") [29]. In its current conception, a replicable biopolymer resides in a lipid container (vesicle), where it replicates by consuming monomer material produced by an autocatalytic metabolism that also produces membrane material. The chemoton employs five metabolic reagents that constitute an autocatalytic cycle, one metabolic precursor for membrane molecules and one for monomers of the genetic systems. With additional nutrients and waste components, the chemoton adds up to 12 components that interact through ten chemical reactions. The subsystems are stoichiometrically coupled. In particular, the metabolic production of genetic, metabolic, and container building blocks are controlled by a differential feedback mechanism that maintains a synchronized growth of the entire system. Mass reaction kinetic studies indicate that the chemoton indeed reproduces [30] but the models assume that, for example, template replication and container division can be achieved without complications. To the best of our knowledge, little experimental work has been done on the chemical realization of the chemoton.

The last decade has seen several proposals for the experimental bottom-up realization of artificial cells and cell-like entities. We illustrate this variety with three proposed integrated designs that differ significantly in their complexity and in the way they treat container, metabolism, and chemical information: the protocells suggested and currently experimentally pursued by the groups of Rasmussen et al., Szostak et al., and Luisi et al. (see Fig. 11.3). (For a concise review of the field of protocell research, see [31])

The *minimal protocell* design of Rasmussen et al. is arguably the simplest proposal for self-replicating, evolving matter to date. The model gains its simplicity

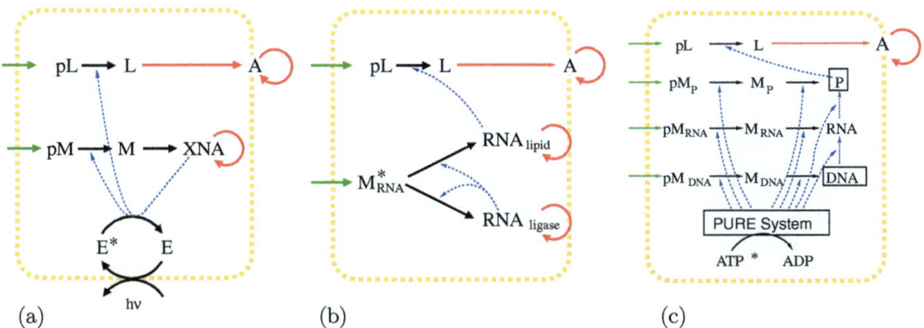

**Fig. 11.3** Currently experimentally pursued bottom-up wet artificial life designs in the schematic language of Rasmussen et al. [32]. **a** The minimal protocell of Rasmussen et al. [33] envisions a catalytic coupling between information and metabolism, **b** Szostak et al.'s (2001) protocell features ribozymatic information molecules, and **c** Luisi et al.'s (2006) semiartificial cell employs a protein kit for the regulation of metabolic activities. Here, L and A denote lipid molecules and the entire aggregate, M and I denote information monomers and templates, and E denotes metabolic molecules. *Black arrows* depict chemical reactions, *red arrows* self-assembly processes, and *dotted blue arrows* catalysis. © MIT Press 2009

through several "unorthodox" simplifications that will be presented in depth in Sect. 11.4. Drawing on simplicity is not a shortcoming but a declared goal of the design, which focuses on the transition from nonliving to living matter "from scratch", that is, from only molecules of nonbiological origin.

Szostak et al. base their protocell on RNA sequences that exhibit catalytic activity – so-called *ribozymes* [34]. Ribozymes are single-stranded RNA strands with the ability to fold into complex three-dimensional structures that exhibit catalytic activities similar to proteins (hence the name ribozyme following the word "enzyme"). The design of Szostak et al. envisions employing two ribozymes encapsulated in semipermeable vesicles: one ribozyme catalyzes the metabolic turnover of lipid precursors into functional lipids to allow for growth of the protocell container, while the second ribozyme supposedly catalyzes the template-directed replication of RNA from shorter oligomers or monomers and should thus allow for the replication of both ribozymes. The hypothetical design is backed up by research on ribozymatic ligases and polymerases [35, 36]. In Szostak et al's protocell, the container membrane has to be permeable to lipid precursors and RNA oligomers while being impermeable to the ribozymes themselves – a property that has been shown experimentally by Mansy et al. [37].

Where the above bottom-up approaches employ only ingredients of nonbiological origin, the protocell design of Luisi et al. hinges on components of biological origin, and can therefore be regarded as a *semiartificial cell*. The approach employs significantly more components and more closely resembles life as we know it today: the team envisions encapsulating the minimal biological proteins necessary for DNA replication, transcription, translation, and protein synthesis into a vesicle. This protein set – characteristically referred to as the PURE system – would thus be able to reproduce itself from nucleotides, amino acids, lipid precursors, ATP and other

**Table 11.1** Recent experimental achievements in encapsulated DNA/RNA replication and protein synthesis in vesicles. Asterisks indicate experiments that employ cell extract

| Year | Achievement | Fig. 11.4 | Refs. |
|------|-------------|-----------|-------|
| 1994 | Template-free enzymatic RNA polymerization | (a) | [39] |
| 1995 | Enzymatic RNA replication | | [40] |
| 1995 | Enzymatic polymerase chain reaction | | [41] |
| 2001 | Transcription and translation * | (c) | [42] |
| 2002 | Enzymatic transcription | (b) | [16] |
| 2004 | Protein expression * | (d) | [43] |
| 2004 | Cascading genetic network transcription * | (e) | [44] |

biomolecules [38]. Although this design is arguably the most complex approach in terms of number of molecular components, the research benefits from the fast accumulation of knowledge in the fields of molecular and synthetic biology. Some recent milestones of vesicle-encapsulated DNA-replication and protein synthesis that use either enzymes or cell extract are summarized in Table 11.1.

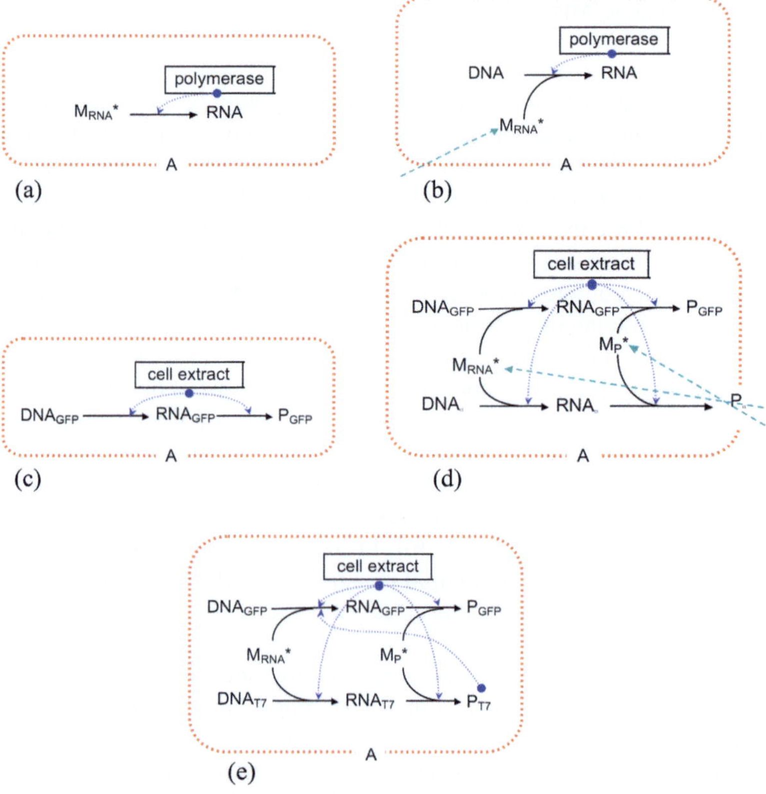


**Fig. 11.4** Several experimental milestones of encapsulated DNA replication and protein expression in vesicles. Schematics are based one the graphical language of Rasmussen et al. [32]. See Fig. 11.3 for the interpretation of the diagrams. © MIT Press 2009. **a** Chakrabarti et al. [39], **b** Monnard and Deamer [16], **c** Yu et al. [42], **d** Noireaux and Libchaber [43], **e** Ishikawa et al. [44]

With the complexity of Luisi's envisioned semiartificial cell, the bottom-up approach stretches toward the top-down research on minimal cells that trims down natural genomes in the quest for the minimal gene set needed for self-replication. However, while the two research communities meet conceptually, they are still separated by at least one order of magnitude (expressed in terms of the total number of genes). Schematics of these achievements are shown in Fig. 11.4.

## 11.4 The Minimal Protocell

In this section, we detail the design and current work on the minimal protocell by Rasmussen et al. [33]. This work is currently being pursued in a combined experimental and theoretical approach at the University of Southern Denmark and the Los Alamos National Laboratory, USA. Theoretical investigations and computer simulations constitute a major part of the protocell assembly project, where they are mostly used (i) to explore design alternatives, (ii) to interpret experimental results, and (iii) to relate them to well-developed theories of, for example, soft condensed matter or minimal replicator systems.

### *11.4.1 Design Principles*

The minimal protocell features three major simplifications to reduce the number and complexity of needed components. First, the minimal protocell design places the functional molecules for information and metabolism at the exterior interface of a lipid container, rather than encapsulating them in the interior volume of a vesicle. Second, rather than encoding enzymes, the information component of the minimal protocell directly affects the metabolism via its electrochemical properties. Third, a unique reaction mechanism is designed to produce all building blocks from appropriate precursors.

Contemporary life and most artificial protocell designs encapsulate their metabolic and information components in the interior of a vesicular container. In contrast, our minimal protocell takes a radically different approach by placing the functional molecules at the *exterior* interface of a lipid container. In order to achieve this, the molecules are decorated with hydrocarbon chains that give them the properties of an amphiphile.

Attachment instead of encapsulation allows not only vesicles but also a variety of other container candidates to be considered, namely oil-in-water emulsion compartments (surfactant-coated oil droplets), water-in-oil emulsions, and reverse micelles. In addition, resource and waste molecules do not need to pass a membranous barrier but can freely diffuse toward and away from the interface. This nullifies the need for pore proteins or other sophisticated regulation mechanisms for substrate uptake and disposal. Because the exterior interface is a free-energy sink for the functional molecules, the simple "sticky gum" design can spontaneously self-assemble in solution.

### *11.4.2 Building Blocks*

Our protocell design is based on three molecular species to realize the three subsystems container, metabolism, and information, plus three molecular species that act as nutrients (some of them are shown in Fig. 11.5). The container of the protocell is formed by fatty acids (a simple single-chained surfactant), the metabolism is carried out by ruthenium bipyridine (a commercial pigment), and the inheritable information is represented in a short nucleic acid strand, most likely DNA. Nucleic and fatty acid esters (namely picolyl esters) are nutrients of the metabolism. The metabolism also consumes a hydrogen donor, which can be regarded as a third nutrient. The detailed metabolic reaction mechanism will be presented in Sect. 11.4.4.

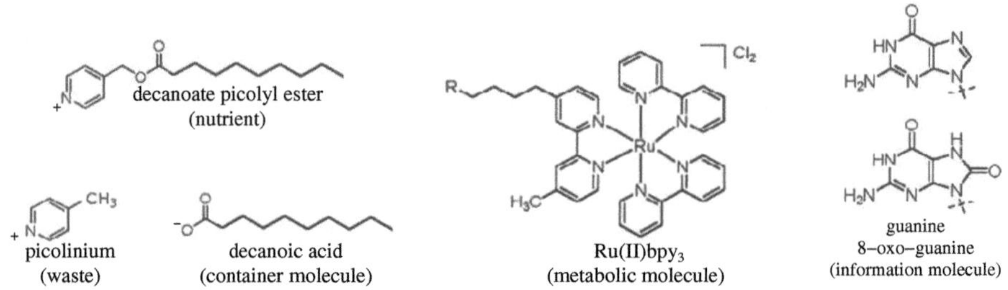

**Fig. 11.5** Components of the minimal protocell. The container of the protocell is formed by decanoic acid and decanoate picolyl ester. Ruthenium(II) bipyridine drives the metabolic turnover of nutrients into functional building blocks. The inheritable information of the protocell is realized in DNA, where 8-oxo-guanine, a derivative of the natural nucleobase guanine, is the main component

### *11.4.3 Life Cycle of the Protocell*

The life cycle of the envisioned protocell is schematically shown in Fig. 11.6, here for a microemulsion container and the photosensitizer covalently bound to the information polymer. The life cycle starts with the spontaneous self-assembly of a microemulsion compartment in aqueous solution – that is, a fatty acid ester droplet (yellow) that is coated by fatty acid surfactants (green). The informational polymer (black and white circles) and photosensitizer (red circle) attach to the water–lipid interface due to their amphiphilic properties (a). Nucleotide double strands are melted on the container surface (b) to allow for the hybridization of piecewise complementary oligomers after the system is cooled below the melting temperature again (c). Exposing the system to light drives the metabolic turnover of fatty acid ester molecules into functional fatty acids, and likewise the ligation reaction of the information oligomers (d). Finally, due to the ongoing activity of the metabolism, the container is envisioned to spontaneously divide into daughter aggregates, each one equipped with a copy of the necessary components (e). Feeding new nutrients to the system completes the life cycle of the minimal protocell.

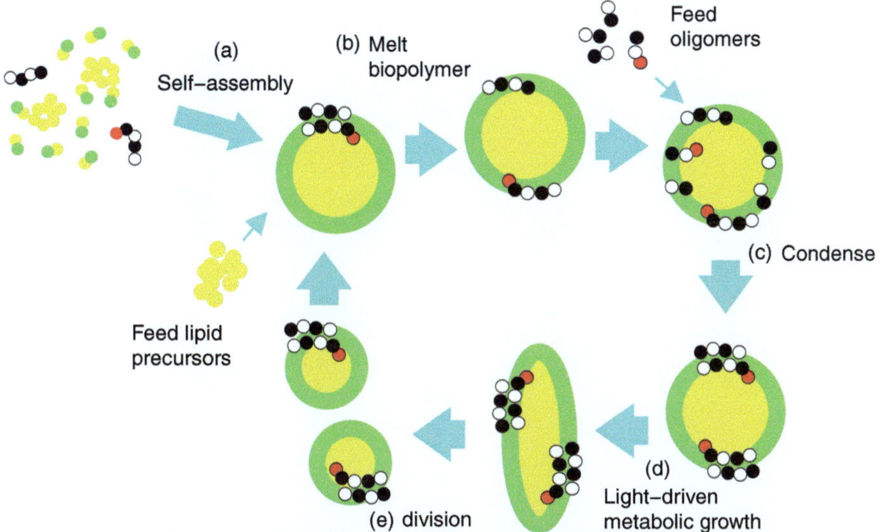



**Fig. 11.6** Life cycle of the minimal protocell (in this case based on a surfactant-coated oil droplet container). **a** Spontaneous self-assembly of the protocells from its components in aqueous solution. Melting (**b**) and condensation (**c**) of information polymers initiates the replication of information. **d** A light-driven metabolism transforms nutrients into new surfactants and forms covalent bonds between the supplied information precursors. **e** The metabolic turnover induces autonomous division and leads to two copies of the original aggregate

In the above setup, the container, information molecule, and metabolism are coupled in various ways. Obviously, both the replication of the container and replication of the genome depend on a functioning metabolism, as the latter provides building blocks for aggregate growth and reproduction. In addition to that, the container also has a catalytic influence on the replication of both the metabolic elements and the genome by co-localizing precursors, photosensitizers, and nucleic acids. Finally, the nucleic acid catalyzes the metabolism, as will be discussed in the next section.

In order to check the performance of the entire envisioned system and the coupling of its components, we have performed spatially resolved, physically grounded simulations in an extended dissipative particle dynamics (DPD) framework [45–47]. DPD is a particle-based mesoscopic simulation method that employs Newton's laws of motion to determine the trajectory of so-called "beads", which represent either individual molecules or volume elements of bulk fluids. Essentially, DPD is a Lagrangian solver of the Navier–Stokes equation that incorporates thermal fluctuations – as such the method can capture the thermodynamic and hydrodynamic properties of complex fluids, oil–water and ternary emulsions being two of them. In contrast to detailed atomistic molecular dynamics simulations, DPD allows system trajectories up to the range of microseconds to be computed on a desktop computer.

We have successfully used the DPD framework to model every step in the life cycle of the protocell, including its self-assembly, nutrient feeding, hybridization, and ligation of a single surface-attached nucleotide strand, as well as a spontaneous

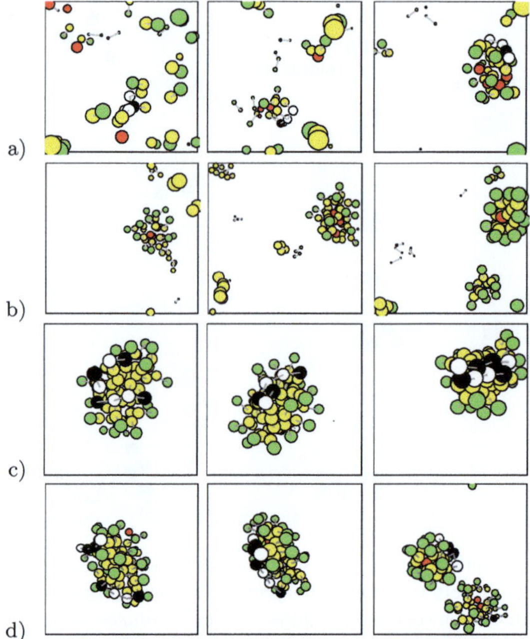

**Fig. 11.7** Critical steps in the life cycle of the minimal protocell. From *top* to *bottom*: **a** spontaneous self-assembly in aqueous solution; **b** precursor feeding, turnover, and induced container replication; **c** interface-associated information replication; and **d** replication of the entire aggregate [46]

division process of the microemulsion compartment (see Fig. 11.7) similar to the experimental setup of Bachman et al. In general, we were able to confirm the viability of every step in the envisioned life cycle or suggest design alternatives. It should be pointed out, however, that the main obstacle to completing a full life cycle in one integrated simulation is the successful replication of information polymers, once more than one copy of the information molecule is present on the container surface. As has been discussed in Sect. 11.2.1, the templates have a significantly stronger tendency to bind to each other than to the complementary oligomers. Naturally, these findings are on a systemic level – simulation results must not be taken as predictive and are constantly compared to the results of our experimental work.

### 11.4.4 The Metabolism of the Protocell

At the heart of the minimal protocell design lies a complex chemical reaction that couples the three subsystems of container, metabolism, and information. The overall reaction is a catabolic breakdown of energy-rich precursors, where the net products are lipid molecules L (fatty acids) and DNA oligomers that, in a second step, undergo template-directed ligation. In the current design of the protocell, the photosensitizer must be provided as a nutrient and does not undergo any metabolic net transformation. Although the overall reaction is catabolic, one step in the reaction mechanism requires the intake of energy (namely light energy).

The net reactants are picolyl esters of the respective acids, which are cleaved into radicals with the aid of an electron. The electron is eventually donated by a hydrogen source, but the electron transfer occurs over two cycles of coupled redox reactions. The downstream cycle employs a photoreaction where the photosensitizer ruthenium(II) trisbipyridine gets excited by visible light – turning it into an electron acceptor for an appropriate reductant. The excited electron – which cannot jump back into its ground state once an electron has been received – now participates in the ester cleavage. The upstream cycle couples the metabolic production to the information stored in the DNA biopolymer: a derivative of the biological nucleotide guanine (8-oxo-guanine) has the correct redox potential to donate an electron to the ruthenium complex in its excited state. This nucleobase is eventually restored by accepting an electron from the hydrogen source – which completes the second redox cycle.

Several undesired side reactions occur in this system:

i. Fluorescence of the ruthenium complex leads back to the ground state – only one of $n$ photons is actually used to drive precursor fragmentation.
ii. Hydrolysis of the ester produces fatty and nucleic acids without participation of the complex reaction.
iii. The photocatalyst slowly degrades and reduces the overall system performance.

Figure 11.8 shows an experimental confirmation of the above mechanism: the graphs show the production of fatty acids from fatty acid esters in the presence

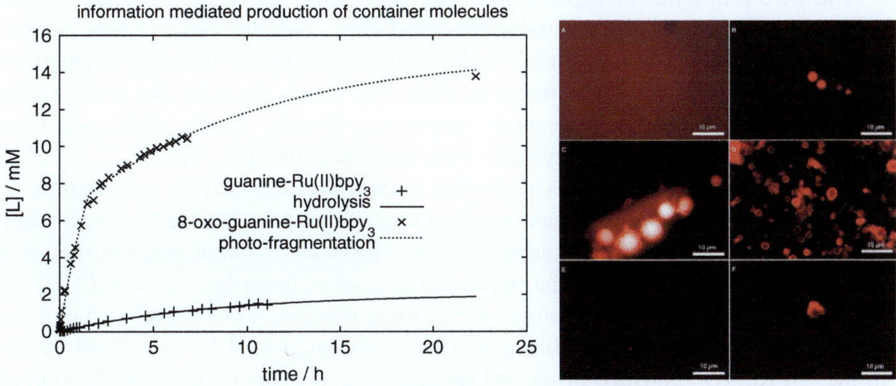


**Fig. 11.8** *Left*: Experimental measurements and fitted reaction kinetics models of the production of fatty acid in the presence of guanine (+ signs and 8-oxo-guanine (x signs). The initial rate of the reaction changes by a factor of more than 25 depending on the presence of absence of oxo-guanine. See text for details. *Right*: Micrographs of the reaction. The *upper four panels* show the conversion in the case of 8-oxo-guanine at times 0, 6, 8, and 24 h. The *lower two panels* show the absence of conversion in the case of guanine after 0 and 24 h. The visible spot is phase-separated precursor that has not been converted into vesicles. Bars correspond to 10 µ m. © American Chemical Society 2009

of the ruthenium complex and either (I) guanine or (II) oxo-guanine [48]. The graph further shows fits to a pseudo-first-order reaction pL $\longrightarrow$ L, which assumes that the electron relay system is a catalyst with kinetics on a faster time scale. In the absence of oxo-guanine (+ signs and solid line), the entire lipid production is due to background hydrolysis (which proceeds with an estimated initial rate of $6.7 \times 10^{-8}$ Ms$^{-1}$). In the presence of oxo-guanine (x signs and dashed line), the total turnover initially proceeds more than 25 times faster. After about 1.6 h, however, the turnover rate decreases by a factor of about five, which might be caused by the appearance of vesicles in the solution. We are currently comparing these experiments to a model of the reaction mechanism. Qualitative analysis of a similar reaction system exposed different kinetic regimes, where only some of them are susceptible to evolution [49].

### 11.4.5 Toward Inheritable Information and Darwinian Evolution

The design of the minimal protocell envisions incorporating metabolically active nucleotides such as 8-oxo-guanine into short DNA strands (5–20 bases in length) and providing complementary oligomers to replicate the information content as described in Sect. 11.2.1. To run the ligation energetically downhill, one of the two oligomers is activated, whereas the other is decorated with a protection group to prevent uncontrolled ligation. This protection group is identical to the one used in the lipid metabolism (picolyl ester), which should allow the replication of information to be controlled with the light-driven reaction system described above. Our hypothesis is that the catalytic activity of oxo-guanine will depend on the nucleotide sequence that the electron has to travel to reach the photosensitizer [50].

The wire properties of DNA are not the only place where variation, and hence selection, enters. In the DPD model presented above, we can observe that the tendency of nucleotides to form stable hybridization complexes critically depends on their base sequence. This constitutes an additional fitness function that arises solely from the geometry of the molecular interactions. Similar results are known from experimental studies of nonenzymatic template replication [51, 52].

Wet-lab experiments on this replicator system are currently in preparation. In parallel, we are developing a mass reaction kinetics model of the replicator system. Unsurprisingly, the theoretical analysis of this replicator system predicts the same issue of product inhibition as the unprotected systems described in Sect. 11.2.1. In order to overcome product inhibition, the local template concentration has to be kept constant. One way to achieve this has been suggested by von Kiedrowski: In the "SPREAD" (surface-promoted replication and exponential amplification of DNA) approach, template molecules are attached to a matrix. The resulting spatial separation prevents the templates from interacting and inhibiting each other [53], leaving all strands available for interaction with oligomers.

Rasmussen and coworkers [54, 55] derive, by means of mass reaction kinetics, that a mechanism similar to SPREAD naturally occurs in the design of the minimal protocell: as the information molecules are attached to the container surface by hydrophobic anchors, division of the protocell container prevents templates from inhibiting each other. In particular, they show that the growth rates of container and information molecules synchronize in the minimal protocell design, thus keeping the local concentration of templates per aggregate constant. This result has since been generalized and shown to hold under very modest conditions, as long as the production rate of container molecules is proportional to the concentration of information templates [56].

Returning to the question posed in the introduction, we can now state that one reason for the predominance of cellular organisms over naked replicators might be found in the enhanced replication abilities of integrated information–metabolism–container systems that are able to actively overcome the limitation of product inhibition and reconstitute exponential growth (thereby overthrowing any competing naked replicators).

## 11.5  The Evolutionary Potential of Protocells

The presented designs differ significantly in the way they use inheritable chemical information in order to control the metabolism of the aggregate (Table 11.2). The minimal protocell exploits the redox potential of nucleobases and the wire property of biopolymers to regulate the turnover rate of a specific metabolic reaction. The ribozymatic protocell of Szostak et al. utilizes the property of RNA to form ternary structures and their possible catalytic activities. Finally, the semiartificial cell of Luisi et al. is based on the very transcription/translation mechanism and subsequent enzymatic regulation that is characteristic of biological life today. With the growing complexity of this mapping from chemical information to metabolic regulation comes an increase in flexibility that is likely to determine the evolutionary potential of each design. Hypothetically assuming that we had experimentally achieved the above designs and were able to replicate and select protocell populations over hundreds or thousands of life cycles, what evolutionary outcome could we hope to observe?

**Table 11.2** Feature comparison of the presented protocell designs in order of their complexity

|  | Metabolism | Division | Information | Evolvability |
| --- | --- | --- | --- | --- |
| Minimal protocell (Rasmussen et al.) | Mainly catabolic | Autonomous or externally driven | Redox-catalytic | Adaptation |
| Ribozyme cell (Szostak et al.) | Anabolic from activated precursors | Externally driven | Ribozymatic | Limited construction |
| Semi-artificial cell (Luisi et al.) | Anabolic from activated precursors | Autonomous or externally driven | Encoding | Universal construction |
| Modern cell | Autocatalytic ana- and catabolism | Autonomous, enzymatic | Encoding | Universal construction |

If the information carrier participates directly in one particular metabolic reaction – as in the case of the minimal protocell – "phenotypic" variability is limited to the rate of that particular reaction. Consequently, evolution is limited to adaptation toward an optimum, for example toward the most effective catalyst. One could imagine adding different photosystems to Rasmussen et al.'s design, such that adaptation becomes multidimensional, but we would still not expect the evolution of this system to generate genuine novelty.

In the ribozymatic protocell, the information carrier also participates directly in the metabolism. However, as ribozymes can fold into versatile shapes with a high possible catalytic specificity, the evolutionary potential appears greater, as information molecules can presumably evolve to promote unforeseen metabolic reaction types, thereby introducing a source for novelty to emerge. When metabolic regulation is based on enzymes – as in the semiartificial protocell – the class of accessible reactions is again expected to increase.

The main difficulty of a ribozymatic information system is that a single molecular species (RNA) needs to perform well both as an information carrier system and as an actuator. This in turn limits the evolutionary potential to only those actuators that can also perform well as templates in the replication process. With the introduction of *encoding information* in the semiartificial cell as well as in all modern life, the storage of information (genome) is physically separated from its action (proteome). In this way, the ability of the information carrier to replicate becomes independent of the information that it actually stores.

**Acknowledgements** The author thanks Steen Rasmussen, Pierre-Alain Monnard, Goran Goranovič, James Boncella, Hans-Joachim Ziock, the members of the FLint Center for Fundamental Living Technology, and the Protocell Assembly team of the Los Alamos National Laboratory for useful discussions.

# References


1. M. Hanczyc, in *Protocells: Bridging Nonliving and Living Matter*, ed. by S. Rasmussen, M. Bedau, L. Chen, D. Deamer, D. Krakauer, N. Packard, P. Stadler (MIT Press, Cambridge, MA, 2008), pp. 3–18
2. J.I. Glass, N. Assad-Garcia, N. Alperovich, S. Yooseph, M.R. Lewis, M. Maruf, C.A. Hutchison III, H.O. Smith, J.C. Venter, Proc. Natl. Acad. Sci. USA **103**(2), 425 (2006)
3. B. Alberts, A. Johnson, J. Lewis, M. Raff, K. Roberts, P. Watson, *Molecular Biology of the Cell* (Garland Science, New York, NY, 2002)
4. P.A. Monnard, in *Prebiotic Evolution and Astrobiology*, ed. by J.T.F. Wong, A. Lazcano (Landes Bioscience, Austin, TX, 2008)
5. D. Sievers, G. von Kiedrowski, Nature **369**, 221 (1994)
6. B.G. Bag, G. von Kiedrowski, Pure Appl. Chem. **68**(11), 2145 (1996)
7. G. von Kiedrowski, Angew. Chem. Int. Ed. **25**(10), 932 (1986)
8. G. von Kiedrowski, B. Wlotzka, J. Helbing, M. Matzen, S. Jordan, Angew. Chem. Int. Ed. **30**(4), 423 (1991)
9. I. Scheuring, E. Száthmáry, J. Theor. Biol. **212**, 99 (2001)
10. P.L. Luisi, P. Waldea, T. Oberholzer, Curr. Opin. Colloid Interface Sci. **4**(1), 33 (1999)

11. D. Deamer, J.P. Dworkin, S.A. Sandford, M.P. Bernstein, L.J. Allamandola, Astrobiology **2**(4) (2002)
12. S.S. Mansy, Int. J. Mol. Sci. **10**, 835 (2009)
13. P.A. Bachmann, P.L. Luisi, J. Lang, J. Am. Chem. Soc. **113**, 8204 (1991)
14. P.A. Bachmann, P. Walde, P.L. Luisi, J. Lang, J. Am. Chem. Soc. **112**, 8200 (1990)
15. K. Suzuki, T. Ikegami, Artif. Life **15**(1), 59 (2009)
16. P.A. Monnard, D. Deamer, Anatom. Record **268**, 196 (2002)
17. M. Hanczyc, J.W. Szostak, Curr. Opin. Chem. Biol. **8**, 660 (2004)
18. T. Baumgart, S.T. Hess, W.W. Webb, Nature **425**, 821 (2003)
19. H. Noguchi, M. Takasu, Biophys. J. **83**, 299 (2002)
20. J. Macía, R.V. Solé, J. Theor. Biol. **245**(3), 400 (2007)
21. R.V. Solé, J. Macía, H. Fellermann, A. Munteanu, J. Sardanyés, S. Valverde, in *Protocells: Bridging Nonliving and Living Matter*, ed. by S. Rasmussen, M. Bedau, L. Chen, D. Deamer, D. Krakauer, N. Packard, P. Stadler (MIT Press, Cambridge, MA, 2008), pp. 213–231
22. P.A. Bachmann, P.L. Luisi, J. Lang, Nature **357**, 57 (1992)
23. S.A. Kauffman, J. Theor. Biol. **119**, 1 (1986)
24. J. Farmer, S. Kauffman, N. Packard, Physica D **22**, 50 (1986)
25. R.J. Bagley, J.D. Farmer, in *Artificial Life II*, ed. by C.G. Langton, C. Taylor, J.D. Farmer, S. Rasmussen (Addison-Wesley, Reading, MA, 1991), pp. 93–140
26. E. Smith, H. Morowitz, Proc. Natl. Acad. Sci. USA **101**(36), 13 168 (2004)
27. H.R. Maturana, F.J. Varela, *Autopoiesis and Cognition: The Realization of the Living* (Reidel, Dordrecht, 1980)
28. P.L. Luisi, Naturwissenschaften **90**, 49 (2003)
29. T. Gánti, *The Principles of Life* (Oxford University Press, Oxford, 2003)
30. A. Munteanu, R.V. Solé, J. Theor. Biol. **240**(3), 434 (2006)
31. S. Rasmussen, M. Bedau, L. Chen, D. Deamer, D. Krakauer, N. Packard, P. Stadler (eds.), *Protocells: Bridging Nonliving and Living Matter* (MIT Press, Cambridge, MA, 2008)
32. S. Rasmussen, M. Bedau, J. McCaskill, N. Packard, in *Protocells: Bridging Nonliving and Living Matter*, ed. by S. Rasmussen, M. Bedau, L. Chen, D. Deamer, D. Krakauer, N. Packard, P. Stadler (MIT Press, Cambridge, MA, 2008), pp. 71–100
33. S. Rasmussen, L. Chen, M. Nilsson, S. Abe, Artif. Life **9**, 269 (2003)
34. W. Szostak, D.P. Bartel, P.L. Luisi, Synthesizing life. Nature **409**, 387–390 (2001)
35. J.A. Doudna, J.W. Szostak, Nature **339**, 519 (1989)
36. D.P. Bartel, P.J. Unrau, Trends Cell Biol. **9**, M9 (1999)
37. S.S. Mansy, J.P. Schrum, M. Krishnamurthy, S. Tobé, D.A. Treco, J.W. Szostak, Nature **454**, 122 (2008)
38. P.L. Luisi, F. Ferri, and P. Stano. Approaches to semi-synthetic minimal cells: a review. Naturwissenschaften, **93**, 1–13 (2006)
39. A.C. Chakrabarti, R.R. Breaker, G.F. Joyce, D.W. Deamer, J. Mol. Evol. **39**, 555 (1994)
40. T. Oberholzer, R. Wick, P.L. Luisi, C.K. Biebricher, Biochem. Biophys. Res. Commun. **207**(1), 250 (1995)
41. T. Oberholzer, M. Albrizio, P.L. Luisi, Chem. Biol. **2**(10), 677 (1995)
42. W. Yu, K. Sato, M. Wakabayashi, T. Nakaishi, E.P. Ko-Mitamura, Y. Shima, I. Urabe, T. Yomo, J. Biosci. Bioeng. **92**, 590 (2001)
43. V. Noireaux, A. Libchaber, Proc. Natl. Acad. Sci. USA **101**(51), 17669 (2004)
44. K. Ishikawa, K. Sato, Y. Shima, I. Urabe, T. Yomo, FEBS Lett. **576**, 387 (2004)
45. H. Fellermann, R. Solé, Philos. Trans. R. Soc. Lond. B **362**(1486), 1803 (2007)
46. H. Fellermann, S. Rasmussen, H.J. Ziock, R. Solé, Artif. Life **13**(4), 319 (2007)
47. H. Fellermann, Physically embedded minimal self-replicating systems – studies by simulation. Ph.D. thesis, University of Osnabrück (2009)
48. M. DeClue, P.A. Monnard, J. Bailey, S. Maurer, G. Colins, H.J. Ziock, S. Rasmussen, J. Boncella, J. Am. Chem. Soc. **131**, 931 (2009)
49. C. Knutson, G. Benkö, T. Rocheleau, F. Mouffouk, J. Maselko, A. Shreve, L. Chen, S. Rasmussen, Artif. Life **14**(2), 189 (2008)

50. Y.A. Berlin, A.L. Burin, M.A. Ratner, Superlattices Microstruct. **28**(4), 241 (2000)
51. G.F. Joyce, Orig. Life Evol. Biosph. **14**, 613 (1984)
52. O.L. Acevedo, L.E. Orgel, J. Mol. Biol. **197**(2), 187 (1987)
53. A. Luther, R. Brandsch, G. von Kiedrowski, Nature **396**, 245 (1998)
54. T. Rocheleau, S. Rasmussen, P.E. Nielson, M.N. Jacobi, H. Ziock, Philos. Trans. R. Soc. Lond. B **362**, 1841 (2007)
55. A. Munteanu, C.S.O. Attolini, S. Rasmussen, H. Ziock, R.V. Solé, Philos. Trans. R. Soc. Lond. B **362**, 1847 (2007)
56. R. Serra, T. Carletti, Artif. Life **13**(2), 123 (2007)

# Chapter 12
# Towards a Minimal System for Cell Division

**Petra Schwille**

**Abstract** We have entered the "omics" era of the life sciences, meaning that our general knowledge about biological systems has become vast, complex, and almost impossible to fully comprehend. Consequently, the challenge for quantitative biology and biophysics is to identify appropriate procedures and protocols that allow the researcher to strip down the complexity of a biological system to a level that can be reliably modeled but still retains the essential features of its "real" counterpart. The virtue of physics has always been the reductionist approach, which allowed scientists to identify the underlying basic principles of seemingly complex phenomena, and subject them to rigorous mathematical treatment. Biological systems are obviously among the most complex phenomena we can think of, and it is fair to state that our rapidly increasing knowledge does not make it easier to identify a small set of fundamental principles of the big concept of "life" that can be defined and quantitatively understood. Nevertheless, it is becoming evident that only by tight cooperation and interdisciplinary exchange between the life sciences and quantitative sciences, and by applying intelligent reductionist approaches also to biology, will we be able to meet the intellectual challenges of the twenty-first century. These include not only the collection and proper categorization of the data, but also their true understanding and harnessing such that we can solve important practical problems imposed by medicine or the worldwide need for new energy sources. Many of these approaches are reflected by the modern buzz word "synthetic biology", therefore I briefly discuss this term in the first section. Further, I outline some endeavors of our and other groups to model minimal biological systems, with particular focus on the possibility of generating a minimal system for cell division.

P. Schwille (✉)
Biophysics/BIOTEC, Technische Universität Dresden, D-01307 Dresden, Germany; Max Planck Institute for Molecular Cell Biology and Genetics, D-01307 Dresden, Germany
e-mail: schwille@biotec.tu-dresden.de

H. Meyer-Ortmanns, S. Thurner (eds.), *Principles of Evolution*,
The Frontiers Collection, DOI 10.1007/978-3-642-18137-5_12,
© Springer-Verlag Berlin Heidelberg 2011

## 12.1 Two Concepts of Synthetic Biology

The first major representation of synthetic biology, which still prevails and dominates the field, is that of microbial engineering. The idea is to identify genetically encoded, well-characterized functional elements that could be used as a tool-box quite analogously to electronic systems, with bacteria as a chassis, on which assembly and systems engineering can be performed. In other words, bacteria with new medically or environmentally interesting properties are generated by combining known genetic features in their genome like a switching circuit. The concept very much relies on digital logic, and has the shortcoming that, of course, not all of these functional elements coming from differently evolved organisms are biologically compatible with each other. Nevertheless, the large growing community, particularly in the US, and the large selection of biological "hardware" and "software" units that are already available display the attractiveness and promise of this approach.

Another driving force behind synthetic biology is the emergence of nanotechnology, in particular the biologically oriented and inspired community among nanotechnologists. Biological building blocks such as proteins and nucleic acids can be considered as molecular machines with complex functionality, already adapted to nanoscopic scales and systems. This renders it plausible to devise bio-tech hybrid devices that display functional features of the biological systems but can be controlled externally by physical parameters such as nanoscopic technical switches. The most prominent among the already successfully employed bio-nano hybrid systems are three-dimensional (3D) nanostructures using single-stranded and double-stranded DNA building blocks. But the real promise lies in the functional integration of large protein machineries that can also accomplish complex tasks such as energy production, waste degradation, pathogen defense, or damage repair outside and inside living organisms. Consequently, the pathway provided by nanotechnology and materials science towards synthetic biology involves not only the functional reconstitution of biological units and circuits in technological environments but also the design of interfaces that can actually be controlled by nonbiological cues.

The second representation of synthetic biology is the more biophysical one, namely the striving for a better understanding of biological, particularly cellular systems, by a so-called "bottom-up biology". Physicist Richard Feynman once formulated the famous phrase "What I cannot create, I do not understand". In a strict sense, following this quote, we would only fully understand a biological system if we were able to make it from scratch. Besides ethical implications that I will not be able to address here, it appears to be a rather hopeless enterprise to make a "modern" cell, let alone a whole organism, in all its complexity. On the other hand, life has arisen from a presumably much simpler subsystem containing unknown and probably no longer existing key molecules. The success of several functional in vitro assays for biological subsystems, functioning in environments of dramatically reduced complexity, suggests that it is indeed possible to reconstitute essential features and distinct modules of the cell from small and physically controllable sets of molecules, and by doing this learn more about the fundamental physical and chemical laws which nature builds the phenomenon of life on. It is this concept of synthetic biology, summarized in the vision of a minimal cell, that I will further discuss with respect to its biophysical implications.

## 12.2 The Concept of a Minimal Cell

There are many motivations for the development of minimal cellular systems, or – to give them a more provocative name – artificial cells. One obvious motivation is to find possible models of how primordial cells could have developed to become the first major organizational units of life, compartmenting biological information, and thus forming the first true individuals set apart from their environment – from then on being subject to the mechanisms of Darwinian evolution. Another motivation is a more technical one, with biomedical implications: if we manage to create functional models of cells, we may in the future be able to replace the real ones that fail or somehow misbehave in our organisms. This would be a completely novel approach to what is presently aimed at and partly achieved by stem cell technology – but with fewer, or at least different, ethical implications. The third, and certainly in the short term most relevant, motivation is the striving for better quantitative analysis and modeling of biological systems – as the quantitative researcher is often frustrated when working with native cells under physiological conditions. Hence, minimal systems of cellular modules have in recent decades helped tremendously to elucidate underlying physical and chemical laws that govern complex phenomena of living matter. In vitro models of the cellular cytoskeleton, along with reconstituted motors, have triggered remarkable studies that led to the recognition of mechanisms of dynamic instability, and resolved the step size of single motor proteins per adenosine triphosphate (ATP) hydrolysis. In vitro models of cellular membranes, on the other hand, have helped to reveal the hypothesized biological functionality of lipid domains, or rafts, in membrane protein recruitment and signaling. With increasing knowledge and identification of proteins that transform cellular membranes, morphology and shape changes that are key to cellular metabolism can be mimicked, and the outsourcing of functionality into organelles can be better understood by revealing and reconstituting the proteome machinery required for organelle assembly and maintenance.

In short, dissecting the cellular interaction network module by module, although it does not give us a complete view of the full system, will at least help us in understanding the principles that might have been assembled and combined in ancient forms of living systems. In fact, by admiring the immense entanglement of cellular networks, with their stunning complexity that raises very little hope for understanding the full system, we have to admit that cells as we know them today do not tell us much about the first physical principles and (bio-)chemical modules that governed their evolution. Among the most remarkable features in modern cells and organisms is not only the general processing and inheriting of genetic information, but also the possibility of adapting to environmental conditions, and the robustness of the biochemical machinery with respect to external and internal disturbances. It is thus well conceivable that nature's solution to a specific biological problem, for example, cell division or generally the budding or fusion of vesicles from and to membranes, is not the most straightforward in terms of underlying physical mechanisms, and could be realized in simpler systems with fewer molecular players. This is particularly important to bear in mind when considering the possibility of generating an artificial cell, or, more modestly, engineering a specific functionality by employing a set of biological devices, for example, proteins.

## 12.3 Minimal Systems for Cell Division: An Attempt

One of the most fundamental transformations of biological systems, and key to the understanding of the origin of life, is cell division, that is, the controlled splitting of a compartment carrying biological information into two daughter compartments. What this compartment was like in early life forms is unknown, although the assumption that it was made up of amphiphilic molecules similar to membrane lipids is quite appealing, due to the large tendency of these molecules to self-organize in aqueous systems into complex structures. So let us assume for the moment that a cell could have evolved from lipid vesicles filled with genetic material. How could the controlled splitting of the vesicles be realized? It is known that many physical factors govern the stability of vesicles, such as surface tension and elasticity of the membrane, as well as the osmotic balance between the inside and the outside. If any of these factors are modified owing to environmental or internal dynamics, for example, ion exchange or accumulation of charged molecules, or the modification of lipids forming the membrane, the vesicle will become unstable and at some point break up into two or more smaller units. This will, however, not guarantee a remotely homogeneous size distribution of the daughter vesicles. Therefore, it is more interesting to speculate about how to better control the shape transformation of a membrane vesicle. Ideally, division into two equally sized units upon a defined cue or switch would have to be achieved. From cell biology, numerous proteins are known that can contribute to such a controlled splitting, most of them somehow related to cytoskeletal elements, because the force to transform membranes is often exerted by molecular motors. Although much is known today about the relationship between motor activity, filament dynamics, and shape changes in cytokinesis, the process of cell division as a whole, particularly in eukaryotic systems, has evolved to a degree of complexity that makes it quite impossible to relate back to possible primordial mechanisms of compartment splitting.

In a nave attempt to find a minimal divisome machinery for the inducible, controlled division of biomolecular compartments, the following set of molecular modules seems to be promising: (1) a membrane vesicle made of lipids, (2) membrane proteins serving as an anchoring machinery for force-inducing molecules, (3) a filamentous system to provide a minimum of stability against bursting, and to ensure points of action to apply the forces for membrane transformation, (4) a mechanism to identify the actual division site, (5) force-inducing molecules, and (6) energy to power the system. The beauty of having ATP as an energy source is that it easily facilitates bistable switches, and thus switchable systems, before and after ATP hydrolysis. Can we now build a system as described, as a potential solution for a minimal, controllable, division machinery for lipid vesicles? The first requirement is to generate vesicles, the second to reconstitute proteins into their membranes, and the third to anchor filaments to the vesicle surface. All of these steps have so far successfully been employed in giant vesicle systems, as will be outlined below.

### 12.3.1 Step One: Modeling Membrane Morphogenesis – Nature's Solution to Compartmentation

Although they certainly do not represent the only possible solutions for efficient compartmentation of biological material, membrane vesicles belong to the most intriguing systems to confine aqueous environments. This is mainly due to the wealth of possible shapes and structures that amphiphilic molecules such as lipids can assume in the presence of water, and the simple transformability of their 3D figures. Giant unilamellar vesicles (GUVs), for example, among the most attractive lipid structures owing to their size, have been studied by membrane and lipid researchers for more than two decades. With comfortable dimensions between single and hundreds of micrometers that are easily accessible to optical imaging and manipulation techniques, they have proven ideal model systems to study membrane morphology and mechanical parameters, such as surface tension, elasticity, and local curvature, relevant to membrane structure and transformations. Moreover, if a selection of different lipid species is chosen, they can be structured two-dimensionally by forming distinct domains, or membrane phases, on their surface. Of particular importance for cell biological research are GUV systems consisting of a ternary mixture of phosphatidylcholine (PC), sphingomyelin, and cholesterol (Fig. 12.1). It can be demonstrated that, dependent on their exact composition, these ternary mixtures exhibit defined domains of different lipid order or fluidity, which can easily be imaged by confocal or wide-field imaging, using fluorescent markers that bind preferentially to a certain lipid environment. The most prominent representations of phases in membranes are the so-called fluid (or liquid disordered) phase, characterized by a relatively low order in the tail region, with correspondingly high lateral mobility of the lipids; the gel phase, displaying high order in the tail region and low mobility of the lipids; and an intermediate phase, the so-called liquid-ordered phase, where relatively high lipid mobility can be observed, in spite of a rather strong local lipid order [1].

Since the development of the raft hypothesis [2] in cell biology, there has also been rising interest from the biological community in better understanding the relevance of local lipid order for the lateral sorting and induction of functionality of membrane proteins. It is now widely accepted that the quantitative representation and local order of specific lipids in membranes of various organelles, in tight concert with the respective proteins inserted or attached to them, contributes to their functionality. In this respect, the above-mentioned liquid-ordered phase is presently

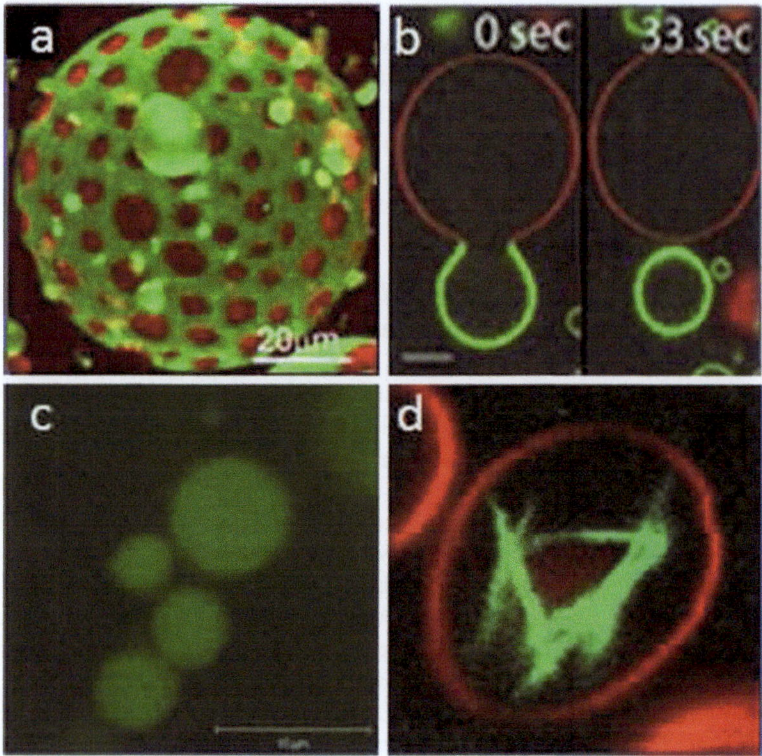


**Fig. 12.1 a** Phase-separating GUV, liquid-ordered (*red*) and liquid-disordered domains (*green*) [1]. **b** Budding of a liquid-ordered (*green*) domain away from the GUV surface [3]. **c** GUVs with active ion channels, pumping $Ca^{2+}$ into the lumen of the vesicle. **d** GUVs with filaments tightly anchored to the surface (**c** and **d**: [4])

supposed to be the most likely organizational form of lipids in these functionally important membrane segments, termed lipid rafts. Since the exact relationships are often too complex and the structural features such as rafts in live cells too small to be resolved quantitatively, minimal systems with reduced complexity, such as GUVs, have paved the way to a more fundamental understanding of lipid–lipid and lipid–protein interactions of physiological importance.

The question for our minimal cell approach is now whether the GUV model system can be worked into a more elaborate model of biomolecular self-organization. The next obvious step would be to include factors that are able to controllably induce transformations, such as division, of these cell-like compartments, and to then combine the transformable compartments with information units that could be reproduced during division.

As the simplest solution to membrane transformations, vesicle shapes can be easily modified through variations of the physical parameters that control shape and size. There have been numerous reports about the splitting of vesicles [5], for example their breakup into two or more daughter vesicles. They can be split if lipid mass is constantly added, thereby destabilizing the surface [6], or if osmotic pressure is increased, potentially also through reactions of encapsulated material, as beautifully demonstrated by the Szostak group [7] with replicating RNA molecules. In our own work, we demonstrated that splitting of phase-separating vesicles can occur along the phase boundaries of liquid-ordered and liquid-disordered membrane domains, if the salt concentration outside the vesicle is slightly altered with respect to that inside [3]. Figure 12.1b shows the budding of the liquid-ordered domain away from the larger part of the vesicle, which is in the liquid-disordered state. The mechanism behind this process is the larger increase of the line tension along the domain boundary with respect to the increase of surface tension, which acts against a budding transition. Although this kind of vesicle transformation or splitting is not very attractive in terms of creating a minimal self-replicating system, due to its very poor local controllability, the presence of local 2D structures or faults within the membrane, as in this case of a phase-separating vesicle, may well promote or recruit the activity of membrane-transforming proteins.

In cells, membranes are constantly being transformed into spatial structures such as lobes, cristae, stacks, tubes, and vesicles. Prominent membrane transformations involve the uptake and release of molecules, their packaging and transport from and to distinct sites, the transformation of whole organelles, and, finally, the large-scale restructuring of the cell membrane during cell division. All of these transformations are tightly regulated and catalyzed by specific protein machineries, presumably triggered through a sophisticated interplay between the local lipid environment and the specifically adapted protein structure and function [8]. Although force-inducing motor proteins, as mentioned above, are often involved in large-scale membrane transformations, most of the intracellular membrane traffic, for example between the Golgi network and the endoplasmic reticulum, actually relies on the recruitment of cytosolic protein machineries, called "coats", prone to cooperatively form higher order structures, thus inducing buds with specific membrane curvature, which can then be easily transformed into 3D structures such as tubes or vesicles. In recent years, evidence has accumulated that this recruitment of these cytoplasmic coats occurs predominantly at specific sites, for example with already existing curvatures through lipid asymmetry, or on domains with higher membrane fluidity. Here, in vitro reconstitution of this machinery onto GUVs, especially with domains, is a valuable tool for better understanding and characterizing the physical parameters governing coat protein recruitment and the subsequent membrane transformation under defined conditions [9]. Another way to induce membrane transformations by soluble factors is to involve enzymes that modify the lipid structure, for example, by removing or changing the head groups of the membrane lipids. The relevance of the local lipid environment in membrane deformations and budding was highlighted in a recent study that compared the transformation of GUV membranes upon activity of sphingomyelinase with the creation of intraluminal vesicles of multivesicular endosomes [10]. On the GUVs, conversion of sphingomyelin into ceramide, by removing the large characteristic head group, resulted in the spontaneous, protein-free budding of small vesicles away from the GUV membrane. In the cells, it was found that cargo is segregated into distinct subdomains on the endosomal membrane, and that the transfer of exosome-associated domains into the lumen of the endosome required the sphingolipid ceramide. Purified exosomes were enriched in ceramide, and the release of exosomes was reduced after the inhibition of neutral sphingomyelinases.

The probably most sophisticated and best-controlled system of membrane transformation, characteristic of higher, eukaryotic, cells, is the deformation by motor protein activity through filaments that are anchored to the membrane. Filaments of the cytoskeleton on one hand stabilize the cell and thus allow it to grow to larger sizes to accommodate more complex biological functions. On the other hand, through their defined attachment sites, they allow for a better definition of where exactly the deformations can occur. Membrane transformation through motor activity, requiring energy input in the form of ATP, can also be studied in cell-free GUV model systems. In vitro experiments with reconstituted filaments and purified proteins or cell extracts has shown that motor proteins, moving along actin or microtubules, exert forces on the membranes to which they are attached. One very nice example is the pulling of membrane tubes by motor proteins from free-standing GUV membranes, by attaching purified kinesin-1 motors to the membranes via micrometer-sized beads. Using the beads, the involved forces could in these geometries be measured directly with optical tweezers [11]. These experiments were, however, performed with filaments and motors acting on the outside of vesicles and could thus not be easily developed into a vesicle division system. The question is thus whether it would be possible to somehow attach the filaments to the membrane inside of the giant vesicles. This will be discussed in the next section.

## 12.3.2 Step Two: Adding Mechanical Stability – Creation of an Artificial Cortex/Cytoskeleton

In cells, the interplay between filaments anchored to membranes and active motor proteins, which induce the required forces to transform these membranes, tightly regulates the division and opens up the control of these processes by providing interfaces for other proteins or protein machineries that activate and deactivate the actual division process. A general requirement to couple the activity of cytoskeletal motors to model membranes is the establishment of a proper interface, that is, the reconstitution of a cytoskeleton- or cortex-like structure on the membrane, through stationary or transient anchors. Although minimal systems based on filaments and molecular motors have been extensively studied over the past decades, involving many beautiful in vitro assays, mainly on microtubule–kinesin systems as mentioned above, the linkage of these systems to membranes has so far not been studied in detail.

Actin, presumably one of the major constitutents of cellular cortices, was polymerized within GUVs quite early on, but without stable attachment between the membrane and the filaments [12]. There are many actin-based superstructures interacting with the plasma membrane at different cellular locations, such as the highly branched, polymerizing actin network at the leading edge of migratory cells, stress fibers that are attached to sites of adhesion, long actin filaments in filopodia, and actin-rich structures found at invaginations in endocytic and phagocytic structures. Nucleation of actin networks is spatially and temporally coordinated by a complex interplay of several proteins: small GTPases (hydrolase enzymes that can bind and hydrolyze guanosine triphosphate, GTP) in concert with lipid second messengers such as phosphatidylinisitol 4,5-biphosphate (PIP2). By in vitro reconstitution on GUVs containing PIP2 using purified components, the minimal requirements of actin-based motility were elucidated. It was found that in addition to actin, an activated Arp2/3 (actin related protein 2 and 3) complex for enhanced nucleation, actin depolymerizing factors, capping proteins, and ATP were required to reconstitute sustained motility [13]. Specifically for this anchoring system, PIP2, in collaboration with the small GTPase Cdc42, binds N-WASP (neuronal Wiskott–Aldrich syndrom protein) and triggers a conformational change that allows binding to and activation of the Arp2/3 complex, which in turn nucleates the formation of a branched actin network. In a recent approach to anchoring an actin network to the outer surface of GUVs, Liu and Fletcher [14] used this combination of N-WASP bound to PIP2 and activated Arp2/3 to polymerize actin. On phase-separated vesicles, N-WASP, Arp2/3, and actin only formed networks on tetramethyl rhodamine (TMR)-PIP2 enriched domains. In analyzing the domain melting temperature, it was found that the actin network on the surface of the membrane can lead to the induction of new domains, and the stabilization of existing domains. Furthermore, the actin network seems to spatially bias the location of domain formation after the temperature is cycled above and below the melting temperature. This work nicely illustrates how a dynamic actin cytoskeleton can organize the cellular membrane, not only by restricting lipid and protein diffusion, but also by actively organizing membrane domains. Recently, an elegant protocol based on reverse emulsions was released, allowing actin polymerization to nucleate and assemble at the inner membrane of a GUV [15].

A completely different approach to tightly anchoring cytoskeletal elements to GUV membranes was chosen by our group [4]. Using porcine total brain lipid extracts rather than synthetic lipids, static linkage of actin filaments through the ankyrin/spectrin machinery bound to functional ion channels in the GUV membrane was accomplished (Fig. 12.1d). The use of membrane fractions maintains the complex lipid composition found in a native brain membrane state, in addition to containing the necessary integral membrane proteins, such as ion channels, for anchoring the cytoskeleton. As attachment machinery, we isolated spectrin and ankyrin, added them to the GUVs and thereby anchored the actin filaments to the inner walls of the porcine GUVs. In this way, assembling a quaternary-protein system to the membrane surface, our work demonstrated the ability to use GUVs as a model "cell-like" compartment, in which multiprotein systems can be reconstituted and examined in the presence of complex lipid mixtures. On the other hand, this assay may signify a critical step towards achieving the requirements for actin-induced division of a minimal cell system. The GUVs containing actin filaments could be visualized either with the lipid dye DiD-C18, or by immunostaining with specific antibodies, targeting proteins known to be integral to brain membranes.

We were able to show that in addition to their functionality as membrane anchoring sites, the activity of the channels was preserved in these model systems (Fig. 12.1c). GUVs were prepared in the presence of ATP and Calcium Green, a sensitive calcium probe that becomes more fluorescent in the presence of $Ca^{2+}$, such that ATP and calcium ended up inside the GUVs. Then the GUVs were washed extensively with buffer to remove the Calcium Green on the outside. Upon addition of 5 mM $CaCl_2$ and 1 mM ATP, the interior of the GUVs became bright fluorescent green, while GUVs grown and washed in the absence of ATP remained at a nearly nondetectable fluorescence level.

To add the filaments to the inside of GUVs, they were grown by electroswelling in the presence of purified actin treated with phalloidine-Alexa 488 to fluorescently label the filaments. Then the GUVs were washed extensively with buffer, in order to remove the majority of the actin filaments on the outside of the GUVs. The resulting GUVs contained actin filaments dispersed throughout their interior (Fig. 12.1d). When GUVs were prepared from highly enriched protein fractions containing spectrin and ankyrin, the filaments within the GUVs were no longer dispersed, but displayed dense packing near the walls of the GUVs. In the majority of cases, the filaments were anchored to the interior wall of the GUV. After essential controls with total brain lipid extracts not including the ion channels, these results indicated the first reconstitution of stably anchored cytoskeleton to the interior walls of GUVs via the spectrin–ankyrin proteins, which bind to functional transmembrane ion channels (such as Na/K ATPase). The spectrin-based membrane skeleton is a network of cytoplasmic structural proteins, first investigated in erythrocytes, that underlies regions of the plasma membrane in diverse cells and tissues. Within this complex architecture, spectrin is thought to connect certain membrane proteins at the cell surface with actin and microtubules in the cytoplasm, and thereby affect the topography and dynamic behavior of these proteins.

### 12.3.3 Step Three: How Should the Division Site Be Defined? Pattern Formation and Self-Organization in Minimal Systems

Having achieved a system composed of membrane and filaments, we still have to solve one of the biggest problems in the quest to reconstitute cell division, namely to achieve a controlled division at a defined site. In other words: how do we tell the vesicle where to split, or better: how to split into two equal halves? Clearly, such a controlled division using a protein-based divisome machinery is one of the most spectacular visions when it comes to converting GUV into minimal cells. Since the simplest divisomes that we know of are the prokaryotic ones, avoiding the double task of nuclear division and cytokinesis, it makes sense to speculate about reconstituting bacterial divisomes in vesicles. One particularly well-studied system is the divisome of *E. coli* bacteria, based around a contractile ring, called the Z ring. This division ring is formed by a set of at least ten proteins that assemble at mid-cell to drive cytokinesis [16]. Initially three proteins, FtsZ, FtsA, and ZipA assemble together, forming a proto-ring, to which the other components are added. To ensure the correct assembly of this ring at mid-cell, and thus to solve the task of controlled splitting into two equal halves mentioned above, two main positioning mechanisms are required: the Min system and nucleoid occlusion, which select the constriction site [17].

The Min is probably the best characterized divisome subsystem in quantitative terms. It consists of MinC, D, and E proteins, which oscillate repeatedly from one pole of the cell to the opposite pole [18]. MinC inhibits FtsZ polymerization and thereby presumably prevents the ring from forming away from the cell center. MinD, a membrane-bound ATPase, activates MinC's inhibitory activity and directs it to the membrane, while MinE activates MinD's ATPase to drive the oscillation. Their combined action results in FtsZ assembly into a ring at mid-cell, where the local time-averaged concentration of MinC inhibitor is lowest. The other spatial regulatory system, nucleoid occlusion, is an additional mechanism to avoid septation at places occupied by the bacterial nucleoid (bulk chromosomal DNA), and therefore acts as a fail-safe system. The relationship between MinD and MinE can be considered a classical energy-consuming self-organized system as already suggested by Turing in 1952 [19], where energy is consumed by the ATPase MinD. When bound to ATP, MinD dimerizes and exposes its membrane targeting sequence (MTS), attaching the previously soluble protein to the inner bacterial membrane. MinE then binds to membrane-bound MinD, stimulating the ATP turnover. Subsequently, both proteins detach from the membrane and become soluble again. The dynamic pattern formation is ascribed to dynamical instability driven by the hydrolysis of ATP [20, 21].

The beauty of the Min system is that its self-organization behavior, leading to the definition of a preferred division site in the bacterial cell, can be easily reconstituted in minimal membrane-protein systems. This was shown by our group using purified and fluorescently-labelled MinD and MinE, and supported membranes made of *E. coli* lipids PE, PG, and cardiolipin [22]. Naturally, the presence of 2D open planes, rather than closed compartments such as cells, prevents the formation of oscillations because of the lack of turning points. Instead, upon addition of ATP, impressive traveling wave patterns on the membrane could be observed (Fig. 12.2a), MinE following MinD at the trailing edge of the wave, and the relationship between wavelength, propagation speed, and diffusional mobility of the proteins in solution being consistent with the oscillations in the cellular setting.

With the reconstitution of the self-organizing Min system in vitro, an important step towards the controlled splitting of membrane compartments has been made. Although the reconstituted Min assay does not yet involve any filaments or other handles to exert forces to the membrane, they provide the unique feature of creating spatiotemporal patterns that can later be used to identify the division site. The next logical steps towards the assembly of a divisome machinery into vesicles are now (1) to observe how the Min waves react to the closing of the membrane surface, by injecting the reaction-diffusion system to vesicles, (2) to find a similar minimal system solution for in vitro assembly of the Z ring as the key element for division, and (3) to find a way of actually constricting the Z ring. These are the topics of ongoing work in our laboratory. As a first promising attempt towards ring assembly

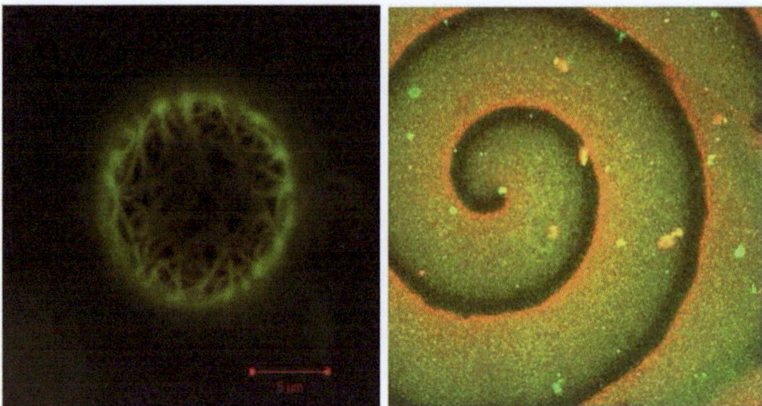

**Fig. 12.2 a** Spiral waves, formed by the self-organization of reconstituted and fluorescently labeled MinD/MinE on planar membranes made of *E. coli* lipids [22]. **b** FtsZ filaments with membrane binding domain and labeled by *yellow* fluorescent protein, reconstituted on a GUV surface (S. Chiantia and S. Arumugam, unpublished work)

in cell-free systems, the attachment of FtsZ filaments to GUV surfaces was realized with a mutant of FtsZ that exhibits a membrane-targeting sequence (Osawa et al. 2008), avoiding the complete FtsA/ZipA membrane anchoring machinery in order to attach the protein to the GUV surface (Fig. 12.2b). Owing to the absence of MinC gradients, the filaments are still isotropically attached to the surface. Nevertheless, the image shows a minimal realization of a prokaryotic filament network on a membrane vesicle, and thus a potentially very interesting study object for a better understanding of the mutual relationship between membranes and membrane-deforming protein machineries.

## 12.4 Outlook

In this chapter, I have discussed the possibility of creating a bottom-up system for cell division, realized within giant unilamellar membrane vesicles. This is set in the context of synthetic biology, as one implementation of minimal cellular systems, which could become extremely valuable for quantitative biosciences such as biophysics. Although we are still far from constructing an artificial cell bottom-up, the realization of artificial cellular modules with defined properties, and achieving specific tasks characteristic to living systems, becomes plausible. Without doubt, these minimal in vitro systems are presently among the most appealing assays to complement cellular work for a better dissection of complex protein networks, and may also help in identifying appropriate models for early stages in the evolution of cells as we know them today. As shown in the last section, minimal systems can also allow us to reconstitute essential patterns of biological self-organization, which may otherwise be hidden behind the complexity of living cells and organisms, to make them accessible to rigorous mathematical modeling. In this way, we hope also to assemble in the future minimal biological systems to mimic processes such as pattern formation or differentiation in developing tissue, which have been subject to physical modeling for decades without the possibility of testing these models in controllable experimental settings. In other words, synthetic biology of minimal systems opens up exciting future perspectives not only for the principle, but also for the quantitative understanding of living systems by modern biological physics.

**Acknowledgments** Financial support of our synthetic biology work in the form of a fellowship of the Max Planck Society (MPG) is gratefully acknowledged.

# References


1. D. Scherfeld, N. Kahya, P. Schwille, Biophys. J. **85**, 3758 (2003)
2. K. Simons, E. Ikonen, Nature **387**, 569 (1997)
3. K. Bacia, P. Schwille, T. Kurzchalia, Proc. Natl. Acad. Sci. USA **102**, 3272 (2005)
4. D. Merkle, N. Kahya, P. Schwille, ChemBioChem **9**, 2673 (2008)
5. P.L. Luisi, P. Stano, S. Rasi, F. Mavelli, Artif. Life **10**, 297 (2004)
6. B. Bozic, S. Svetina, Eur. Biophys. J. **33**, 565 (2004)
7. I.A. Chen, R.W. Roberts, J.W. Szostak, Science **305**, 1474 (2004)
8. H.T. McMahon, J.L. Gallop, Nature **438**, 590 (2005)
9. P. Sens, L. Johannes, P. Bassereau, Curr. Opin. Cell Biol. **20**, 1 (2008)
10. K. Trajkovic, C. Hsu, S. Chiantia, L. Rajendran, D. Wenzel, F. Wieland, P. Schwille, B. Brügger, M. Simons, Science **319**, 1244 (2008)
11. A. Roux, G. Cappello, J. Cartaud, J. Prost, B. Goud, P. Bassereau, Proc. Natl. Acad. Sci. USA **99**, 5394 (2002)
12. L. Limozin, E. Sackmann, Phys. Rev. Lett. **89**, 168103 (2002)
13. V. Delatour, E. Helfer, D. Didry, K.H.D. Le, J.F. Gaucher, M.F. Carlier, G. Romet-Lemonne, Biophys. J. **94**, 4890 (2008)
14. A.P. Liu, D.A. Fletcher, Biophys. J. **92**, 697 (2007)
15. L.L. Pontani, J. van der Gucht, G. Salbreux, G. Heuvingh, J.F. Joanny, C. Sykes, Biophys. J. **96**, 192 (2009)
16. M. Vicente, A.I. Rico, R. Martinez-Arteaga, J. Mingorance, J. Bacteriol. **188**, 19 (2006)
17. X.C. Yu, W. Margolin, Mol. Microbiol. **32**, 315 (1999)
18. D.M. Raskin, P.A. de Boer, Proc. Natl. Acad. Sci. USA **96** 4971 (1999)
19. A.M. Turing, Philos. Trans. R. Soc. Lond. B **237**, 37 (1952)
20. M. Howard, A.D. Rutenberg, S. de Vet, Phys. Rev. Lett. **87**, 278102 (2001)
21. K. Kruse, M. Howard, W. Margolin, Mol. Microbiol. **63**, 1279 (2007)
22. M. Loose, E. Fischer-Friedrich, J. Ries, K. Kruse, P. Schwille, Science **320**, 789 (2008)

# Part IV
# From Cells to Societies

# Chapter 13
# Bacterial Games

Erwin Frey and Tobias Reichenbach

**Abstract** Microbial laboratory communities have become model systems for studying the complex interplay between nonlinear dynamics of evolutionary selection forces, stochastic fluctuations arising from the probabilistic nature of interactions, and spatial organization. Major research goals are to identify and understand mechanisms that ensure viability of microbial colonies by allowing for species diversity, cooperative behavior and other kinds of "social" behavior. A synthesis of evolutionary game theory, nonlinear dynamics, and the theory of stochastic processes provides the mathematical tools and conceptual framework for a deeper understanding of these ecological systems. We give an introduction to the modern formulation of these theories and illustrate their effectiveness, focusing on selected examples of microbial systems. Intrinsic fluctuations, stemming from the discreteness of individuals, are ubiquitous, and can have important impact on the stability of ecosystems. In the absence of speciation, extinction of species is unavoidable, may, however, take very long times. We provide a general concept for defining survival and extinction on ecological time scales. Spatial degrees of freedom come with a certain mobility of individuals. When the latter is sufficiently high, bacterial community structures can be understood through mapping individual-based models, in a continuum approach, onto stochastic partial differential equations. These allow progress using methods of nonlinear dynamics such as bifurcation analysis and invariant manifolds. We conclude with a perspective on the current challenges in quantifying bacterial pattern formation, and how this might have an impact on fundamental research in nonequilibrium physics.

E. Frey (✉)
Arnold Sommerfeld Center for Theoretical Physics and Center for NanoScience,
Ludwig-Maximilians-Universität München, D-80333 München, Germany
e-mail: frey@lmu.de

H. Meyer-Ortmanns, S. Thurner (eds.), *Principles of Evolution*,
The Frontiers Collection, DOI 10.1007/978-3-642-18137-5_13,
© Springer-Verlag Berlin Heidelberg 2011

## 13.1 Introduction

Microbial systems are complex assemblies of large numbers of individuals, interacting competitively under multifaceted environmental conditions. Bacteria often grow in complex, dynamical communities, pervading the earth's ecological systems, from hot springs to rivers and the human body [1]. As an example, in the latter case, they can cause a number of infectious diseases, such as lung infection by *Pseudomonas aeruginosa*. Bacterial communities, quite generically, form biofilms [1, 2], that is, they arrange into a quasi-multicellular entity with strong interactions. These interactions include competition for nutrients, cooperation by providing various kinds of public goods essential for the formation and maintenance of the biofilm [3], communication through the secretion and detection of extracellular substances [4, 5], chemical warfare [6], and, finally, physical forces. The ensuing complexity of bacterial communities has conveyed the idea that they constitute "social groups", where the coordinated action of individuals leads to various kinds of system-level functionalities [7].

Since additionally microbial interactions can be manipulated in a multitude of ways, many researchers have turned to microbes as the organisms of choice to explore fundamental problems in ecology and evolutionary dynamics [6, 8, 9]. Much effort is currently being devoted to qualitative and quantitative understanding of basic mechanisms that maintain the *diversity* of microbial populations. Hereby, within exemplary models, the formation of dynamic spatial patterns has been identified as a key promoter [10–13]. In particular, the crucial influence of self-organized patterns on biodiversity has been demonstrated in recent experimental studies [6] employing three bacterial strains that display cyclic competition. The latter is metaphorically described by the game "rock–paper–scissors", where rock blunts scissors, scissors cut paper, and paper wraps rock in turn. For the three bacterial strains, and for low microbes motility, cyclic dominance leads to the stable coexistence of all three strains through self-formation of spatial patterns. In contrast, stirring the system, as can also result from high motilities of the individuals, destroys the spatial structures, which results in the takeover of one subpopulation and the extinction of the others after a short transition. There is also an ongoing debate in sociobiology about how *cooperation* within a population emerges in the first place and how it is maintained in the long run. Microbial communities again serve as versatile model systems for exploring these questions [8, 9]. In those systems, cooperators are producers of a common good, usually a metabolically expensive biochemical product. Hence a successfully cooperating collective of microbes permanently runs the risk of being undermined by nonproducing strains ("cheaters"), who save themselves the metabolically costly production for biofilm formation [3, 14]. As partial resolutions to this puzzling dilemma, recent studies emphasize nonlinear benefits [8] and population bottlenecks in permanently regrouping populations [9].

This chapter is intended as an introduction to some of the theoretical concepts that are useful in deepening our understanding of these systems. We will start with an introduction to the language of game theory and after a short discussion of "strategic games" quickly move to "evolutionary game theory". The latter is the natural framework for the evolutionary dynamics of populations consisting of multiple interacting species, where the success of a given individual depends on the behavior of the surrounding ones. It is most naturally formulated in the language of nonlinear dynamics, where the game theory terms "Nash equilibrium" or "evolutionary stable strategy" map onto "fixed points" of ordinary nonlinear differential equations. Illustrations of these concepts are given in terms of two-strategy games and the cyclic Lotka–Volterra model, also known as the "rock–paper–scissors" game. Before embarking on the theoretical analysis of the role of stochasticity and space we give, in the short Sect. 13.3, some examples of game-theoretical problems in biology, mainly taken from the field of microbiology.

A deterministic description of populations of interacting individuals in terms of nonlinear differential equations lacks some important features of actual ecological systems. The molecular processes underlying the interaction between individuals are often inherently stochastic and the number of individuals is always discrete. As a consequence, there are random fluctuations in the composition of the population, which can have an important impact on the stability of ecosystems. In the absence of speciation, extinction of species is unavoidable, may, however, take very long times. Sect. 13.4 starts with some elementary, but very important, notes on extinction times, culminating in a general concept for defining survival and extinction on ecological time scales. These ideas are then illustrated for the "rock-scissors-paper" game.

Cyclic competition of species, as metaphorically described by the children's game "rock–paper–scissors", is an intriguing motif of species interactions. Laboratory experiments on populations consisting of different bacterial strains of *E. coli* have shown that bacteria can coexist if a low mobility enables the segregation of the different strains and thereby the formation of patterns [6]. In Sect. 13.5 we analyze the impact of stochasticity as well as individuals' mobility on the stability of diversity as well as the emerging patterns. Within a spatially extended version of the May–Leonard model [15] we demonstrate the existence of a sharp mobility threshold [13], such that diversity is maintained below, but jeopardized above that value. Computer simulations of the ensuing stochastic cellular automaton show that entangled rotating spiral waves accompany biodiversity. In our final section we conclude with a perspective on the current challenges in quantifying bacterial pattern formation and how this might also have an impact on fundamental research in nonequilibrium physics.

## 13.2 The Language of Game Theory

### 13.2.1 Strategic Games and Social Dilemmas

Classical game theory [16] describes the behavior of rational players. It attempts to mathematically capture behavior in strategic situations, in which an individual's success in making choices depends on the choices of others. A classical example of a strategic game is the *prisoner's dilemma*. It can be formulated as a kind of a *public good game*, where a cooperator provides a benefit $b$ to another individual, at a cost $c$ to itself (with $b - c > 0$). In contrast, a defector refuses to provide any benefit and hence does not pay any costs. For the selfish individual, irrespective of whether the partner cooperates or defects, defection is favorable, as it avoids the cost of cooperation, exploits cooperators, and ensures not being exploited. However, if all individuals act rationally and defect, everybody is, with a gain of 0, worse off compared to universal cooperation, where a net gain of $b - c > 0$ would be achieved. This unfavorable outcome of the game, where both play "defect", is called *Nash equilibrium* [17]. The prisoner's dilemma therefore describes, in its most basic form, the fundamental problem of establishing cooperation. It is summarized in the following payoff matrix (for the column player):

$$\begin{array}{c|cc} \mathbf{P} & \text{Cooperator (C)} & \text{Defector (D)} \\ \hline C & b - c & -c \\ D & b & 0 \end{array} \qquad (13.1)$$

This scheme can be generalized to include other basic types of social dilemmas [18, 19]. Namely, two cooperators that meet are both *rewarded* a payoff $\mathcal{R}$, while two defectors obtain a *punishment* $\mathcal{P}$. When a defector encounters a cooperator, the first exploits the second, gaining the *temptation* $\mathcal{T}$, while the cooperator only gets the *sucker's payoff* $\mathcal{S}$. Social dilemmas occur when $\mathcal{R} > \mathcal{P}$, such that cooperation is favorable in principle, while temptation to defect is large: $\mathcal{T} > \mathcal{S}$, $\mathcal{T} > \mathcal{P}$. These interactions may be summarized by the payoff matrix

$$\begin{array}{c|cc} \mathbf{P} & \text{Cooperator (C)} & \text{Defector (D)} \\ \hline C & \mathcal{R} & \mathcal{S} \\ D & \mathcal{T} & \mathcal{P} \end{array} \qquad (13.2)$$

Variation of the parameters $\mathcal{T}$, $\mathcal{P}$, $\mathcal{R}$, and $\mathcal{S}$ yields four distinct types of games. The *prisoner's dilemma* arises if the temptation $\mathcal{T}$ to defect is larger than the reward $\mathcal{R}$, and if the punishment $\mathcal{P}$ is larger than the sucker's payoff $\mathcal{S}$. As we have already seen above, in this case, defection is the best strategy for the selfish player. Within the three other types of games, defectors are not always better off. For the *snowdrift game* the temptation $\mathcal{T}$ is still higher than the reward $\mathcal{R}$ but the sucker's payoff $\mathcal{S}$ is larger than the punishment $\mathcal{P}$. Therefore, now actually cooperation is favorable when meeting a defector, but defection pays off when encountering a cooperator, and a rational strategy consists of a mixture of cooperation and defection. The snowdrift game derives its name from the potentially cooperative interaction present when two drivers are trapped behind a large pile of snow, and each driver must decide whether to clear a path. Obviously, then the optimal strategy is the opposite of the opponent's (cooperate when your opponent defects and defect when your opponent cooperates). Another scenario is the *coordination game*, where mutual agreement is preferred: either all individuals cooperate or defect as the reward $\mathcal{R}$ is higher than the temptation $\mathcal{T}$ and the punishment $\mathcal{P}$ is higher than sucker's payoff $\mathcal{S}$. Lastly, the scenario of *by-product mutualism* (also called *harmony*) yields cooperators fully dominating defectors since the reward $\mathcal{R}$ is higher than the temptation $\mathcal{T}$ and the sucker's payoff $\mathcal{S}$ is higher than the punishment $\mathcal{P}$.

## *13.2.2 Evolutionary Game Theory*

Strategic games are thought to be a useful framework in economic and social settings. In order to analyze the behavior of biological systems, the concept of rationality is not meaningful. Evolutionary game theory (EGT), as developed mainly by Maynard Smith and Price [20, 21], does not rely on rationality assumptions but on the idea that evolutionary forces such as natural selection and mutation are the driving forces of change. The interpretation of game models in biology is fundamentally different from strategic games in economics or social sciences. In biology, strategies are considered to be inherited programs that control the individual's behavior. Typically one looks at a population composed of individuals with different strategies who interact generation after generation in game situations of the same type. The interactions may be described by deterministic rules or stochastic processes, depending on the particular system under study. The ensuing dynamic process can then be viewed as an iterative (nonlinear) map or a stochastic process (with either discrete or continuous time). This naturally puts evolutionary game theory in the context of nonlinear dynamics and the theory of stochastic processes. We will see later on how a synthesis of the two approaches helps us to understand the emergence of complex spatio-temporal dynamics.

In this section, we focus on a *deterministic* description of *well-mixed* populations. The term "well-mixed" signifies systems where the individual's mobility (or diffusion) is so large that one may neglect any spatial degrees of freedom and assume that every individual is interacting with everyone at the same time. This is a mean-field picture, where interactions are given in terms of the average number of individuals playing a particular strategy. Frequently, this situation is visualized as an "urn model", where two individuals from a population are randomly selected to play with each other according to some specified game-theoretical scheme. The term "deterministic" means that we are seeking a description of populations where the number of individuals $N_i(t)$ playing a particular strategy $A_i$ is macroscopically large such that stochastic effects can be neglected.

### 13.2.2.1 Pairwise Reactions and Rate Equations

In the simplest setup the interaction between individuals playing different strategies can be represented as a reaction process characterized by a set of rate constants. For example, consider a game where three strategies $\{A, B, C\}$ cyclically dominate each other, as in the "rock–paper–scissors" game: $A$ invades $B$, $B$ outperforms $C$, and $C$ in turn dominates over $A$, schematically drawn in Fig. 13.1.

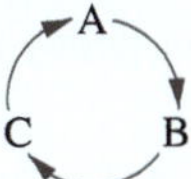


**Fig. 13.1** Illustration of cyclic dominance of three states $A$, $B$, and $C$: $A$ invades $B$, $B$ outperforms $C$, and $C$ in turn dominates over $A$

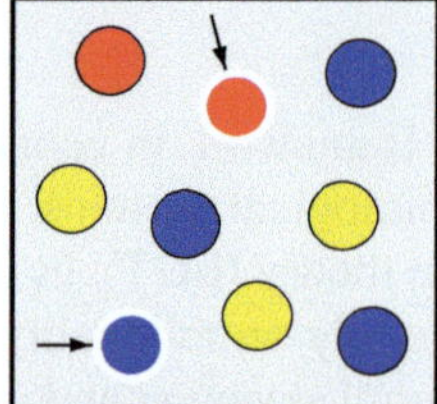 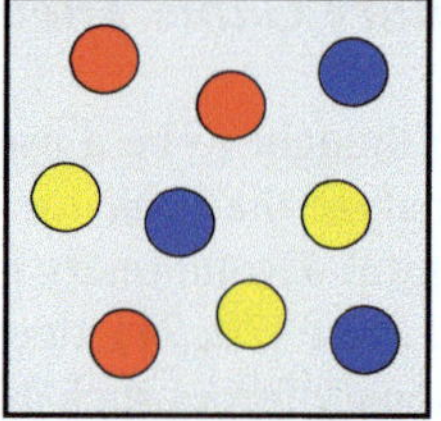

**Fig. 13.2** The urn model describes the evolution of well-mixed finite populations. Here, as an example, we show three species as *yellow* (*A*), *red* (*B*), and *blue* (*C*) spheres. At each time step, two randomly selected individuals are chosen (indicated by *arrows* in the *left picture*) and interact with each other according to the rules of the game, resulting in an updated composition of the population (*right picture*)

In an evolutionary setting, the game may be played according to an urn model as illustrated in Fig. 13.2: at a given time $t$ two individuals from a population with constant size $N$ are randomly selected to play with each other (react) according to the reaction scheme

$$
\begin{aligned}
A + B &\xrightarrow{k_A} A + A \,, \\
B + C &\xrightarrow{k_B} B + B \,, \\
C + A &\xrightarrow{k_C} C + C \,,
\end{aligned}
\tag{13.3}
$$

where $k_i$ are rate constants, that is, probabilities per unit time. This interaction scheme is termed a *cyclic Lotka–Volterra model*.[1] It is equivalent to a set of chemical reactions, and in the deterministic limit of a well-mixed population one obtains *rate equations* for the frequencies $(a, b, c) = (N_A, N_B, N_C)/N$:

$$
\begin{aligned}
\partial_t a &= a(k_A b - k_C c) \,, \\
\partial_t b &= b(k_B c - k_A a) \,, \\
\partial_t c &= c(k_C a - k_B b) \,.
\end{aligned}
\tag{13.4}
$$

Here the right-hand sides give the balance of "gain" and "loss" processes. The phase space of the model is the simplex $S_3$, where the species' densities are constrained by $a + b + c = 1$. There is a constant of motion for the rate equations (13.5), namely the quantity $\rho := a^{k_B} b^{k_C} c^{k_A}$ does not evolve in time [24]. As a consequence, the phase portrait of the dynamics, shown in Fig. 13.3, yields neutrally stable cycles with fixed $\rho$ around the reactive fixed point $F$. This implies that the deterministic dynamics is oscillatory with the amplitude and frequency determined by the initial composition of the population.

---

[1] The two-species Lotka–Volterra equations describe a predator–prey system where the per capita growth rate of the prey decreases linearly with the number of predators present. In the absence of

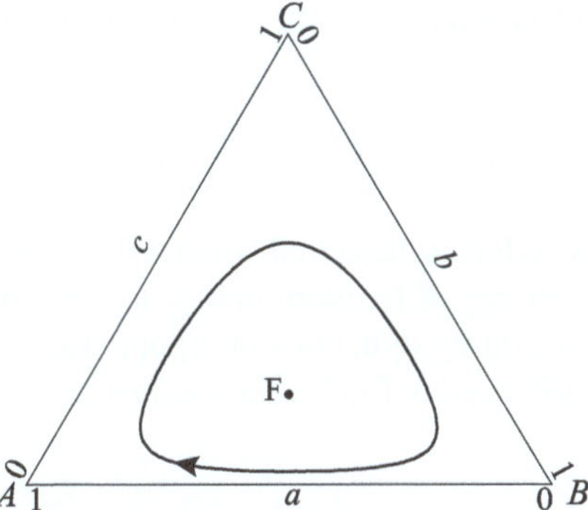



**Fig. 13.3** The three-species simplex for reaction rates $k_A = 0.2$, $k_B = 0.4$, $k_C = 0.4$. Since there is a conserved quantity, the rate equations predict cyclic orbits of constant $\rho = a^{k_B} b^{k_C} c^{k_A}$; $F$ signifies the neutrally stable reactive fixed point

### 13.2.2.2 The Concept of Fitness and Replicator Equations

Another line of thought to define an evolutionary dynamics, often taken in the mathematical literature of evolutionary game theory [21, 24], introduces the concept of fitness and then assumes that the per capita growth rate of a strategy $A_i$ is given by the surplus in its fitness with respect to the average fitness of the population. We will illustrate this reasoning for two-strategy games with a payoff matrix given by (13.2). Let $N_A$ and $N_B$ be the number of individuals playing strategy $A$ (cooperator) and $B$ (defector), respectively, in a population of size $N = N_A + N_B$. Then the relative abundances of strategies $A$ and $B$ are given by

$$a = \frac{N_A}{N} , \quad b = \frac{N_B}{N} = (1 - a) . \tag{13.5}$$

The "fitness" of a particular strategy $A$ or $B$ is defined as a constant *background fitness*, set to 1, plus the *average payoff* obtained from playing the game:

$$f_A(a) := 1 + \mathcal{R}a + \mathcal{S}(1 - a) , \tag{13.6}$$

$$f_B(a) := 1 + \mathcal{T}a + \mathcal{P}(1 - a) . \tag{13.7}$$

In order to mimic an evolutionary process one is seeking a dynamics which guarantees that individuals using strategies with a fitness larger than the average fitness increase while those using strategies with a fitness below average decline in number. This is, for example, achieved by choosing the per capita growth rate, $\partial_t a / a$, of individuals playing strategy $A$ proportional to their surplus in fitness with respect to the average fitness of the population:

$$\bar{f}(a) := a f_A(a) + (1 - a) f_B(a) . \tag{13.8}$$

---

prey, predators die, but there is a positive contribution to their growth which increases linearly with the amount of prey present [22, 23].

The ensuing ordinary differential equation is known as the *standard replicator equation* [21, 24]

$$\partial_t a = \left[ f_A(a) - \bar{f}(a) \right] a \ . \tag{13.9}$$

Lacking a detailed knowledge of the actual "interactions" of individuals in a population, there is, of course, plenty of freedom in how to write down a differential equation describing the evolutionary dynamics of a population. Indeed, there is another set of equations frequently used in EGT, called *adjusted replicator equations*, which reads

$$\partial_t a = \frac{f_A(a) - \bar{f}(a)}{\bar{f}(a)} a \ . \tag{13.10}$$

The correct form to be used in an actual biological setting may be neither of these standard formulations. Typically, some knowledge about the molecular mechanisms is needed to formulate a realistic dynamics. As we will learn in Sect. 13.3, the functional form of the payoff depends on the microbes' metabolism and is, in general, a nonlinear function of the relative abundances of the various strains in the population.

One may also criticise the assumption of constant population size made in evolutionary game theory. The internal evolution of different traits and the dynamics of the species population size are in fact not independent [25]. Species typically coevolve with other species in a changing environment and a separate description of both evolutionary and population dynamics is in general not justified. In particular, not only does a species' population dynamics affect the evolution within each species, as considered for example by models of density-dependent selection [26], but population dynamics is also biased by the internal evolution of different traits. One visual example of this coupling is provided by biofilms, which permanently grow and shrink. In these microbial structures diverse strains live, interact, and outcompete each other while simultaneously affecting the population size [14]. A proper combined description of the total temporal development should therefore be solely based on isolated birth and death events, as recently suggested by Melbinger et al. [27]. Such an approach offers also a more biological interpretation of evolutionary dynamics than common formulations such as the Fisher–Wright or Moran process [28–31]: fitter individuals prevail due to higher birth rates and not by winning a tooth-and-claw struggle where the birth of one individual directly results in the death of another.

### 13.2.3 Nonlinear Dynamics of Two-Player Games

This section is intended to give a concise introduction to elementary concepts of nonlinear dynamics [32]. We illustrate those for the evolutionary dynamics of two-player games characterized in terms of the payoff matrix (13.2) and the ensuing replicator dynamics

$$\partial_t a = a(f_A - \bar{f}) = a(1-a)(f_A - f_B) \,. \tag{13.11}$$

This equation has a simple interpretation: the first factor, $a(1-a)$, is the probability of $A$ and $B$ meeting and the second factor, $f_A - f_B$, is the fitness advantage of $A$ over $B$. Inserting the explicit expressions for the fitness values one finds

$$\partial_t a = a(1-a)\big[\mu_A(1-a) - \mu_B a\big] =: F(a)\,, \tag{13.12}$$

where $\mu_A$ is the relative benefit of $A$ playing against $B$ and $\mu_B$ is the relative benefit of $B$ playing against $A$:

$$\mu_A := \mathcal{S} - \mathcal{P}\,, \qquad \mu_B := \mathcal{T} - \mathcal{R}\,. \tag{13.13}$$

Equation (13.12) is a one-dimensional nonlinear first-oder differential equation for the fraction $a$ of players $A$ in the population, whose dynamics is most easily analyzed graphically. The sign of $F(a)$ determines the increase or decrease of the dynamic variable $a$; see the right half of Fig. 13.4. The intersections of $F(a)$ with the $a$-axis (zeros) are *fixed points*, $a^*$. Generically, these intersections have a finite slope $F'(a^*) \neq 0$; a negative slope indicates a stable fixed point and a positive slope, an unstable fixed point. Depending on some *control parameters*, here $\mu_A$ and $\mu_B$, the first or higher order derivatives of $F$ at the fixed points may vanish. These special parameter values mark "threshold values" for changes in the flow behavior (*bifurcations*) of the nonlinear dynamics. We may now classify two-player games as illustrated in Fig. 13.4.

For the prisoner's dilemma $\mu_A = -c < 0$ and $\mu_B = c > 0$ and hence players with strategy $B$ (defectors) are always better off (compare the payoff matrix). Both players playing strategy $B$ is a Nash equilibrium. In terms of the replicator equations this situation corresponds to $F(a) < 0$ for $a \neq 0$ and $F(a) = 0$ at $a = 0, 1$, such that $a^* = 0$ is the only stable fixed point. Hence the term "Nash equilibrium"

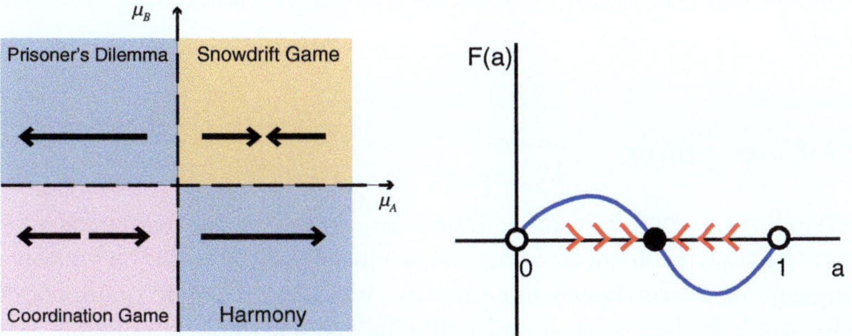


**Fig. 13.4** Classification of two-player games. *Left*: The *black arrows* in the control parameter plane $(\mu_A, \mu_B)$ indicate the flow behavior of the four different types of two-player games. *Right*: Graphically the solution of a one-dimensional nonlinear dynamics equation, $\partial_t a = F(a)$, is simply read off from the signs of the function $F(a)$; illustration for the snowdrift game

translates into the "stable fixed point" of the replicator dynamics (nonlinear dynamics).

For the snowdrift game both $\mu_A > 0$ and $\mu_B > 0$, such that $F(a)$ can change sign for $a \in [0, 1]$. In fact, $a^*_{\text{int}} = \mu_A/(\mu_A + \mu_B)$ is a stable fixed point while $a^* = 0, 1$ are unstable fixed points; see the right panel of Fig. 13.4. Inspection of the payoff matrix tells us that it is always better to play the opposite strategy to your opponent. Hence there is no Nash equilibrium in terms of pure strategies $A$ or $B$. This corresponds to the fact that the boundary fixed points $a^* = 0, 1$ are unstable. There is, however, a Nash equilibrium with a *mixed strategy*, where a rational player would play strategy $A$ with probability $p_A = \mu_A/(\mu_A + \mu_B)$ and strategy $B$ with probability $p_B = 1 - p_A$. Hence, again, the term "Nash equilibrium" translates into the "stable fixed point" of the replicator dynamics (nonlinear dynamics).

For the coordination game, there is also an interior fixed point at $a^*_{\text{int}} = \mu_A/(\mu_A + \mu_B)$, but now it is unstable, while the fixed points at the boundaries $a^* = 0, 1$ are stable. Hence we have *bistability*: for initial values $a < a^*_{\text{int}}$ the flow is towards $a = 0$ whereas it is towards $a = 1$ otherwise. In the terminology of strategic games there are two Nash equilibria. The game harmony corresponds to the prisoner's dilemma with the roles of $A$ and $B$ interchanged.

## 13.3 Games in Microbial Metapopulations

Two of the most fundamental questions that challenge our understanding of evolution and ecology are the origin of *cooperation* [4, 5, 8, 9, 33–36] and *biodiversity* [6, 7, 37–39]. Both are ubiquitous phenomena yet conspicuously difficult to explain since the fitness of an individual or the whole community depends in an intricate way on a plethora of factors, such as spatial distribution and mobility of individuals, secretion and detection of signaling molecules, toxin secretion leading to inter-strain competition and changes in environmental conditions. It is fair to say that we are still a long way from a full understanding, but the versatility of microbial communities makes their study a worthwhile endeavor with exciting discoveries still ahead of us.

### *13.3.1 Cooperation*

Understanding the conditions that promote the emergence and maintenance of cooperation is a classic problem in evolutionary biology [21, 40, 41]. It can be stated in the language of the prisoner's dilemma. By providing a public good, cooperative behavior would be beneficial for all individuals in the whole population. However, since cooperation is costly, the population is at risk from invasion by "selfish" individuals (cheaters), who save the cost of cooperation but can still obtain the benefit of cooperation from others. In evolutionary theory many principles have been proposed to overcome this dilemma of cooperation: repeated interaction [35, 40], punishment [35, 42], or kin discrimination [14, 43]. All of these principles share one fundamental feature: they are based on some kind of selection mechanism. Similar to the old debate between "selectionists" and "neutralists" in evolutionary theory [44], there is an alternative. Owing to random fluctuations, a population initially composed of both cooperators and defectors may (with some probability) become fixed in a state of cooperators only [45].

There have been an increasing number of experiments using microorganisms trying to shed new light on the problem of cooperation [8, 9, 33, 34]. Here, we will briefly discuss a recent experiment on "cheating in yeast" [8]. Budding yeast prefers to use the monosaccharides glucose and fructose as carbon sources. If it has to grow on sucrose instead, the disaccharide must first be hydrolyzed by the enzyme invertase. Since a fraction of approximately $1 - \epsilon = 99\%$ of the produced monosaccharides diffuses away and is shared with neighboring cells, it constitutes a public good available to the whole microbial community. This makes the population susceptible to invasion by mutant strains that save the metabolic cost of producing invertase. One is now tempted to conclude from what we have discussed in the previous sections that yeast is playing the prisoner's dilemma game. The cheater strains should take over the population and the wild-type strain should become extinct. But, this is not the case. Gore and collaborators [8] show that the dynamics is rather described as a snowdrift game, in which cheating can be profitable but is not necessarily the best strategy if others are cheating too. The explanation given is that the growth rate as a function of glucose is highly concave and, as a consequence, the fitness function is *nonlinear* in the payoffs[2]

$$f_C(x) := \left[\epsilon + x(1 - \epsilon)\right]^\alpha - c \,, \tag{13.14}$$

$$f_D(a) := \left[x(1 - \epsilon)\right]^\alpha \,, \tag{13.15}$$

with $\alpha \approx 0.15$ determined experimentally. The ensuing phase diagram Fig. 13.5 as a function of capture efficiency $\epsilon$ and metabolic cost $c$ shows an altered intermediate regime with a bistable phase portrait, that is, the hallmark of a snowdrift game as discussed in the previous section. This explains the experimental observations. The lesson to be learned from this investigation is that defining a payoff function is not a trivial matter, and a naive replicator dynamics fails to describe biological reality. It is, in general, necessary to take a detailed look at the nature of the biochemical processes responsible for the growth rates of the competing microbes.

---

[2] Note that $\epsilon$ is the fraction of carbon source kept by cooperators solely for themselves and $x(1-\epsilon)$ is the amount of carbon source shared with the whole community. Hence, the linear growth rates of cooperators and defectors would be $\epsilon + x(1 - \epsilon) - c$ and $x(1 - \epsilon)$, respectively, where $c$ is the metabolic cost for invertase production.

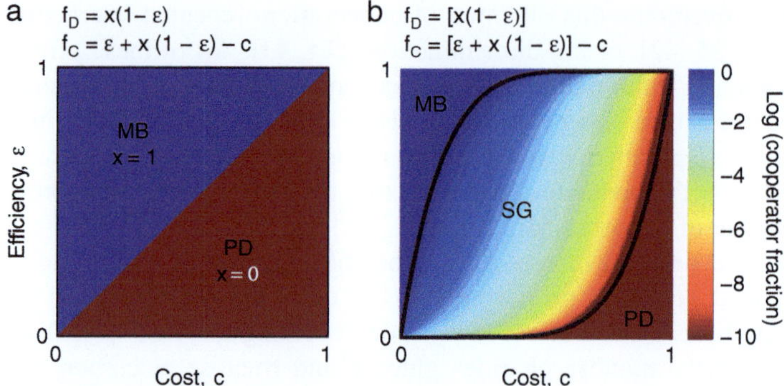


**Fig. 13.5** Game theory models of cooperation in sucrose metabolism of yeast. **a** Phase diagram resulting from fitness functions $f_C$ and $f_D$ linear in the payoffs. This model leads to fixation of cooperators ($x = 1$) at low cost and/or high efficiency of capture ($\epsilon > c$, implying that the game is mutually beneficial (MB)) but fixation of defectors ($x = 0$) for high cost and/or low efficiency of capture ($\epsilon < c$, implying that the game is the prisoner's dilemma (PD)). **b** A model of cooperation with experimentally measured concave benefits yields a central region of parameter space that is a snowdrift game (SG), thus explaining the coexistence that is observed experimentally ($\alpha = 0.15$). Adapted from [8]

## 13.3.2 Pattern Formation

Investigations of microbial pattern formation have often focused on one bacterial strain [46–48]. In this respect, it has been found that bacterial colonies on substrates with a high nutrient level and intermediate agar concentrations, representing "friendly" conditions, grow in simple compact patterns [49]. When instead the level of nutrient is lowered, when the surface on which bacteria grow possesses heterogeneities, or when the bacteria are exposed to antibiotics, complex, fractal patterns are observed [46, 50, 51]. Other factors that affect the self-organizing patterns include motility [52], the kind of bacterial movement, for example, swimming [53], swarming, or gliding [54, 55], as well as chemotaxis and external heterogeneities [56]. Another line of research has investigated patterns of multiple co-evolving bacterial strains. As an example, recent studies looked at growth patterns of two functionally equivalent strains of *Escherichia coli* and showed that, due to fluctuations alone, they segregate into well-defined, sector like regions [47, 57].

## 13.3.3 The Escherichia coli Col E2 System

Several colibacteria such as *Escherichia coli* are able to produce and secrete specific toxins called colicins that inhibit growth of other bacteria. Kerr and coworkers [6] have studied three strains of *E. coli*, one of which is able to produce the toxin Col E2 that acts as a DNA endonuclease. This poison-producing strain (C) kills a sensitive strain (S), which outgrows the third, resistant one (R), as resistance bears certain

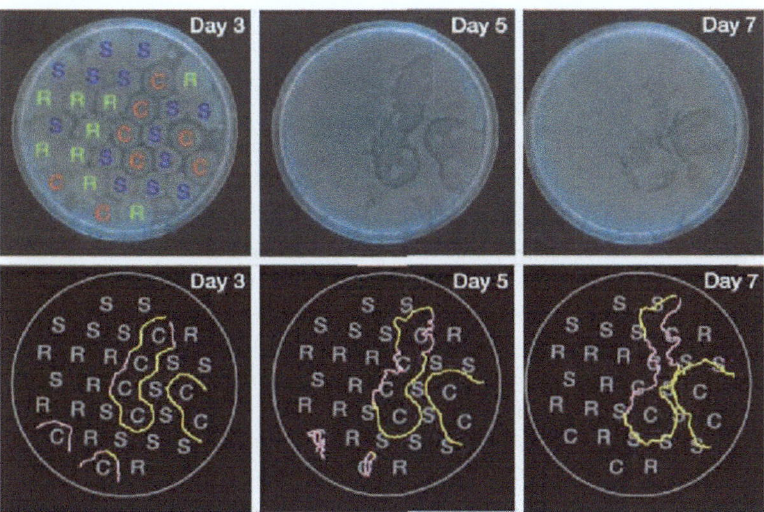

**Fig. 13.6** The three strains of the *E. coli* Col E2 system evolve into spatial patterns on a Petri dish. The competition of the three strains is cyclic (of "rock–paper–scissors" type) and therefore nonequilibrium in nature, leading to dynamic patterns. The pictures have been modified from [6]

costs. The resistant bacteria grow faster than the poisonous ones, as the latter are resistant and produce poison, which is yet an extra cost. Consequently, the three strains of *E. coli* display cyclic competition, similar to the children's game "rock–paper–scissors".

When placed on a Petri dish, all three strains coexist, arranging in time-dependent spatial clusters dominated by one strain. In Fig. 13.6, snapshots of these patterns monitored over several days are shown. Sharp boundaries between different domains emerge, and all three strains coevolve at comparable densities. The patterns are dynamic: Owing to the nonequilibrium character of the species' interactions, clusters dominated by one bacterial strain cyclically invade each other, resulting in an endless hunt of the three species on the Petri dish. The situation changes considerably when the bacteria are placed in a flask with additional stirring. Then, only the resistant strain survives, the two others dying out after a short transient time.

These laboratory experiments thus provide intriguing experimental evidence for the importance of spatial patterns for the maintenance of biodiversity. In this respect, many further questions regarding the spatio-temporal interactions of competing organisms under different environmental conditions lie ahead. Spontaneous mutagenesis of single cells can lead to enhanced fitness under specific environmental conditions or due to interactions with other species. Moreover, interactions with other species may allow unfit, but potentially pathogenic bacteria to colonize certain tissues. Additionally, high concentrations of harmless bacteria may help pathogenic ones to nest on tissues exposed to extremely unfriendly conditions. Information about bacterial pattern formation arising from bacterial interaction may therefore allow mechanisms to avoid pathogenic infection to be developed.

## 13.4 Stochastic Dynamics in Well-Mixed Populations

The machinery of biological cells consists of networks of molecules interacting with each other in a highly complex manner. Many of these interactions can be described as chemical reactions, where the intricate processes that occur during the encounter of two molecules are reduced to reaction rates, that is, probabilities per unit time. This notion of stochasticity carries over to the scale of microbes in a manifold of ways. There is *phenotypic noise* . Owing to fluctuations in transcription and translation, phenotypes vary even in the absence of genetic differences between individuals and despite constant environmental conditions [58, 59]. In addition, phenotypic variability may arise due to various external factors such as cell density, nutrient availability, and other stress conditions. A general discussion of phenotypic variability in bacteria may be found in recent reviews [60–63]. There is *interaction noise*. Interactions between individuals in a given population, as well as cell division and cell death, occur at random points in time (following some probability distribution) and lead to discrete steps in the numbers of the different species. Then, as noted long ago by Delbrück [64], a deterministic description, as discussed in the previous section, breaks down for small copy numbers. Finally, there is *external noise* due to spatial heterogeneities or temporal fluctuations in the environment. In this section we will focus on *interaction noise*, whose role for extinction processes in ecology has recently been recognized to be very important, especially when the deterministic dynamics exhibits neutral stability [65–67] or weak stability [45, 68]. After a brief and elementary discussion of extinction times we will introduce a general concept for defining survival and extinction on ecological time scales. The concept of extinction will be illustrated for the stochastic dynamics of the cyclic Lotka–Volterra model [66].

### *13.4.1 Extinction Times and Classification of Coexistence Stability*

For a deterministic system, given an initial condition, the outcome of the evolutionary dynamics is certain. However, processes encountered in biological systems are often stochastic. For example, consider the degradation of a protein or the death of an individual bacterium in a population. To a good approximation it can be described as a stochastic event that occurs with a probability per unit time (rate) $\lambda$, known as a stochastic *linear death process*. Then the population size $N(t)$ at time $t$ becomes a random variable, and its time evolution becomes a set of integers $\{N_\alpha\}$ changing from $N_\alpha$ to $N_\alpha - 1$ at particular times $t_\alpha$; this is also called a realization of the stochastic process. Now it is no longer meaningful to ask for *the* time evolution of a particular population, as one would do in a deterministic description in terms of a rate equation, $\partial_t N = -\lambda N$. Instead one studies the time evolution of an ensemble of systems or tries to understand the distribution of times $\{t_\alpha\}$. A central quantity in this endeavor is the probability $P(N, t)$ of finding a population of size $N$ given that at some time $t = 0$ there was some initial ensemble of populations. Assuming that the stochastic process is Markovian, its dynamics is given by the following master equation:

$$\partial_t P(N,t) = \lambda(N+1)P(N+1,t) - \lambda N P(N,t) \,. \tag{13.16}$$

A master equation is a "balance equation" for probabilities. The right-hand side simply states that there is an increase in $P(N,t)$ if in a population of size $N+1$ an individual dies with rate $\lambda$, and a decrease in $P(N,t)$ if in a population of size $N$ an individual dies with rate $\lambda$. Master equations can be analyzed by standard tools from the theory of stochastic processes [69, 70].

A quantity of central interest is the average extinction time $T$, that is, the expected time for the population to reach the state $N = 0$. This state is also called an *absorbing state* since (for the linear death process considered here) there are processes leading into but not out of this state. The expected extinction time $T$ can be obtained using rather elementary concepts from probability theory. Consider the probability $Q(t)$ that a given individual is still alive at time $t$ conditioned on that it was alive at some initial time $t = 0$. Since an individual will be alive at time $t + dt$ if it was alive at time $t$ and did not die within the time interval $[t, t + dt]$ we immediately obtain the identity

$$Q(t + dt) = Q(t)(1 - \lambda t) \quad \text{with} \quad Q(0) = 1 \,. \tag{13.17}$$

The ensuing differential equation (in the limit $dt \to 0$), $\dot{Q} = -\lambda Q$ is solved by $Q(t) = \mathrm{e}^{-\lambda t}$. This identifies $\tau = 1/\lambda$ as the expected waiting time for a particular individual to die. We conclude that the waiting times for the population to change by one individual is distributed exponentially and its expected value is $\tau_N = \tau/N$ for a population of size $N$; note that each individual in a population has the same chance of dying. Hence we can write for the expected extinction time for a population with initial size $N_0$

$$T = \tau_{N_0} + \tau_{N_0-1} + \cdots + \tau_1 = \sum_{N=1}^{N_0} \frac{\tau}{N} \approx \tau \int_1^{N_0} \frac{1}{N}\,dN = \tau \ln N_0 \,. \tag{13.18}$$

We have learned that for a system with a "drift" towards the absorbing boundary of the state space the expected time to reach this boundary scales, quite generically, logarithmically in the initial population size, $T \sim \ln N_0$. Note that within a deterministic description, $\dot{N} = -\lambda N$, the population size would exponentially decay to zero but never reach it, $N(t) = N_0 \mathrm{e}^{-t/\tau}$. This is, of course, flawed in two ways. First, the process is not deterministic and, second, the population size is not a real number. Both features are essential to understand the actual dynamics of a population at low copy numbers of individuals.

Now we would like to contrast the linear death process with a "neutral process", where death and birth events balance each other, that is, where the birth rate $\mu$ exactly equals the death rate $\lambda$. In a deterministic description one would write

$$\partial_t N(t) = -(\lambda - \mu)N(t) = 0 \tag{13.19}$$

and conclude that the population size remains constant at its initial value. In a stochastic description, one starts from the master equation

$$\partial_t P(N, t) = \lambda(N+1)P(N+1, t) + \lambda(N-1)P(N-1, t) - 2\lambda N P(N, t) \,. \quad (13.20)$$

Though this could be solved exactly using generating functions it is instructive to derive an approximation valid in the limit of a large population size, that is, $N \gg 1$. This is most easily done by simply performing a second-order Taylor expansion without worrying too much about the mathematical validity of such an expansion. With

$$(N \pm 1)P(N \pm 1, t) \approx NP(N, t) \pm \partial_N\big[NP(N, t)\big] + \frac{1}{2}\partial_N^2\big[NP(N, t)\big]$$

one obtains

$$\partial_t P(N, t) = \lambda \partial_N^2\big[NP(N, t)\big] \,. \quad (13.21)$$

Measuring the population size in units of the initial population size at time $t = 0$ and defining $x = N/N_0$, this becomes

$$\partial_t P(x, t) = D\partial_x^2\big[xP(x, t)\big] \quad (13.22)$$

with the "diffusion constant" $D = \lambda/N_0$. This implies that all time scales in the problem scale as $t \sim D^{-1} \sim N_0$; this is easily seen by introducing a dimensionless time $\tau = Dt$, resulting in a rescaled equation

$$\partial_\tau P(x, \tau) = \partial_x^2\big[xP(x, \tau)\big] \,. \quad (13.23)$$

Hence for a (deterministically) "neutral dynamics", the extinction time, that is, the time to reach the absorbing state $N = 0$, scales, also quite generically, linearly in the initial system size $T \sim N_0$.

Finally, there are processes such as the snowdrift game where the deterministic dynamics drives the population towards an interior fixed point well separated from the absorbing boundaries $x = 0$ and $x = 1$. In such a case, starting from an initial state in the vicinity of the interior fixed point, the stochastic dynamics has to overcome a finite barrier in order to reach the absorbing state. This is reminiscent of a chemical reaction with an activation barrier, which is described by an Arrhenius law. Hence we expect that the extinction time scales exponentially in the initial population size $T \sim \exp N_0$.

These simple arguments on the dependence of the mean extinction time $T$ of competing species on the system size $N$ can now be used to define a general framework to distinguish neutral from selection-dominated evolution. For a selection-dominated parameter regime, instability leads to steady decay of a species, and therefore to fast extinction [13, 71, 72]: The mean extinction time $T$ increases only logarithmically with the population size $N$, $T \sim \ln N$, and a larger system size does not ensure much longer coexistence. This behavior can be understood by noting that a species disfavored by selection decreases at a constant rate. Consequently, its population size decays exponentially in time, leading to a logarithmic dependence of the extinction time on the initial population size. In contrast, stable existence of a species induces $T \sim \exp N$, such that extinction takes an astronomically long time for large populations [45, 71, 72]. In this regime, extinction stems from large fluctuations that cause sufficient deviation from the (deterministically) stable coexistence. These large deviations are exponentially suppressed and hence the time until a rare extinction event occurs scales exponentially with the system size $N$. Then coexistence is maintained on ecologically relevant time scales, which typically lie below $T$. An intermediate situation, that is, when $T$ has a power-law dependence on $N$, $T \sim N^{\alpha}$, signals dominant influences of stochastic effects and corresponds to neutral evolution. Here the extinction time grows considerably, though not exponentially, with increasing population size. Large $N$ therefore clearly prolongs coexistence of species but can still allow extinction within biologically reasonable time scales. Summarizing these considerations, we have proposed a quantitative classification of a coexistence's stability in the presence of absorbing states, which is presented in Table 13.1 [13].

The strength of this classification lies in that it only involves quantities that are directly measurable (for example through computer simulations), namely the mean extinction time and the system size. Therefore, it is generally applicable to stochastic processes, for example, incorporating additional internal population structure such as individuals' age or sex, or where individuals' interaction networks are more complex, such as lattices, scale-free networks, or fractal ones. In these situations, it is typically impossible to infer analytically, from the discussion of fixed points stability, whether the deterministic population dynamics yields a stable or unstable coexistence. However, based on the scaling of extinction time $T$ with system size $N$, differentiating stable from unstable diversity according to the above classification is feasible. In Sect. 13.5, we will follow this line of thought and fruitfully apply the above concept to the investigation of a "rock–paper–scissors" game on a two-dimensional lattice, where individuals' mobility is found to mediate between stable and unstable coexistence.

**Table 13.1** Classification of coexistence stability

---

*Stability:* If the mean extinction time $T$ increases faster than any power of the system size $N$, meaning $T/N^{\alpha} \to \infty$ in the asymptotic limit $N \to \infty$ and for any value of $\alpha > 0$, we refer to the coexistence as stable. In this situation, typically, $T$ increases exponentially in $N$.

*Instability:* If the mean extinction time $T$ increases more slowly than any power in the system size $N$, meaning $T/N^{\alpha} \to 0$ in the asymptotic limit $N \to \infty$ and for any value of $\alpha > 0$, we refer to the coexistence as unstable. In this situation, typically, $T$ increases logarithmically in $N$.

*Neutral stability:* Neutral stability lies between stable and unstable coexistence. It emerges when the mean extinction time $T$ increases proportional to some power $\alpha > 0$ of the system size $N$, meaning $T/N^{\alpha} \to \mathcal{O}(1)$ in the asymptotic limit $N \to \infty$.

---

## 13.4.2 Cyclic Three-Strategy Games

As we have learned in the previous section, the coexistence of competing species is, owing to unavoidable fluctuations, always transient. Here we illustrate this for the cyclic Lotka–Volterra model ("rock–paper–scissors" game) introduced in Sect. 13.2.2 on evolutionary game theory as a mathematical description of nontransitive dynamics. Like the original Lotka–Volterra model, the deterministic dynamics of the "rock–paper–scissors" game yields oscillations along closed, periodic orbits around a coexistence fixed point. These orbits are neutrally stable due to the existence of a conserved quantity $\rho$. If noise is included in such a game, it is clear that eventually only one of the three species will survive [66, 73–75] (Dobrinevski and Frey). However, it is far from obvious which species will most likely win the contest. Intuitively, one might think, at a first glance, that it pays for a given strain to have the highest reaction rate and hence strongly dominate its competitors. As it turns out, however, the exact opposite strategy is the best [76]. One finds what could be called a "law of the weakest": When the interactions between the three species are (generically) asymmetric, the "weakest" species (i.e., the one with the smallest reaction rate) survives at a probability that tends to one in the limit of a large population size, while the other two are guaranteed to go extinct.

The reason for this unexpected behavior is illustrated in Fig. 13.7, showing a deterministic orbit and a typical stochastic trajectory. For asymmetric reaction rates, the fixed point is shifted from the center Z of the phase space (simplex) towards one of the three edges. All deterministic orbits are changed in the same way, squeezing in the direction of one edge. In Fig. 13.7 reaction rates are chosen such that the distance $\lambda_A$ to the $a$-edge of the simplex, where $A$ would win the contest, is smallest. The important observation here is that for simple geometric reasons $\lambda_A$ is smallest because the reaction rate $k_A$ is smallest! Intuitively, the absorbing state that

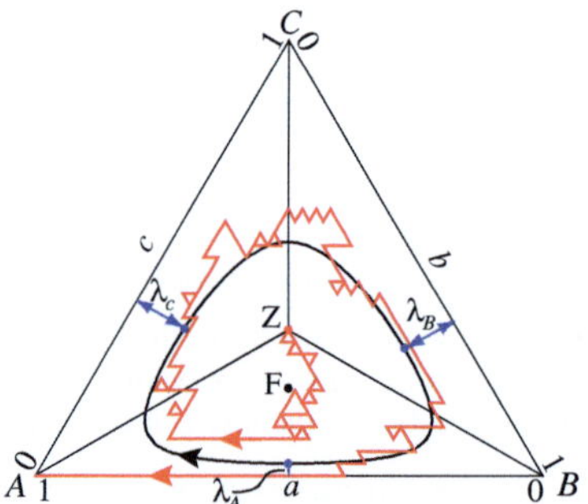


**Fig. 13.7** The phase space $S_3$. We show the reactive fixed point F, the center Z, and a stochastic trajectory (*red*). It eventually deviates from the "outermost" deterministic orbit (*black*) and reaches the absorbing boundary. $\lambda_A$, $\lambda_B$ and $\lambda_C$ (*blue/dark gray*) denote the distances of the "outermost" orbit to the boundaries. Parameters are $(k_A, k_B, k_C) = (0.2, 0.4, 0.4)$ and $N = 36$. Figure adapted from [76]

is reached from this edge has the highest probability of being hit, as the distance $\lambda$ from the deterministic orbit towards this edge is shortest. Indeed, this behavior can be validated by stochastic simulations complemented by a scaling argument [76].

## 13.5 Spatial Games with Cyclic Dominance

*Spatial distribution of individuals*, as well as their *mobility*, are common features of real ecosystems that often come paired [77]. On all scales of living organisms, from bacteria residing in soil or on Petri dishes, to the largest animals living in savannas – such as elephants – or in forests, populations' habitats are spatially extended and *individuals interact locally* within their neighborhood. Field studies as well as experimental and theoretical investigations have shown that the locality of the interactions leads to the self-formation of complex spatial patterns [6, 11, 77–90]. Another important property of most individuals is *mobility*. For example, bacteria swim and tumble, and animals migrate. As motile individuals are capable of enlarging their district of residence, mobility may be viewed as a mixing, or stirring mechanism that "counteracts" the locality of spatial interactions.

### *13.5.1 The Role of Mobility in Ecosystems*

The interplay between mobility and spatial separation on the spatio-temporal development of populations is one of the most interesting and complex problems in theoretical ecology [13, 77–79, 81, 83]. If *mobility is low*, locally interacting populations can exhibit involved spatio-temporal patterns, such as traveling waves [91], and, for example, lead to the self-organization of individuals into spirals in myxobacteria aggregation [91] and insect host–parasitoid populations [11]. In contrast, *high mobility* results in well-mixed systems where the spatial distribution of the populations is irrelevant [13]. In this situation, spatial patterns no longer form: The system adopts a spatially uniform state, which therefore drastically differs from the low-mobility scenario. Pioneering work on the role of mobility in ecosystems was performed by Levin [10], who investigated the dynamics of a population residing in two coupled patches: Within a deterministic description, he identified a critical value for the individuals' mobility between the patches. Below the critical threshold, all subpopulations coexisted, while only one remained above that value. Later, more realistic models of many patches, partially spatially arranged, were also studied; see for example [11, 81, 82, 92] and references therein. These works shed light on the formation of patterns, in particular traveling waves and spirals. However, patch models have been criticized for treating the space in an "implicit" manner (i.e., in the form of coupled habitats without internal structure) [38]. In addition, the above investigations were often restricted to deterministic dynamics and thus did not address the spatio-temporal influence of noise. To overcome these limitations, Durrett and Levin [37] proposed considering interacting particle systems, that is, stochastic spatial models with populations of discrete individuals distributed on lattices. In this realm, studies have mainly focused on numerical simulations and on deterministic reaction–diffusion equations, or coupled maps [12, 37–39, 83, 88, 93–95].

### 13.5.2 Cyclic Dominance in Ecosystems

An intriguing motif of the complex competitions in a population, promoting species diversity, is constituted by three subpopulations exhibiting cyclic dominance, also called nontransitive competition. This basic motif is metaphorically described by the "rock–paper–scissors" game, where rock crushes scissors, scissors cut paper, and paper wraps rock. Such nonhierarchical, cyclic competitions, where each species outperforms another but is also itself outperformed by another one, have been identified in different ecosystems such as coral reef invertebrates [96], rodents in the high-Arctic tundra in Greenland [97], lizards in the inner Coast Range of California [98], and microbial populations of colicinogenic *E. coli* [6, 99]. As we have discussed in Sect. 13.3, in the latter situation it has been shown that spatial arrangement of quasi-immobile bacteria on a Petri dish leads to the stable coexistence of all three competing bacterial strains, with the formation of irregular patterns. In stark contrast, when the system is well-mixed, there is spatial homogeneity, resulting in the takeover of one subpopulation and the extinction of the others after a short transient.

### 13.5.3 The May–Leonard Model

In ecology competition for resources has been classified [100] into two broad groups, *scramble* and *contest*. Contest competition involves direct interaction between individuals. In the language of evolutionary game theory the winner in the competition replaces the loser in the population (Moran process). In contrast, scramble competition involves rapid use of limiting resources without direct interaction between the competitors. The May–Leonard model [15] of cyclic dominance between three subpopulations $A$, $B$, and $C$ dissects the nontransitive competition between these into a contest and a scramble step. In the contest step an individual of subpopulation $A$ outperforms a $B$ through "killing" (or "consuming"), symbolized by the ("chemical") reaction $AB \rightarrow A\oslash$, where $\oslash$ denotes an available empty space. In the same way, $B$ outperforms $C$, and $C$ beats $A$ in turn, closing the cycle. We refer to these contest interactions as selection and denote the corresponding rate by $\sigma$. In the scramble step, which mimics a finite carrying capacity, each member of a subpopulation is allowed to reproduce only if an empty space is available, as described by the reaction $A\oslash \rightarrow AA$ and analogously for $B$ and $C$. For all subpopulations, these reproduction events occur with rate $\mu$, such that the three subpopulations equally compete for empty space. To summarize, the reactions that define the May–Leonard model (selection and reproduction) read

$$\begin{aligned}
AB &\xrightarrow{\sigma} A\oslash\,, & A\oslash &\xrightarrow{\mu} AA\,, \\
BC &\xrightarrow{\sigma} B\oslash\,, & B\oslash &\xrightarrow{\mu} BB\,, \\
CA &\xrightarrow{\sigma} C\oslash\,, & C\oslash &\xrightarrow{\mu} CC\,.
\end{aligned} \tag{13.24}$$

Let $a$, $b$, and $c$ denote the densities of subpopulations $A$, $B$, and $C$, respectively. The overall density $\rho$ then reads $\rho = a + b + c$. As every lattice site is at most occupied by one individual, the overall density (as well as densities of each subpopulation) varies between 0 and 1, that is, $0 \le \rho \le 1$. With this notation, the rate equations for the reactions (13.24) are given by

$$\begin{aligned}
\partial_t a &= a\left[\mu(1 - \rho) - \sigma c\right]\,, \\
\partial_t b &= b\left[\mu(1 - \rho) - \sigma a\right]\,, \\
\partial_t c &= c\left[\mu(1 - \rho) - \sigma b\right]\,.
\end{aligned} \tag{13.25}$$

The phase space of the model is organized by fixed point and invariant manifolds. Equations (13.25) possess four absorbing fixed points. One of these (unstable) is associated with the extinction of all subpopulations, $(a_1^*, b_1^*, c_1^*) = (0, 0, 0)$. The others are heteroclinic points (i.e., saddle points underlying the heteroclinic orbits) and correspond to the survival of only one subpopulation, $(a_2^*, b_2^*, c_2^*) = (1, 0, 0)$, $(a_3^*, b_3^*, c_3^*) = (0, 1, 0)$, and $(a_4^*, b_4^*, c_4^*) = (1, 0, 0)$, shown in blue in Fig. 13.8. In addition, there exists a *reactive* fixed point, indicated in red in Fig. 13.8, where all three subpopulations coexist (at equal densities), namely $(a^*, b^*, c^*) = \frac{\mu}{3\mu+\sigma}(1, 1, 1)$.

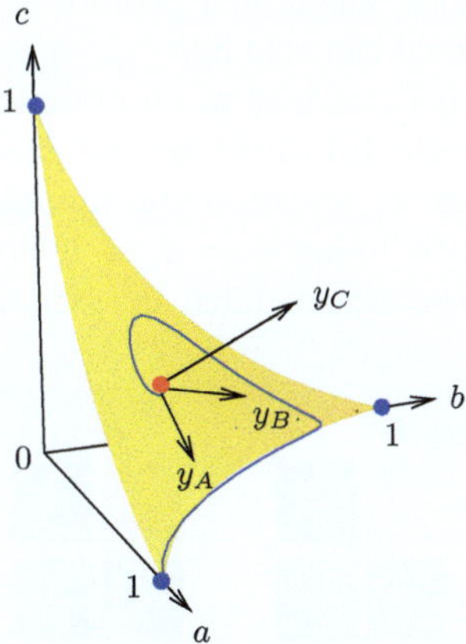


**Fig. 13.8** The phase space of the May–Leonard model. It is spanned by the densities $a$, $b$, and $c$ of species $A$, $B$, and $C$. On an invariant manifold (*yellow*), the flows obtained as solutions of the rate equations (13.25) (an example trajectory is shown in *blue*) initially in the vicinity of the reactive fixed point (*red*) spiral outwards, approaching the heteroclinic cycle that connects three trivial fixed points (*blue*). Adapted from [101]

For a nonvanishing selection rate, $\sigma > 0$, Leonard and May [15] showed that the reactive fixed point is unstable, and the system asymptotically approaches the boundary of the phase space (given by the planes $a = 0$, $b = 0$, and $c = 0$). There, they observed *heteroclinic orbits*: the system oscillates between states where nearly only one subpopulation is present, with rapidly increasing cycle duration. While mathematically fascinating, this behavior was recognized to be unrealistic [15]. For instance, as discussed in Sect. 13.4, the system will, due to finite-size fluctuations, always reach one of the absorbing fixed points in the vicinity of the heteroclinic orbit, and then only one population survives.

### 13.5.4 The Spatially Extended May–Leonard Model

As discussed above, in the experiments by the Kerr group [6] crucial influence of self-organized patterns on biodiversity has been demonstrated, employing three bacterial strains that display cyclic competition. Here, from theoretical studies, we show that cyclic competition of species can lead to highly nontrivial spatial patterns as well as counterintuitive effects on biodiversity. To this end we analyze the stochastic spatially extended version of the May–Leonard model [13], as illustrated in Fig. 13.9. We adopt an interacting particle description where individuals of all subpopulations are arranged on a lattice. Let $L$ denote the linear size of a two-dimensional square lattice (i.e., the number of sites along one edge), such that the total number of sites is $N = L^2$. In this approach, each site of the grid is either occupied by one individual or empty, meaning that the system has a finite carrying capacity, and the reactions are then only allowed between *nearest neighbors*.

In addition, we endow the individuals with a certain form of mobility. Namely, at rate $\epsilon$ all individuals can exchange their position with a nearest neighbor. With that same rate $\epsilon$, any individual can also hop onto a neighboring empty site. These "microscopic" exchange processes lead to an effective diffusion of the individuals described by a macroscopic diffusion constant $D = \epsilon/2L^2$. For simplicity, we consider equal reaction rates for selection and reproduction, and, without loss of generality, set the time unit by fixing $\sigma = \mu \equiv 1$. From the phase portrait of the May–Leonard model it is to be expected that an asymmetry in the parameters yields

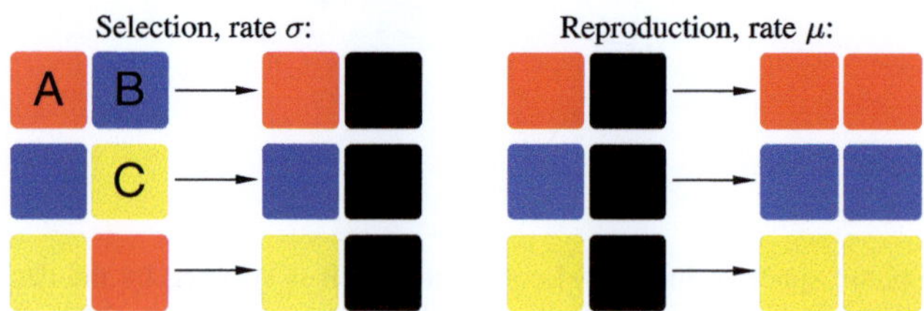


**Fig. 13.9** Individuals on neighboring sites may react with each other according to the rules of cyclic dominance (selection: contest competition), or individuals may give birth to new individuals if they happen to be next to an empty site (reproduction: scramble competition)

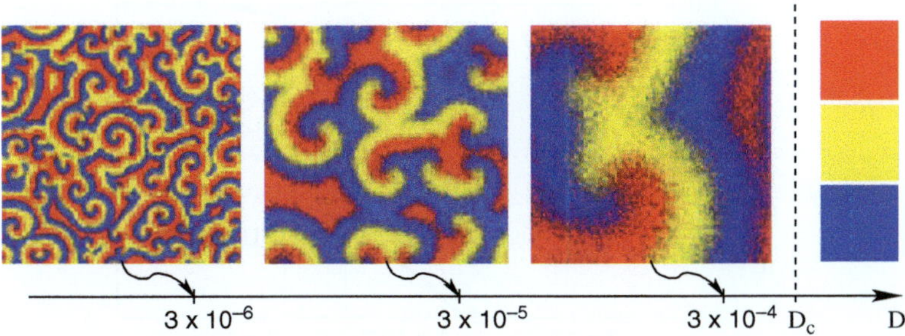



**Fig. 13.10** Snapshots obtained from lattice simulations are shown of typical states of the system after long temporal development (i.e., at time $t \sim N$) and for different values of $D$ (each color, *blue*, *yellow* and *red*, represents one of the species and *black dots* indicate *empty spots*). With increasing $D$ (from *left* to *right*), the spiral structures grow, and outgrow the system size at the critical mobility $D_c$: then, coexistence of all three species is lost and uniform populations remain (*right*). Figure adapted from [13]

only qualitative, not quantitative, changes in the system's dynamics. The length scale is chosen such that the linear dimension of the lattice is the basic length unit, $L \equiv 1$. With this choice of units the diffusion constant measures the fraction of the entire lattice area explored by an individual in one unit of time.

Typical snapshots of the steady states are shown in Fig. 13.10.[3] When the mobility of the individuals is low, one finds that all species coexist and self-arrange by forming patterns of moving spirals. With increasing mobility $D$, these structures grow in size, and they disappear for large enough $D$. In the absence of spirals, the system adopts a uniform state where only one species is present; the others have died out. Which species remains is subject to a random process, all species having an equal chance of surviving in the symmetric model defined above.

The transition from the reactive state containing spirals to the absorbing state with only one subpopulation left is a nonequilibrium phase transition [102]. One way to characterize the transition is to ask how the extinction time $T$, that is, the time for the system to reach one of its absorbing states, scales with system size $N$. In our analysis of the role of stochasticity in Sect. 13.4 we have found the following classification scheme. If $T \sim N$, the stability of coexistence is marginal. Conversely, longer (shorter) waiting times scaling with higher (lower) powers of $N$ indicate stable (unstable) coexistence. These three scenarios can be distinguished by computing the probability $P_{\text{ext}}$ that two species have gone extinct after a waiting time $t \sim N$:

$$P_{\text{ext}} = \text{Prob}\left[\text{only one species left after time } T \sim N\right] . \tag{13.26}$$

---

[3] You may also want to haven a look at the movies posted on http://www.theorie.physik. uni-muenchen.de/lsfrey/research/fields/biological_physics/2007_004/. There is also a Wolfram demonstration project that can be downloaded from the internet: http://demonstrations.wolfram. com/BiodiversityInSpatialRockPaperScissorsGames/.

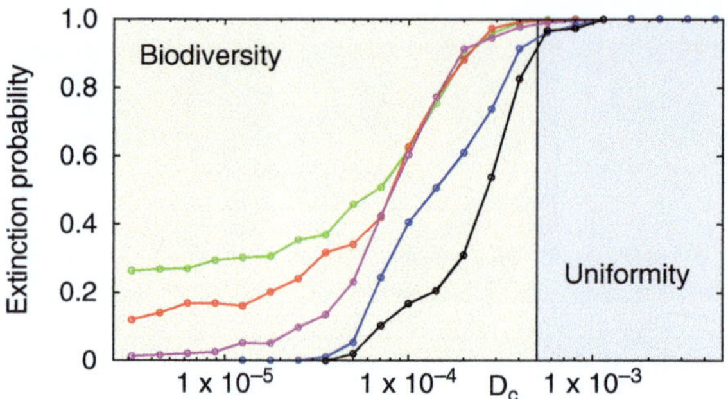


**Fig. 13.11** The extinction probability $P_{ext}$ that, starting with randomly distributed individuals on a square lattice, the system has reached an absorbing state after a waiting time $T \sim N$. $P_{ext}$ is shown as a function of the mobility $D$ (and $\sigma = \mu = 1$) for different system sizes: $N = 20 \times 20$ (*green*), $N = 30 \times 30$ (*red*), $N = 40 \times 40$ (*purple*), $N = 100 \times 100$ (*blue*), and $N = 200 \times 200$ (*black*). As the system size increases, the transition from stable coexistence ($P_{ext} = 0$) to extinction ($P_{ext} = 1$) sharpens at a critical mobility $D_c \approx (4.5 \pm 0.5) \times 10^{-4}$. Figure adapted from [13]

In Fig. 13.11, the dependence of $P_{ext}$ on the mobility $D$ is shown for a range of different system sizes $N$.

As the system size increases, a sharpened transition emerges at a critical value $D_c = (4.5 \pm 0.5) \times 10^{-4}$. Below $D_c$, the extinction probability $P_{ext}$ tends to zero as the system size increases, and coexistence is stable in the sense defined in Sect. 13.4. In contrast, above the critical mobility, the extinction probability approaches one for large system size, and coexistence is unstable. As a central result, agent-based simulations show that there is a critical threshold value for the individuals' diffusion constant, $D_c$, such that a low mobility, $D < D_c$, guarantees coexistence of all three species, while a high mobility, $D > D_c$, induces extinction of two of them, leaving a uniform state with only one species [13].

### 13.5.5 Pattern Formation and Reaction–Diffusion Equations

The emergence of spatial patterns, their form, and characteristic features can be understood by employing a continuum approach that maps the agent-based model to a set of stochastic partial differential equations (SPDEs) (often referred to as Langevin equations) [102]:

$$\partial_t a(\boldsymbol{r}, t) = D\Delta a(\boldsymbol{r}, t) + \mathcal{A}_A[\boldsymbol{a}] + \mathcal{C}_A[\boldsymbol{a}]\xi_A\,,$$
$$\partial_t b(\boldsymbol{r}, t) = D\Delta b(\boldsymbol{r}, t) + \mathcal{A}_B[\boldsymbol{a}] + \mathcal{C}_B[\boldsymbol{a}]\xi_B\,,$$
$$\partial_t c(\boldsymbol{r}, t) = D\Delta c(\boldsymbol{r}, t) + \mathcal{A}_C[\boldsymbol{a}] + \mathcal{C}_C[\boldsymbol{a}]\xi_C\,, \tag{13.27}$$

where $\boldsymbol{a} = (a, b, c)$ and $\Delta$ denotes the Laplacian operator. The first term describes the diffusive motion of each of the individual agents with a macroscopic diffusion constant $D$. The reaction terms $\mathcal{A}_i[\boldsymbol{a}]$ derived in a Kramers–Moyal expansion [101] are identical – as they must be – to the corresponding nonlinear drift term in the diffusion–reaction equation $\boldsymbol{F}[\boldsymbol{a}] = \mathcal{A}[\boldsymbol{a}]$, which describes coevolution of different species in the absence of spatial degrees of freedom and with a large number of interacting individuals. Noise arises because processes are stochastic *and* population size $N$ is finite. While noise resulting from the competition processes (reactions) scales as $1/\sqrt{N}$, noise originating from hopping (diffusion) only scales as $1/N$. In summary, this gives (multiplicative) Gaussian white noise $\xi_i(\boldsymbol{r}, t)$ characterized by the correlation matrix

$$\langle \xi_i(\boldsymbol{r}, t)\xi_j(\boldsymbol{r}', t') \rangle = \Delta_{ij}\delta(\boldsymbol{r} - \boldsymbol{r}')\delta(t - t') \tag{13.28}$$

and amplitudes depending on the system's configuration:

$$\mathcal{C}_A = \frac{1}{\sqrt{N}}\sqrt{a(\boldsymbol{r}, t)\big[\mu(1 - \rho(\boldsymbol{r}, t)) + \sigma c(\boldsymbol{r}, t)\big]}\,,$$

$$\mathcal{C}_B = \frac{1}{\sqrt{N}}\sqrt{b(\boldsymbol{r}, t)\big[\mu(1 - \rho(\boldsymbol{r}, t)) + \sigma a(\boldsymbol{r}, t)\big]}\,,$$

$$\mathcal{C}_C = \frac{1}{\sqrt{N}}\sqrt{c(\boldsymbol{r}, t)\big[\mu(1 - \rho(\boldsymbol{r}, t)) + \sigma b(\boldsymbol{r}, t)\big]}\,. \tag{13.29}$$

The strength of such a continuum description is that it is *generic*, that is, the form of the equations does not depend on, for example, the precise form of the lattice or the shape and size of individuals' neighborhood as long as it is local. It is the interplay between diffusion, mixing the system locally on a certain length scale, and the reaction kinetics, whose features are encoded by the phase portrait of the well-mixed system, which gives rise to the observed complex dynamics. The stochastic reaction–diffusion equations can be solved numerically. Figure 13.12 shows the outcome of such a simulation starting from an inhomogeneous initial condition (and using periodic boundary conditions) [13], and compares the results obtained by agent-based simulations and deterministic diffusion–reaction equations. The comparison of those snapshots reveals a remarkable coincidence of the patterns obtained from agent-based simulations and the continuum approach. As shown in [101, 102], these similarities in patterns are actually fully quantitative and the spatio-temporal correlations functions for the population densities are almost identical.

The approach of mapping the interacting particle system to the SPDEs, (13.28), yields extremely insightful results, as it enables the application of bifurcation theory [103]. Determining the bifurcations that the nonlinear functions $\mathcal{A}_i[\boldsymbol{a}]$ exhibit defines universality classes for the emerging patterns. Namely, in the vicinity of bifurcations, the behavior is described by generic normal forms, characterizing each bifurcation type. The resulting universality classes have already been widely studied in the physical and mathematical community, mostly by investigating deterministic partial differential equations, see for example [104–106] for reviews, as well as references therein. Although specific models for competing populations will not yield SPDEs that are identical to the general equations studied there, their

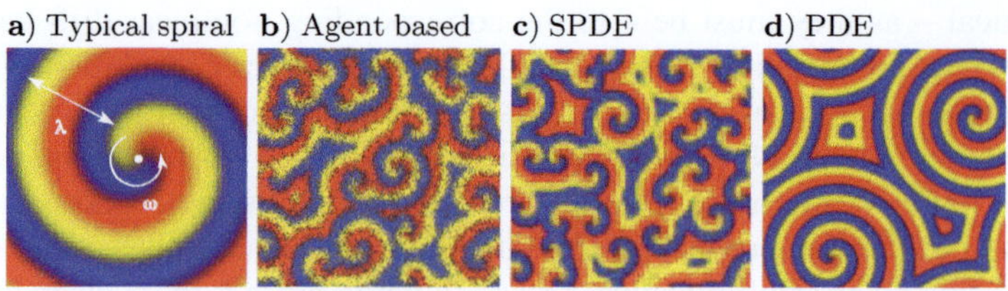


**Fig. 13.12** Spiral patterns. **a** Schematic drawing of a spiral with wavelength λ. It rotates around the origin at a frequency $\omega$. **b** Agent-based simulations for $D < D_c$, when all three species coexist, show entangled, rotating spirals. **c** Stochastic partial differential equations show similar patterns to agent-based simulations. **d** Spiral pattern emerging from the dynamics of the deterministic diffusion–reaction equation starting from a spatially inhomogeneous initial state. Parameters are $\sigma = \mu = 1$ and $D = 1 \times 10^{-5}$. Figure adapted from [13]

bifurcation behavior may coincide with an equation that has already been investigated. Consequently, the specific SPDE falls into that universality class, and generic results may be transferred. In the present case of a spatially extended May–Leonard model, projecting the deterministic version of the diffusion–reaction equation (13.28) onto the *reactive manifold M* one obtains [13, 101, 102]

$$\partial_t z = D\Delta^2 z + (c_1 - i\omega)z - c_2(1 + ic_3)|z|^2 z \,. \tag{13.30}$$

Here, we recognize the celebrated *complex Ginzburg–Landau equation* (CGLE), whose properties have been extensively studied in the past [104, 105]. In particular, it is known that in two dimensions the latter gives rise to a broad range of coherent structures, including spiral waves whose velocity, wavelength, and frequency can be computed analytically. Remarkably, the results for the spirals' velocities, wavelengths, and frequencies agree extremely well with those obtained from the agent-based simulations [13, 101, 102].

Thus the formulation of the spatial game theoretical model in terms of stochastic diffusion–reaction equations has enabled us to reach a comprehensive understanding of the resulting out-of-equilibrium and nonlinear phenomena. Employing a mapping of the diffusion–reaction equation onto the reactive manifold of the nonlinear dynamics indicated that the dynamics of the coexistence regime is in the same "universality class" as the complex Ginzburg–Landau equation. This fact reveals the generality of the phenomena discussed in this chapter. In particular, the emergence of an entanglement of spiral waves in the coexistence state, the dependence of spirals' size on the diffusion rate, and the existence of a critical value of the diffusion above which coexistence is lost are robust phenomena. This means that they do not depend on the details of the underlying spatial structure: While, for specificity, we have (mostly) considered square lattices, other two-dimensional topologies (e.g., hexagonal or other lattices) will lead to the same phenomena, too. Also the details of the cyclic competition have no qualitative influence, as long as the underlying rate equations exhibit an unstable coexistence fixed point and can be recast in the universality class of the Hopf bifurcations. It remains to be explored what kind of mathematical structure corresponds to a broader range of game-theoretical problems.

In this chapter, we have mainly focused on the situation where the exchange rate between individuals is relatively high, which leads to the emergence of regular spirals in two dimensions. However, we have seen that when the exchange rate is low (or vanishes) stochasticity strongly affects the structure of the ensuing spatial patterns. In this case, the (continuum) description in terms of SPDEs breaks down. In this situation, the quantitative analysis of the spatio-temporal properties of interacting particle systems requires the development of other analytical methods, for example, relying on field-theoretic techniques [95]. Fruitful insights into this regime have already been gained by pair approximations or larger-cluster approximations [88, 107–109]. These researchers investigated a set of coupled nonlinear differential equations for the time evolution of the probability of finding a cluster of a certain size in a particular state. While such an approximation improves when large clusters are considered, unfortunately the effort required to solve their coupled equations of motion also drastically increases with the size of the clusters. In addition, the use of these cluster mean-field approaches becomes problematic in the proximity of phase transitions (near an extinction threshold), where the correlation length diverges. Investigations along these lines represent a major future challenge in the multidisciplinary field of complexity science.

The cyclic "rock–paper–scissors" model as discussed in this section can be generalized in manifold ways. The model with asymmetric rates turns out to be in the same universality class as the one with symmetric rates [110]. Qualitative changes in the dynamics, however, emerge when the interaction network between the species is changed. For example, consider a system where each agent can interact with its neighbors according to the following scheme:

$$
\begin{aligned}
AB &\xrightarrow{1} AA \\
BC &\xrightarrow{1} BB \\
CA &\xrightarrow{1} CC
\end{aligned} \quad (13.31)
\qquad
\begin{aligned}
AB &\xrightarrow{\sigma} A\oslash \\
BC &\xrightarrow{\sigma} B\oslash \\
CA &\xrightarrow{\sigma} C\oslash
\end{aligned} \quad (13.32)
\qquad
\begin{aligned}
A\oslash &\xrightarrow{\mu} AA \\
B\oslash &\xrightarrow{\mu} BB \\
C\oslash &\xrightarrow{\mu} CC
\end{aligned} \quad (13.33)
$$

Reactions (13.31) describe *direct dominance* in a Moran-like manner, where an individual of one species is consumed by another from a more predominant species, and the latter immediately reproduces. Cyclic dominance appears as $A$ consumes $B$ and reproduces, while $B$ preys on $C$ and $C$ feeds on $A$ in turn. Reactions (13.32) encode some kind of *toxicity*, where one species kills another, leaving an empty site $\oslash$. These reactions occur at a rate $\sigma$, and are decoupled from *reproduction*, (13.33), which happens at a rate $\mu$. Note that reactions (13.31) and (13.33) describe two different mechanisms of reproduction, both of which are important for ecological systems: In (13.31), an individual reproduces when having consumed prey, as a result of thus increased fitness. In contrast, in reactions (13.33) reproduction depends solely on the availability of empty space. As can be inferred from Fig. 13.13, the spatio-temporal patterns sensitively depend on the strength $\sigma$ of the *toxicity* effect. Actually, as can be shown analytically [111], there is an Eckhaus instability, that

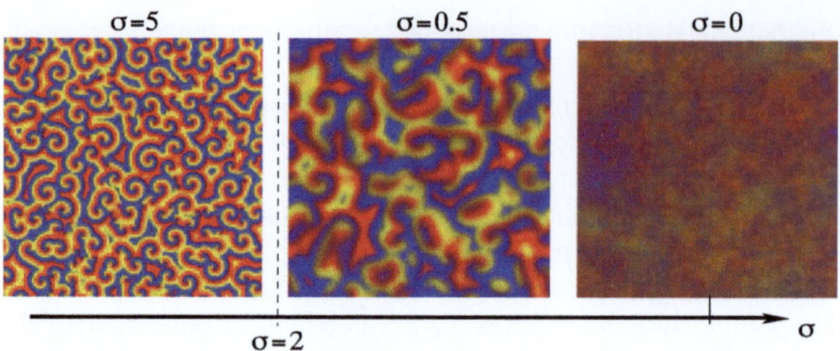


**Fig. 13.13** Snapshots of the biodiverse state for $D = 1 \times 10^{-5}$. **a** For large rates $\sigma$, entangled and stable spiral waves form. **b** A convective (Eckhaus) instability occurs at $\sigma_E \approx 2$; below this value, the spiral patterns blur. **c** At the bifurcation point $\sigma = 0$, only very weak spatial modulations emerge; we have amplified them by a factor of two for better visibility. The snapshots stem from numerical solution of an appropriate SPDE with initially homogeneous densities $a = b = c = 1/4$. Figure adapted from [109]

is, a convective instability: a localized perturbation grows but travels away. The instabilities result in the blurring seen in Fig. 13.13.

It remains to be explored how more complex interaction networks with an increasing number of species and with different types of competition affect the spatio-temporal pattern formation process. Research along these lines is summarized in a recent review [88].

## 13.6 Conclusions and Outlook

In this contribution we have given an introduction to evolutionary game theory. The perspective we have taken was that, starting from agent-based models, the dynamics may be formulated in terms of a hierarchy of theoretical models. First, if the population size is large and the population is well-mixed, a set of ordinary differential equations can be employed to study the system's dynamics and ensuing stationary states. Game-theoretical concepts of "equilibria" then map to "attractors" of the nonlinear dynamics. Setting up the appropriate dynamic equations is a nontrivial matter if one is aiming at a realistic description of a biological system. For instance, as nicely illustrated by a recent study on yeast [8], a linear replicator equation might not be sufficient to describe the frequency dependence of the fitness landscape. We suppose that this is the rule rather than the exception for biological systems such as microbial populations. Second, for well-mixed but finite populations, one has to account for stochastic fluctuations. Then there are two central questions: (i) What is the probability of a certain species going extinct or becoming fixated in a population? (ii) How long does this process take? These questions have to be answered by employing concepts from the theory of stochastic processes. Since most systems have absorbing states, we have found it useful to classify the stability of a given dynamic system according to the scaling of the expected extinction time with population size. Third, and finally, taking into account finite mobility of individuals in an explicit spatial model makes a description in terms of stochastic partial differential equations necessary. These Langevin equations describe the interplay between reactions, diffusion, and noise, which give rise to a plethora of new phenomena. In particular, spatio-temporal patterns or, more generally, spatio-temporal correlations, may emerge, which can dramatically change the ecological and evolutionary stability of a population. For nontransitive dynamics, such as the "rock–paper–scissors" game played by some microbes [6], there is a *mobility threshold*, which demarcates regimes of maintenance and loss of biodiversity [13]. Since, for the "rock–paper–scissors" game, the nature of the patterns and the transition was encoded in the flow of the nonlinear dynamics on the reactive manifold, one might hope that a generalization of the outlined approach might be helpful in classifying a broader range of game-theoretical problems and identify some "universality classes".

What are the ideal experimental model systems for future studies? We think that microbial populations will play a major role since interactions between different strains can be manipulated in a multitude of ways. In addition, experimental tools such as microfluidics and various optical methods allow for easy manipulation and observation of these systems, from the level of an individual up to the level of a whole population. Bacterial communities represent complex and dynamic ecological systems. They appear in the form of free-floating bacteria as well as biofilms in nearly all parts of our environment, and are highly relevant for human health and disease [25]. Spatial patterns arise from heterogeneities of the underlying "landscape" or are self-organized by the bacterial interactions and play an important role in maintaining species diversity [7]. Interactions include, among others, competition for resources and cooperation by sharing of extracellular polymeric substances. Another aspect of interactions is chemical warfare. As we have discussed, some bacterial strains produce toxins such as colicin, which acts as a poison to sensitive strains, while other strains are resistant [6]. Stable coexistence of these different strains arises when they can spatially segregate, resulting in self-organizing patterns. There is a virtually inexhaustible complexity in the structure and dynamics of microbial populations. The recently proposed term "socio-microbiology" [112] expresses this notion in a most vivid form. Investigating the dynamics of those complex microbial populations is a challenging interdisciplinary endeavor, which requires the combination of approaches from molecular microbiology, experimental biophysical methods, and theoretical modeling. The overall goal would be to explore how collective behavior emerges and is maintained or destroyed in finite populations under the action of various kinds of molecular interactions between individual cells. Both biology and physics communities will benefit from this line of research.

Stochastic interacting particle systems are a fruitful testing ground for understanding generic principles in nonequilibrium physics. Here biological systems have been a wonderful source of inspiration for the formulation of new models. For example, MacDonald [113], looking for a mathematical description for mRNA translation into proteins managed by ribosomes, which bind to the mRNA strand and step forward codon by codon, formulated a nonequilibrium one-dimensional transport model, nowadays known as the totally asymmetric simple exclusion process. This model has led to significant advances in our understanding of phase transitions and the nature of stationary states in nonequilibrium systems [114, 115]. Searching for simplified models of epidemic spreading without immunization, Harris [116] introduced the contact process. In this model infectious individuals can either heal themselves or infect their neighbors. As a function of the infection and recovery rate it displays a phase transition from an active to an absorbing state, that is, the epidemic disease may either spread over the whole population or vanish after some time. The broader class of absorbing-state transitions has recently been reviewed [117]. Another well-studied model is the voter model, where each individual has one of two opinions and may change it by imitation of a randomly chosen neighbor. This process mimics opinion making in a naive way [118]. Actually, it was first considered by Clifford and Sudbury [119] as a model for the competition of species and only later named voter model by Holley and Liggett [120]. It has been shown rigorously that on a regular lattice there is a stationary state where two "opinions" coexist in systems with spatial dimensions where the random walk is not recurrent [118, 121]. A question of particular interest is how opinions or strategies may spread in a population. In this context it is important to understand the coarsening dynamics of interacting agents. For a one-dimensional version of the "rock–paper–scissors" game, Frachebourg and collaborators [122, 123] have found that, starting from some random distribution, the species organize into domains that undergo (power law) coarsening until finally one species takes over the whole lattice. If this model is generalized to account for species mobility and multiple occupation of each site, several distinct pathways to extinction emerge, ranging from annihilating propagating waves to intermittent dynamics arising from heteroclinic orbits [124] (Rulands, et al.). When mutation is included the coarsening process is counteracted and in an interesting interplay between equilibrium and nonequilibrium processes a reactive stationary state emerges [125]. Yet another endeavor in nonequilibrium dynamics is to find global variables that provide a characterization of the system. Entropy production has been proposed as a useful observable [126, 127], and different principles governing its behavior have been suggested [128, 129], although problems arise from the different definitions of entropy employed and the different approaches to nonequilibrium dynamics [127, 130, 131]. Recent investigations of the "rock–paper–scissors" model with mutations show that entropy production can indeed characterize the behavior of population dynamics models. At a critical point the dynamics exhibits a transition from large, limit-cycle-like oscillations to small, erratic oscillations. It is found that the entropy production peaks very close to this critical point and tends to zero upon deviating from it [132]. One may hope that, in a similar manner, entropy production may yield valuable information about other models in evolutionary game theory.

**Acknowledgements** We are indebted to Benjamin Andrae, Maximilian Berr, Jonas Cremer, Alexander Dobrinevsky, Anna Melbinger, Mauro Mobilia, Steffen Rulands, and Anton Winkler, with whom we had the pleasure to work on game theory. They have contributed with a multitude of creative ideas and through many insightful discussions have shaped our understanding of the topic. Financial support from the German Excellence Initiative via the program "Nanosystems Initiative Munich" and the German Research Foundation via the SFB TR12 "Symmetries and Universalities in Mesoscopic Systems" is gratefully acknowledged.

# References


1. L. Hall-Stoodley, J.W. Costerton, P. Stoodley, Nat. Rev. Microbiol. **2**, 95 (2004)
2. T.J. Battin, W.T. Sloan, S. Kjelleberg, H. Daims, I.M. Head, T.P. Curtis, L. Eberl, Nat. Rev. Microbiol. **5**, 76 (2007)
3. G.J. Velicer, Trends Microbiol. **11**, 330 (2003)
4. J.B. Xavier, K.R. Foster, Proc. Natl. Acad. Sci. USA **104**, 876 (2007)
5. C.D. Nadell, J.B. Xavier, S.A. Levin, K.R. Foster, PLoS Biol. **6**, e14 (2008)
6. B. Kerr, M.A. Riley, M.W. Feldman, B.J.M. Bohannan, Nature **418**, 171 (2002)
7. J.B. Xavier, E. Martinez-Gracia, K.R. Foster, Am. Nat. **174**, 1 (2009)
8. J. Gore, H. Youk, A. van Oudenaarden, Nature **459**, 253 (2009)
9. J.S. Chuang, O. Rivoire, S. Leibler, Science **323**, 272 (2009)
10. S.A. Levin, Am. Nat. **108**, 207 (1974)
11. M.P. Hassell, H.N. Comins, R.M. May, Nature **353**, 255 (1991)
12. R. Durrett, S. Levin, Theor. Popul. Biol. **46**, 363 (1994)
13. T. Reichenbach, M. Mobilia, E. Frey, Nature **448**, 1046 (2007)
14. S.A. West, A.S. Griffin, A. Gardner, S.P. Diggle, Nat. Rev. Microbiol. **4**, 597 (2006)
15. R.M. May, W.J. Leonard, SIAM J. Appl. Math. **29**, 243 (1975)
16. M.J. Osborne, *An Introduction to Game Theory* (Oxford University Press, Oxford, 2004)
17. J.F. Nash, Proc. Natl. Acad. Sci. USA **36**, 48 (1950)
18. R.M. Dawes, Annu. Rev. Psychol. **31**, 169 (1980)
19. R. Axelrod, W.D. Hamilton, Science **211**, 1390 (1981)
20. J. Maynard Smith, G.R. Price, Nature **246**, 15 (1973)
21. J. Maynard Smith, *Evolution and the Theory of Games* (Cambridge University Press, Cambridge, 1982)
22. A.J. Lotka, J. Am. Chem. Soc. **42**, 1595 (1920)
23. V. Volterra, Mem. Accad. Lincei **2**, 31 (1926)
24. J. Hofbauer, K. Sigmund, *Evolutionary Games and Population Dynamics* (Cambridge University Press, Cambridge, 1998)
25. M.E. Hibbing, C. Fuqua, M.R. Parsek, S.B. Peterson, Nat. Rev. Microviol. **8**, 15 (2010)
26. J. Roughgarden, Ecology **3**, 453 (1971)
27. A. Melbinger, J. Cremer, E. Frey, Evolutionary game theory in growing populations. Phys. Rev. Lett. **105**, 178101 (2010)
28. S. Wright, Proc. Natl. Acad. Sci. USA **31**, 382 (1945)
29. S. Wright, *Evolution and the Genetics of Populations* (Chicago University Press, Chicago, IL, 1969)
30. W.J. Ewens, *Mathematical Population Genetics*, 2nd edn. (Springer, New York, NY, 2004)
31. P.A. Moran, *The Statistical Processes of Evolutionary Theory* (Clarendon, Oxford, 1964)
32. S.H. Strogatz, *Nonlinear Dynamics and Chaos* (Westview, Boulder, CO, 1994)
33. D. Greig, M. Travisano, Proc. R. Soc. Lond. B **271**, S25 (2004)
34. A. Buckling, F. Harrison, M. Vos, M.A. Brockhurst, A. Gardner, S.A. West, A. Griffin, FEMS Microbiol. Ecol. **62**, 135 (2007)
35. R.L. Trivers, Q. Rev. Biol. **46**, 35 (1971)
36. C.M. Waters, B.L. Bassle, Annu. Rev. Cell Dev. Biol. **21**, 319 (2005)
37. R. Durrett, S. Levin, J. Theor. Biol. **185**, 165 (1997)
38. R. Durrett, S. Levin, Theor. Popul. Biol. **53**, 30 (1998)
39. T.L. Czárán, R.F. Hoekstra, L. Pagie, Proc. Natl. Acad. Sci. USA **99**, 786 (2002)
40. R. Axelrod, *The Evolution of Cooperation* (Basic Books, New York, NY, 1984)
41. M.A. Nowak, Science **314**, 1560 (2006)
42. T. Yamagishi, J. Pers. Soc. Psychol. **51**, 110 (1986)
43. W.D. Hamilton, *Narrow Roads of Gene Land: Evolution of Social Behaviour* (Oxford University Press, Oxford, 1996)
44. J.F. Crow, M. Kimura, *An Introduction to Population Genetics* (Blackburn Press, Caldwell, NJ, 2009)

45. J. Cremer, T. Reichenbach, E. Frey, New J. Phys. **11**, 093029 (2009)
46. E. Ben-Jacob, I. Cohen, H. Levine, Adv. Phys. **49**, 395 (2000)
47. O. Hallatschek, P. Hersen, S. Ramanathan, D.R. Nelson, Proc. Natl. Acad. Sci. USA **104**, 19926 (2007)
48. C.J. Ingham, E.B. Jacob, BMC Microbiol. **8**, 36 (2008)
49. A.T. Henrici, *The Biology of Bacteria: The Bacillaceae*, 3rd edn. (Heath, Lexington, MA 1948)
50. E. Ben-Jacob, I. Cohen, I. Golding, D.L. Gutnick, M. Tcherpakov, D. Helbing, I.G. Ron Open, Physica A **282**, 247 (2000)
51. M. Matsushita, H. Fujikawa, Physica A **168**, 498 (1990)
52. R. Rudner, O. Martsinkevich, W. Leung, E.D. Jarvis, Mol. Microbiol. **27**, 687 (1998)
53. M. Eisenbach, Mol. Microbiol. **4**, 161 (1990)
54. J. Henrichsen, Acta Pathol. Microbiol. Scand. B **80**, 623 (1972)
55. J. Henrichsen, L.O. Froholm, K. Bovre, Acta Pathol. Microbiol. Scand. B **80**, 445 (1972)
56. S. Park, P.M. Wolanin, E.A. Yuzbashyan, H. Lin, N.C. Darnton, J.B. Stock, P. Silberzan, R. Austin, Proc. Natl. Acad. Sci. USA **100**, 13910 (2003)
57. O. Hallatschek, D.R. Nelson, Theor. Popul. Biol. **73**, 158 (2008)
58. H.H. McAdams, A. Arkin, Trends Genet. **15**, 65 (1999)
59. M. Kaern, T.C. Elston, W.J. Blake, J.J. Collins, Nat. Rev. Microbiol. **6**, 451 (2005)
60. J.W. Veening, W.K. Smits, O.P. Kuipers, Annu. Rev. Microbiol. **62**, 193 (2008)
61. W.K. Smits, O.P. Kuipers, J.W. Veening, Nat. Rev. Microbiol. **4**, 259 (2006)
62. D. Dubnau, R. Losick, Mol. Microbiol. **61**, 564 (2006)
63. M. Leisner, K. Stingl, E. Frey, B. Maier, Curr. Opin. Microbiol. **11**, 553 (2008)
64. M. Delbrück, J. Chem. Phys. **8**, 120 (1940)
65. A. Traulsen, J.C. Claussen, C. Hauert, Phys. Rev. Lett. **95**, 238701 (2005)
66. T. Reichenbach, M. Mobilia, E. Frey, Phys. Rev. E **74**, 051907 (2006)
67. A. Traulsen, J.C. Claussen, C. Hauert, Phys. Rev. E **74**, 011901 (2006)
68. J. Cremer, T. Reichenbach, E. Frey, Eur. Phys. J. B **63**, 373 (2008)
69. N.G. van Kampen, *Stochastic Processes in Physics and Chemistry* (North-Holland, Amsterdam, 1981)
70. C.W. Gardiner, *Handbook of Stochastic Methods* (Springer, Berlin, Heidelberg, 2007)
71. T. Antal, I. Scheuring, Bull. Math. Biol. **68**, 1923 (2006)
72. C. Taylor, Y. Iwasa, M.A. Nowak, J. Theor. Biol. **243**, 245 (2006)
73. M. Ifti, B. Bergersen, Eur. Phys. J. E **10**, 241 (2003)
74. M. Ifti, B. Bergersen, Eur. Phys. J. B **37**, 101 (2004)
75. A. Dobrinevski, E. Frey, Extinction in neutrally stable stochastic Lotka-Volterra models, Submitted [arXiv:1001.5235]
76. M. Berr, T. Reichenbach, M. Schottenloher, E. Frey, Phys. Rev. Lett. **102**, 048102 (2009)
77. R.M. May, *Stability and Complexity in Model Ecosystems*, 2nd edn. (Princeton University Press, Princeton, NJ, 1974)
78. J.D. Murray, *Mathematical Biology*, 3rd edn. (Springer, Berlin, Heidelberg, 2002)
79. A.M. Turing, Philos. Trans. R. Soc. Lond. B **237**, 37 (1952)
80. M.A. Nowak, R.M. May, Nature **359**, 826 (1992)
81. M.P. Hassell, H.N. Comins, R.M. May, Nature **370**, 290 (1994)
82. B. Blasius, A. Huppert, L. Stone, Nature **399**, 354 (1999)
83. A.A. King, A. Hastings, Theor. Popul. Biol. **64**, 431 (2003)
84. G. Szabo, C. Hauert, Phys. Rev. Lett. **89**, 118101 (2002)
85. C. Hauert, M. Doebeli, Nature **428**, 643 (2004)
86. T.M. Scanlon, K.K. Caylor, I. Rodriguez-Iturbe, Nature **449**, 209 (2007)
87. S. Kefi, M. Rietkerk, C.L. Alados, Y. Pueyo, V.P. Papanastasis, A. ElAich, P.C. de Ruiter, Nature **449**, 213 (2007)
88. G. Szabó, G. Fáth, Phys. Rep. **446**, 97 (2007)
89. M. Perc, A. Szolnoki, G. Szabó, Phys. Rev. E **75**, 052102 (2007)
90. M.A. Nowak, *Evolutionary Dynamics* (Belknap Press, Cambridge, MA, 2006)

91. O.A. Igoshin, R. Welch, D. Kaiser, G. Oster, Proc. Natl. Acad. Sci. USA **101**, 4256 (2004)
92. A. McKane, D. Alonso, Bull. Math. Biol. **64**, 913 (2002)
93. E. Liebermann, C. Hauert, M.A. Nowak, Nature **433**, 312 (2005)
94. M. Mobilia, I.T. Georgiev, U.C. Täuber, Phys. Rev. E **73**, 040903(R) (2006)
95. M. Mobilia, I.T. Georgiev, U.C. Täuber, J. Stat. Phys. **128**, 447 (2007)
96. J.B.C. Jackson, L. Buss, Proc. Natl. Acad. Sci. USA **72**, 5160 (1975)
97. O. Gilg, I. Hanski, B. Sittler, Science **302**, 866 (2001)
98. B. Sinervo, C.M. Lively, Nature **380**, 240 (1996)
99. B.C. Kirkup, M.A. Riley, Nature **428**, 412 (2004)
100. A.J. Nicholson, Aust. J. Zool. **2**, 9 (1954)
101. T. Reichenbach, M. Mobilia, E. Frey. J. Theor. Biol. **254**, 368 (2008)
102. T. Reichenbach, M. Mobilia, E. Frey. Phys. Rev. Lett. **99**, 238105 (2007)
103. S. Wiggins, *Introduction to Applied Nonlinear Dynamical Systems and Chaos* (Springer, Berlin, Heidelberg, 1990)
104. M.C. Cross, P.C. Hohenberg, Rev. Mod. Phys. **65**, 851 (1993)
105. I.S. Aranson, L. Kramer, Rev. Mod. Phys. **74**, 99 (2002)
106. W. van Saarloos, Phys. Rep. **386**, 29 (2003)
107. K. Tainaka, Phys. Rev. E **50**, 3401 (1994)
108. K. Sato, N. Konno, T. Yamaguchi, Mem. Muroran Inst. Technol. **47**, 109 (1997)
109. G. Szabó, A. Szolnoki, R. Izsak, J. Phys. A Math. Gen. **37**, 2599 (2004)
110. M. Peltomaki, M. Alava, Phys. Rev. E **78**, 031906 (2008)
111. T. Reichenbach, E. Frey, Phys. Rev. Lett. **101**, 058102 (2008)
112. M.R. Parsek, E.P. Greenberg, Trends Microbiol. **13**, 27 (2005)
113. C.T. MacDonald, J.H. Gibbs, A.C. Pipkin, Biopolymers **6**, 1 (1968)
114. G.M. Schütz, *Exactly Solvable Models for Many-Body Systems Far from Equilibrium*, Vol. 19 of *Phase Transitions and Critical Phenomena* (Academic, San Diego, CA, 2001), pp. 1–251
115. M. Mobilia, T. Reichenbach, H. Hinsch, T. Franosch, E. Frey, Banach Center Publ. **80**, 101 (2008); arXiv:cond-mat/0612516
116. T.E. Harris, Ann. Probab. **2**, 969 (1974)
117. H. Hinrichsen, Adv. Phys. **49**, 815 (2000)
118. C. Castellano, S. Fortunato, V. Loreto, Rev. Mod. Phys. **81**, 591 (2009)
119. P. Clifford, A. Sudbury, Biometrika **60**, 581 (1973)
120. R. Holley, T.M. Liggett, Ann. Probab. **6**, 198 (1978)
121. T.M. Liggett, *Stochastic Interacting Systems: Contact, Voter and Exclusion Processes* (Springer, Berlin, Heidelberg, 1999)
122. L. Frachebourg, P.L. Krapivsky, E. Ben-Naim, Phys. Rev. E **54**, 6186 (1996)
123. L. Frachebourg, P.L. Krapivsky, E. Ben-Naim, Phys. Rev. Lett. **77**, 2125 (1996)
124. S. Rulands, T. Reichenbach, E. Frey, Three-fold way to extinction in populations of cyclically competing species, J. Stat. Mech. L01003 (2011)
125. A. Winkler, T. Reichenbach, E. Frey, Phys. Rev. E **81**, 060901(R) (2010)
126. D.-Q. Jiang, M. Qian, M.-P. Qian, *Mathematical Theory of Nonequilibrium Steady States* (Springer, Berlin, Heidelberg, 2004)
127. F. Schlögl, Z. Phys. **198**, 559 (1967)
128. E.T. Jaynes, Ann. Rev. Phys. Chem. **3**, 579 (1980)
129. P. Glansdorff, I. Prigogine, *Thermodynamic Theory of Structure, Stability and Fluctuations* (Wiley-Interscience, New York, NY, 1971)
130. S. Goldstein, J.L. Lebowitz, Physica D **193**, 53 (2004)
131. U. Seifert, Phys. Rev. Lett. **95**, 040602 (2005)
132. B. Andrae, J. Cremer, T. Reichenbach, E. Frey, Phys. Rev. Lett. **104**, 218102 (2010)

# Chapter 14
# Darwin and the Evolution of Human Cooperation

Karl Sigmund and Christian Hilbe

**Abstract** Humans are characterized by a high propensity for cooperation. The emergence and stability of this trait is one of the hottest topics in evolutionary game theory, and leads to a wide variety of models offering a rich source of complex dynamics based on social interactions. This chapter offers an overview of different approaches to this topic (such as kin selection, group selection, direct and indirect reciprocity) and relates it to some of the views that Darwin expressed over 150 years ago. It turns out that, in many cases, Darwin displayed a remarkably lucid intuition of the major issues affecting the complex mechanisms promoting the evolution of cooperation.

## 14.1 Darwin on Complexity

Charles Darwin (1809–1882), who was celebrated especially in 2009, which marked the 200th anniversary of his birth and the 150th anniversary of the publication of *On the Origin of Species*, was one of the first scientists to become fascinated by the many facets of complexity, be it complexity of organs or complexity of interactions. His approach to explaining the emergence of complex structures by the effect of natural selection operating on inheritable variations is almost universally accepted today. But in addition to this gradual adaptation through ceaseless trial and error, a few big events shaped the course of evolution. Major transitions brought about genomes, eukaryotic cells, multicellular organisms, sexual reproduction, and animal societies. These major transitions in evolution [1] introduced new levels of organization, and even produced new units of selection. In particular, the evolution of cooperation is widely recognized as one of the most important areas for Darwinian theory.

The mathematical tools for the analysis of social interactions, and in particular the interplay of competition and collaboration, are provided by game theory. It

K. Sigmund (✉)
Faculty of Mathematics, University of Vienna, A-1090 Wien, Austria
e-mail: karl.sigmund@univie.ac.at

H. Meyer-Ortmanns, S. Thurner (eds.), *Principles of Evolution*,
The Frontiers Collection, DOI 10.1007/978-3-642-18137-5_14,
© Springer-Verlag Berlin Heidelberg 2011

originated in the 1940s, and was initially meant to analyze rational behavior in economic interactions. In particular, it assumed that the players are rational – they could figure out the consequences of their action, and choose whichever action optimized their expected payoff, taking into account that the co-players would similarly decide upon their strategy. In the hands of evolutionary biologists such as William D. Hamilton and John Maynard Smith, this assumption of rationality, which was hardly applicable to humans, let alone to other biological organisms, was thrown overboard [2]. In evolutionary game theory, the players were the members of a population. Strategies were no longer the outcome of rational decision making, but behavioral programs, and the payoff was not measured in monetary units or derived from some other preference scale, but simply indicated Darwinian fitness, that is, reproductive success. The applications of evolutionary game theory quickly spread through all fields of behavioral sciences [3–5].

It is unlikely that Darwin would have felt at ease with the formal aspects of evolutionary game theory. He often regretted that he "had not proceeded far enough to know something of the great leading principles of mathematics" and wrote that "persons thus endowed seem to possess an extra sense". But he would certainly have enjoyed the investigation of the feedback loops characteristic of evolutionary dynamics. The feedback is based on the fact that the success of a strategy depends on what the other players are doing, and hence on the relative frequencies of the strategies in a population. This success, or payoff, in turn determines how quickly the strategy spreads, provided that it corresponds to an innate program that is inherited by the offspring. Thus evolutionary game theory is very similar to population ecology [4]. Just as the densities of diverse types of predators and prey, for instance, affect growth rates, which, in turn, determine the densities in the following generations, so the frequencies of behavioral types determine their growth rates and hence the future frequencies.

Darwin, with his staggering ecological intuition, delighted in describing "how plants and animals [...] are bound together by a web of complex relations". He relished in figuring out how, "if certain insectivorous birds were to increase in Paraguay", a species of flies would decrease; and how – since these flies parasitize newborn cattle – "that decrease would cause cattle to become abundant; which would certainly greatly alter the vegetation; how this again would largely affect the insects; and this again the insectivorous birds; and so on in ever increasing circles of complexity."

We shall presently see that in a similar vein, different types of cooperation-related behavior can influence each other "in ever increasing circles of complexity".

## 14.2 The Riddle of Cooperation

At first sight, it seems surprising that cooperation, which is so obviously a good thing, should present any problems. If two individuals profit from cooperating, why shouldn't they? But let us be more specific, and consider a "game" where two players independently can decide between two options: to send a gift to the other player, or not; that is, to cooperate (play **C**) or to defect (play **D**). Let us assume that the gift confers a benefit $b$ to the recipient, at a cost $c$ to the donor, with $0 < c < b$. If both players cooperate, each receives a payoff $b - c > 0$, whereas both receive nothing if they both defect. But the socially desirable outcome of mutual cooperation has an Achilles' heel. It is inconsistent in the sense that if the other player duly plays **C**, then one can improve one's own payoff by playing **D**. From the viewpoint of an individual player, it is always better to defect, no matter what the co-player is going to decide. This can be seen from the following payoff matrix, which describes the payoff for player I (who can chose between the rows **C** and **D**, whereas the co-player can chose between the columns **C** and **D**):

$$\begin{array}{c|cc} * & \mathbf{C} & \mathbf{D} \\ \hline \mathbf{C} & b - c & -c \\ \mathbf{D} & b & 0 \end{array} \tag{14.1}$$

The only consistent outcome is for both players to play **D**: consistent in the sense that both players cannot improve their payoff by unilaterally switching strategy. "Unilaterally" is the crucial word: if their actions were correlated, that is, if they had to chose the same move, then cooperation would be the obvious outcome. But they are independent actors. This is just the point about being "individuals". If we assume the viewpoint of "methodological individualism", that is, that societies are based on individual decisions, we find that altruistic behavior – helping others at a cost to oneself – poses a problem indeed.

We stress that we need not assume that the players are rational, and able to predict and optimize their payoff. The outcome is just the same if we consider a population of **C**-players and **D**-players randomly meeting and playing the game. The defectors would always do better, have a higher payoff, which means more offspring, and hence they would spread and ultimately eliminate the cooperators.

This game is an example of a social dilemma : self-interested motives lead to self-defeating moves. There are quite a few other social dilemmas, but we shall stick with this one, and its generalization, the so-called prisoner's dilemma. This describes any symmetric interaction between two players, with two strategies each, having a payoff matrix

$$\begin{array}{c|cc} * & \mathbf{C} & \mathbf{D} \\ \hline \mathbf{C} & R & S \\ \mathbf{D} & T & P \end{array} \tag{14.2}$$

with $T > R > P > S$. The prisoner's dilemma game encapsulates the tug-of-war between the common interest ($R$ is larger than $P$) and the selfish interest (**D** dominates **C**).

## 14.3 Kin Selection

There exist two major ways, in theoretical biology, to come to grips with the problem of cooperation. The first is known as kin selection theory. Clearly, a huge part of biological cooperation occurs within families. This is an immediate corollary of the Darwinian struggle for survival. Genes that promote their own spreading (by enhancing the survival and the fecundity of their carriers) become necessarily more frequent than those that do not. Just as parents programmed to help their children have an obvious advantage in passing along their genetic program, so siblings programmed to help each other will also have an advantage. More precisely, a gene causing you to help your brother will help to spread itself: for it is, with a high probability, carried by your brother too.

This approach was developed in William D. Hamilton's inclusive fitness theory [6], although earlier geneticists had anticipated the basic idea. For instance, R.A. Fisher used the approach in explaining the spread of bright warning colors among distasteful caterpillars. This appears at first sight a suicidal advertisement policy, but a bird who swallows such a lurid caterpillar will most likely not do it again. If this saves the life of the caterpillar's siblings (as will most probably be the case, since they travel in family groups), then the victim's demise has not been in vain. The genes for bright colors will be passed on through the siblings, and can spread. The famous geneticist Haldane is said to have quipped, in a similar vein: "I am ready to lie down my life to save two of my brothers, or eight of my cousins" [7]. Why two? Why eight? Obviously, there is some theory behind it. It is based on the quantification of relatedness.

The *coefficient of relatedness* between two players can be defined in various ways. Here, we simply assume that it measures *relatedness by descent*: this is the probability $\rho$ that a recently mutated gene (or allele, to use the correct term), if it is carried by one player, is also carried by the other. Of course, any two humans are related, if we go back to primordial Eve. But we do not share all our genes. A mutation occurring in the body of your grandfather produces an allele that will be found with probability $1/2$ in his children and with probability $1/4$ in his grandchildren (Fig. 14.1). Under usual circumstances (e.g., no parental inbreeding), the coefficient of relatedness between two siblings is $1/2$; between you and your nephew it is $1/4$, and between two cousins it is $1/8$. You can view your relatives as watered-down copies of yourself. The coefficient $\rho$ measures the amount of dilution. The higher $\rho$, the more your genetic interests coincide.

Suppose now that the coefficient of relatedness with your co-player is $\rho$. You can view any increase in the co-player's fitness as an increase of your own fitness, discounted by the factor $\rho$. The payoff matrix then turns into

$$
\begin{array}{c|cc}
* & \mathbf{C} & \mathbf{D} \\
\hline
\mathbf{C} & (b-c)(1+\rho) & b\rho - c \\
\mathbf{D} & b - c\rho & 0
\end{array}
\tag{14.3}
$$

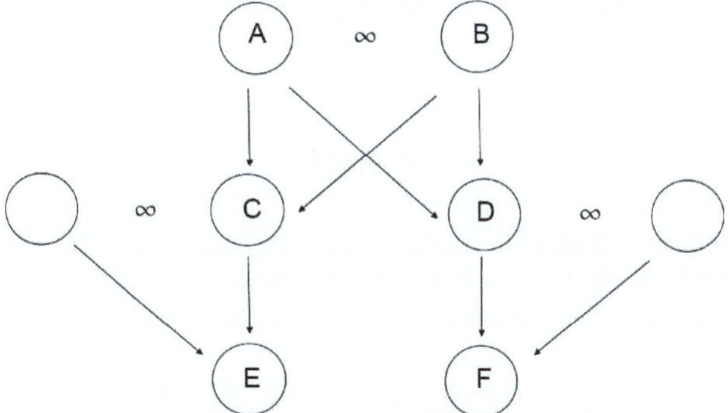


**Fig. 14.1** Coefficients of relatedness in a small family. A and B are the parents, C and D their offspring, E and F their grandchildren. Individuals have two copies (alleles) of each gene, one inherited from the mother, the other from the father. The probability that a specific allele of A is passed to C is 1/2. The probability that a specific allele of C comes from A is 1/2. The degree of relatedness between A and C is 1/2. The siblings C and D can both have inherited a newly mutated allele, either from A (probability $1/2 \times 1/2 = 1/4$) or from B, hence their degree of relatedness is $1/4 + 1/4 = 1/2$. The degree of relatedness between F and C (nephew and uncle) is $1/2 \times 1/2 \times 1/2 + 1/2 \times 1/2 \times 1/2 = 1/4$, the degree of relatedness between E and F (two cousins) is 1/8

If the coefficient of relatedness satisfies $\rho > c/b$, the elements in the first row are larger than the corresponding elements of the second row. So in this case, no matter what your co-player does, it is better to chose the first row, that is, to cooperate. This is known as Hamilton's rule.

## 14.4 Relatedness and Assortment

This "selfish gene" view, elaborated as the theory of kin selection, has been developed to a considerable extent within the last 50 years. A more "modern" version of Hamilton's rule is based on *relatedness by assortment* [8, 9]. Let us denote by $P(\mathbf{C} \mid \mathbf{C})$ the probability that a **C**-player meets with a **C**-player, and by $P(\mathbf{C} \mid \mathbf{D})$ the probability that a **D**-player meets a **C**-player. As the notation suggests, these expressions can be viewed as conditional probabilities. Let us consider

$$r := P(\mathbf{C} \mid \mathbf{C}) - P(\mathbf{C} \mid \mathbf{D}) \, ,$$

and denote by $p$ the frequency of **C**-players in the population. From

$$P(\mathbf{C} \mid \mathbf{D})P(\mathbf{D}) = P(\mathbf{D} \mid \mathbf{C})P(\mathbf{C})$$

(both expressions measure the probability that one player defects and the second cooperates), we deduce

$$P(\mathbf{C} \mid \mathbf{D}) = p[1 - P(\mathbf{D} \mid \mathbf{C}) + P(\mathbf{C} \mid \mathbf{D})] = p(1 - r)$$

and

$$P(\mathbf{C} \mid \mathbf{C}) = r + P(\mathbf{C} \mid \mathbf{D}) = r + (1 - r)p \;.$$

In the prisoner's dilemma game, the expected payoff for cooperators is given by $P(\mathbf{C} \mid \mathbf{C})R + [1 - P(\mathbf{C} \mid \mathbf{C})]S$, and that for defectors by $P(\mathbf{C} \mid \mathbf{D})T + [1 - P(\mathbf{C} \mid \mathbf{D})]P$. Cooperators do better than defectors if and only if

$$r > \frac{P - S + p(T - R - P + S)}{R - S + p(T - R - P + S)} \;.$$

In the special case of the gift-giving game, we have $T - R = P - S$ and hence the inequality reduces to

$$r > \frac{c}{b} \;,$$

which is just Hamilton's rule, but this time for relatedness by assortment rather than relatedness by descent.

Using $r$ instead of $\rho$ has several advantages. In particular, relatedness by descent $\rho$ is not easy to define properly. In the case of a recently mutated gene, it reduces to the same expression as $r$, in principle, but how old is a "recently mutated gene"? Moreover, there exist certainly situations where the positive assortment between cooperators is due to other factors than common descent. Somewhat confusingly, there is a trend to subsume all such situations under the heading of kin selection, even if family ties play no role at all [10]: in the most extreme examples, the two players could be from different species [11]. Moreover, it should be stressed that for games involving more than two players, the relevant "relatedness" becomes a complex expression involving the probabilities for triplets, etc. [9].

## 14.5 Darwin on Kin Selection

The fact that "kin selection" is a semantically overstretched expression does not take anything away from the importance of family ties for promoting cooperation. Many of the most remarkable examples of altruistic behavior occur among social insects, such as bees or ants, where the family ties are extremely tight. Typically, all workers in a beehive or an anthill are sisters, the daughters of one queen, who herself is often the product of intensive inbreeding. The degree of relatedness, in that case, is so high that one can view an insect state as a "super-organism".

Darwin, who had overlooked (like everyone else) the contemporary work by Mendel, did not have a clear idea of how traits could be passed on from one generation to the next. The notion of a gene was unknown to him. Nevertheless, he had as good a grasp of the principles of kin selection as was possible in his time. This can be seen in the following quotes:

> One special difficulty [...] at first appeared to me insuperable, and actually fatal to my whole theory. I allude to the sterile females in insect-communities [who] differ widely [...] from both males and fertile females, and yet, from being sterile, cannot propagate their kind.

"Natural selection may be applied to the family, as well as to the individual ... Thus a well-flavored vegetable is cooked, but the horticulturist sows seeds of the same stock and confidently expects to get nearly the same variety. Or, on the topic of a particularly tasty beef: "the animal has been slaughtered, but the breeder goes with confidence to the same family. Thus I believe it has been with social insects: a slight modification in structure, or instinct, correlated with the sterile condition ... has been advantageous ... consequently the [related] fertile males and females of the same community flourished, and transmitted to their fertile offspring a tendency to produce sterile members having the same modification."

Such modifications can gradually build up to impressive feats of altruism and self-sacrifice. For instance, so-called honeypot ants are worker ants that spend all their life clinging to the wall of a subterranean chamber, their bodies exclusively used for storing nutrient. And worker bees are ready to sting intruders and thus to perform a suicide attack in order to defend their hive. It seems that such acts of self-immolation can only be explained by indirect fitness benefits.

## 14.6 Political Animals

In his zoology, Aristotle classified humans together with ants and bees as "social animals". The similarities between anthills, beehives, and human cities have often since been taken up by other authors, most famously by Bernard de Mandeville in his *Fable of the Bees*, written more than 300 years ago. Some of the parallels are striking indeed: the division of labor, the ceaseless bustle and exchange, the hierarchical organization, etc. Nevertheless, it has become clear that human sociality is very different from that of hymenoptera or termites. Basically, humans have not given up reproduction in favor of a few highly privileged individuals. Most humans can and do reproduce, in contrast to social insects, where the job is delegated to specialized queens and consorts. As a consequence, the degree of relatedness in a beehive is vastly higher than in a city. While it is clear that a lot of human cooperation occurs within the family, and that a tendency to nepotism is nearly universal, there also exist many instances of close collaboration among unrelated individuals. Hence kin selection, while certainly important, cannot be the only cause behind human cooperation.

It is well known that Darwin, in his *Origins of Species*, avoided touching on the human species, except in one sentence: "Light will be thrown on the origin of man and his history". Clearly, he expected his readers to see the light for themselves. It took Darwin more than a dozen years before he spelled out his conclusions, in *The Descent of Man, and Selection in Relation to Sex* and *The Expression of the*

*Emotions in Man and Animals*. The reason for the long delay was caused by Darwin's carefulness in marshaling his facts, but it may also have been due, in part, to hesitation. Darwin had to brace himself for the storm of bigotry, outrage, and ridicule expected to arise. (Among Darwin's files, a particularly voluminous one is devoted entirely to caricatures, ditties, and abuse heaped on him and his family.)

Since human cooperation among nonrelatives cannot be explained by indirect fitness benefits, there must be other reasons, causing direct fitness benefits, reasons which are not genetic but economic in nature. The economist Adam Smith had them in mind when, in *The Theory of Moral Sentiments* (written long before his *Wealth of Nations*), he spoke of "our propensity to trade, barter and truck". Darwin, who was well versed in the writings of economists such as Adam Smith or Robert Malthus, took this up when he wrote: "The small strength and speed of man, his want of natural weapons, etc., are more than counterbalanced ... by his social qualities, which led him to give and receive aid from his fellow men." To give and to receive: this economic viewpoint anticipated the second major approach to the evolution of cooperation.

## 14.7 Reciprocal Altruism

The theory of reciprocal altruism was first developed in a landmark paper by Trivers in 1971 [12]. It defined reciprocal altruism as "the trading of altruistic acts in which benefit is larger than cost, so that over a period of time both parties enjoy a net gain." In the simplest model, this can be described by a repeated prisoner's dilemma game: the same two players meet in round after round. In that case, they are not obliged to choose the same option (**C** or **D**) in every round. In particular, they can use conditional strategies, and decide whether to cooperate or defect according to the past behavior of their co-player. The most natural strategy, in this context, is to reciprocate good with good, and bad with bad. Many experiments have shown that a large majority of humans are conditional cooperators, and want to play **C** if the co-player also chooses **C**. In the context of repeated games, they can base their decision on the past behavior of their co-player. The simplest such strategy is "tit for tat" (**TFT**): it prescribes playing **C** in the first round, and from then on using whichever move the co-player used in the previous round.

If $w$ is the probability that the same two players will engage in a further round of the game, then the expected number of rounds is given by $1/(1 - w)$. If we consider only the two strategies **TFT** and **AllD** (i.e., defect in every round), then the payoff matrix is given by

$$
\begin{array}{c|cc}
* & \textbf{TFT} & \textbf{AllD} \\
\hline
\textbf{TFT} & (b - c)/(1 - w) & -c \\
\textbf{AllD} & b & 0
\end{array}
\tag{14.4}
$$

If $w > c/b$, that is, if the expected number of rounds is sufficiently large, then it does not pay, against a **TFT**-player, to play **AllD**: the advantage gained in the first round cannot make up for the handicap of turning the co-player into a defector. On the other hand, it does not pay, against an **AllD**-player, to use **TFT**. The best policy is to do whatever the other player does. This means that the evolutionary dynamics is bistable. If most players in the population use one of the strategies, then it is best to adopt it, too. If $w$ is close to 1, however, then the contest between **TFT** and **AllD** is rigged in favor of the former: its basin of attraction is much larger [13, 14].

Nevertheless, **TFT** has some weaknesses. For instance, unconditional cooperators (**AllC**-players) can enter a population of **TFT**-players by neutral drift. Both strategies do exactly as well, everybody cooperates in every round, and therefore, selection does not act in one direction or the other. This means that, by sheer chance, a sizable number of **AllC**-players can build up. But once this happens, **AllD** players can invade, since they can exploit the **AllC**-players. The state of a population consisting of **TFT**-, **AllC**- and **AllD**-players can be described by a point $(x_1, x_2, x_3)$ on the unit simplex (since $x_1 + x_2 + x_3 = 1$). The dynamics is interesting, see Fig. 14.2 [15]. Let us first note that the vertices are fixed points. So are the points on the edge where **AllD** is missing ($x_3 = 0$). Suppose now that a small minority of **AllD** is introduced, by some random event. If the frequency of **TFT**-players is sufficiently large, the invader is immediately eliminated again. If the frequency is too small, the defectors take over. But if the frequency of **TFT**-players is in the middle range, not too small and not too large, the **AllD**-players will, at first, prosper and grow. However, by preying on the unconditional **AllD**-players, they destroy the basis of their own success and will, eventually, be eliminated. The state, then, consists again of a mixture of conditional and unconditional altruists, as before, but now the frequency of **TFT**-players is so large that **AllD**-players can no longer invade. They have to wait (figuratively speaking) until neutral drift has again reduced the number

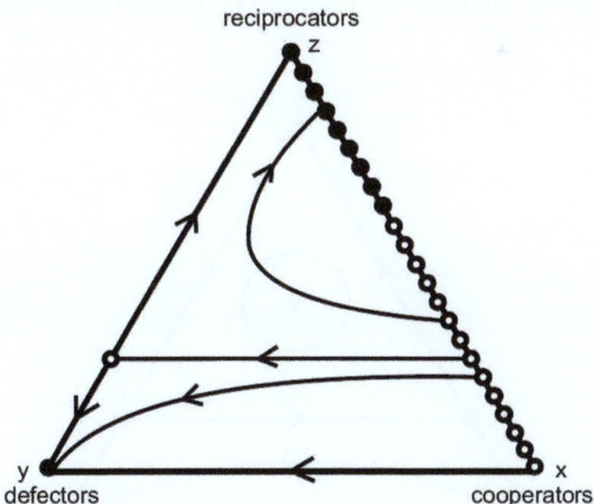


**Fig. 14.2** The good, the bad, and the reciprocator, for the repeated prisoner's dilemma game in the absence of errors. A horizontal line $z = c/wb$ divides the state space. For smaller values of the frequency $z$ of reciprocators, the **AllD**-players win; for larger values, they are eliminated. Here and in all other figures, full circles correspond to stable rest points, and empty circles to unstable rest points

of **TFT**-players in favor of unconditional defectors. If they invade too often, this will not happen. Hence, cooperative societies are more stable if they are challenged more often.

Another weakness of **TFT**-societies is that they are very vulnerable to errors [16]. If two **TFT**-players interact and one of them defects by mistake, this will cause a long vendetta. If such errors are taken into account, the dynamics of a population consisting of **TFT**-, **AllC**-, and **AllD**-players is highly unstable. Either **AllD** eliminates the other two strategies, or else all three will endlessly oscillate (Fig. 14.3).

There are other conditional strategies that do not have the defects of **TFT**. This holds in particular for **Pavlov**. Players using that strategy start with a cooperative move and from then on cooperate if and only if their co-player, in the previous round, chose the same move as themselves. This means that a **Pavlov** player repeats the former move if it yielded a positive income, and switches to the other move if not. As a consequence, **Pavlov** is error-correcting [17]. If two **Pavlov**-players are engaged in a repeated prisoner's dilemma game and one of them inadvertently defects, then in the next round both play **D** and afterwards resume mutual cooperation. Moreover, **AllC**-players cannot subvert a **Pavlov**-population by neutral drift. On the other hand, **Pavlov** cannot invade an **AllD**-population. It needs **TFT** to pave the way [17, 18].

So far, we have implicitly assumed that strategies are inherited from parent to offspring. For the hard-wired behavior of social insects, this is reasonable enough, but for humans, it is absurd to assume that **TFT**-players breed true. Fortunately, we can use the machinery of evolutionary game theory even if strategies are transmitted not through inheritance but through learning. If we assume that humans have a propensity to preferentially imitate more successful strategies, then we can apply the same dynamics to "memes" rather than genes. The role of "mutations", in this context, is provided by the random adoption of behavioral alternatives. Whether in

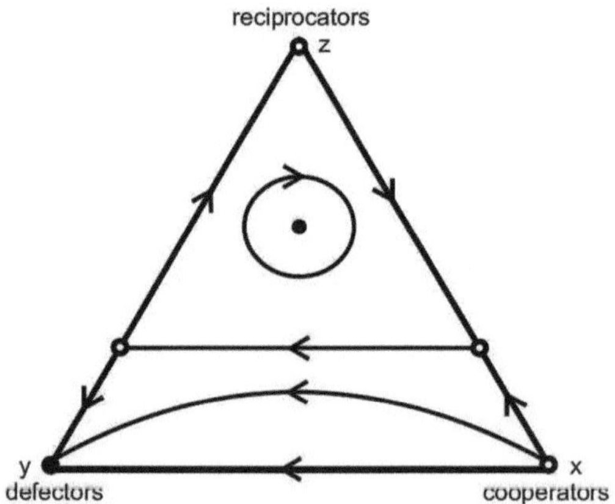


**Fig. 14.3** The good, the bad, and the reciprocator, for the repeated prisoner's dilemma with errors in implementation. If $z$ is below a threshold, defectors win; if $z$ is above the threshold, all three strategies co-exist, their frequencies oscillating periodically

the context of selection–mutation or of imitation–exploration, we are led to the same process of trial-and-error, and hence to the same dynamics [4, 14].

## 14.8 Indirect Reciprocity

Not every help is directed at a recipient able to return that help, but it may well be that an act of help is returned not by the recipient but by a third party. The idea of an indirect, or "generalized" reciprocity can already be found in Trivers's seminal paper of 1971. In direct reciprocity, if Alice provides help to Bob, then Bob should return help to Alice. In indirect reciprocity, if Alice helps Bob, then the help can be returned to Alice by some third party, for instance Charlie (Fig. 14.4). This seems a more subtle form of reciprocation. Direct reciprocity works on the principle that "I'll scratch your back if you scratch mine", indirect reciprocity on the principle "I'll scratch your back if you scratch somebody's". In direct reciprocity, I use my experience with someone. In indirect reciprocity, I also use the experience of others. This is cognitively much more demanding, but both direct and indirect reciprocity are forms of conditional cooperation: the willingness to assist those who are willing to provide assistance.

The evolutionary biologist Richard Alexander, who coined the term "indirect reciprocity", stressed that it "involves reputation and status, and results in everyone in the group continually being assessed and reassessed". He argued that it represents the *Biological Basis of Moral Systems* [19] (the title of his book). Indirect reciprocity requires a high degree of information. Whether Alice provides or refuses help to Bob can be either directly observed by third parties, or learned through gossip from others. This forms the basis for Charlie's decision on whether or not to help Alice, in turn. Charlie in effect acts upon a moral judgment on whether Alice deserves to be helped or not.

In the very simplest model, we can assume that every player has a binary reputation, which can be either G (for good) or B (for bad). Individuals meet randomly, as potential donors or recipients, and the donors can confer a benefit $b$ to the the recipient at a cost $c$ for themselves. Donors who provide help obtain reputation G, and those who refuse obtain B. The discriminating strategy consists in giving help to

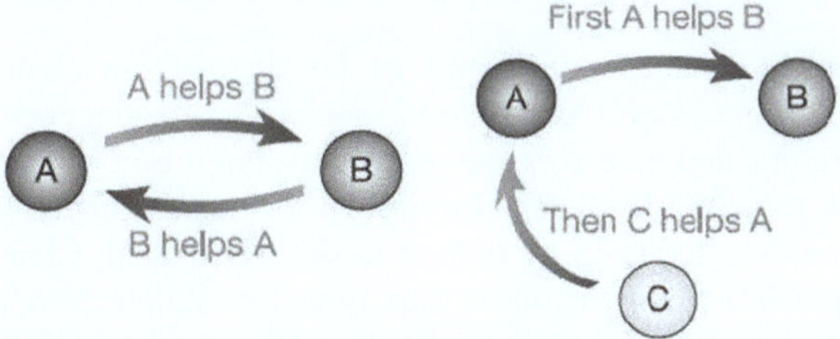


**Fig. 14.4** Direct versus indirect reciprocity. On the *left*, player A gives help to B and B returns the help to A. On the *right*, it is player C who returns the help to A

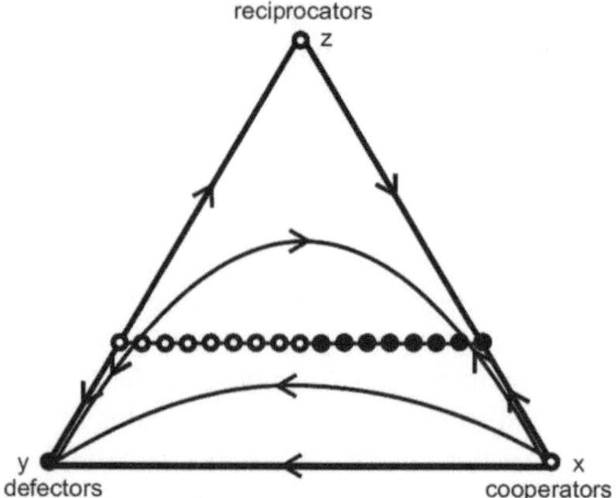


**Fig. 14.5** The good, the bad, and the reciprocator, for indirect reciprocity. If the information on the co-player's past behavior is sufficiently large, defectors, cooperators, and reciprocators coexist. The *horizontal line* of fixed points includes a stable mixture of **AllC**-players and reciprocators, but random shocks can lead, in the long run, to a population consisting only of **AllD**-players

G-recipients, and withholding help from B-recipients. This discriminating strategy is what, in the simpler situation of direct reciprocity (repeated games between the same two players) corresponds to **TFT**. In indirect reciprocity, it may be that players meet only once. Reputation takes the place of repetition.

Let us consider a population consisting of discriminators, as well as **AllC**- and **AllD**-players. It turns out that if the degree of information (i.e., the probability $q$ of knowing the other player's reputation) is sufficiently large, then the population will either evolve towards fixation of the all-out defectors, or towards a mixture of discriminating and undiscriminating cooperators (Fig. 14.5). This mixture can eventually be subverted by neutral drift, in a manner reminiscent but not quite similar to the direct reciprocity case. If we assume that each player, in time, becomes better informed about the co-players, then the mixture of discriminating and undiscriminating altruists actually becomes a stable attractor for the evolutionary dynamics.

## 14.9 Competition of Moral Systems

The discriminating strategy considered so far displays an element of paradox [20, 21]. Indeed, it is certainly useful to the society when cooperation is channeled towards cooperators, and defectors are kept at bay, but it is costly to the discriminator. By refusing to help a B-player, discriminators acquire themselves the B-label, and are therefore less likely to be helped in the next round. Clearly, it would be better to distinguish between justifiable and unjustifiable defection, but this requires more sophisticated rules for assessing what is good and what is bad.

The very rudimentary moral system considered up to now is called SCORING: according to this system, it is always bad to refuse to help. The STANDING rule

**Table 14.1** The assessment specifies which image to assign to the potential donor of an observed interaction ("good → bad" means "a *good* player helps a *bad* player", "bad ↛ good" means "a *bad* player refuses to help a *good* player", etc.)

| Situation/strategy | SCORING | STANDING | JUDGING |
| --- | --- | --- | --- |
| Good → good | Good | Good | Good |
| Good → bad | Good | Good | Bad |
| Bad → good | Good | Good | Good |
| Bad → bad | Good | Good | Bad |
| Good ↛ good | Bad | Bad | Bad |
| Good ↛ bad | Bad | Good | Good |
| Bad ↛ good | Bad | Bad | Bad |
| Bad ↛ bad | Bad | Good | Bad |

seems more reasonable: for this rule, it is bad to refuse help a to good player, but not to refuse help to a bad player. A more sterner version is named JUDGING: it views, additionally, help to bad players as a bad behavior (Table 14.1).

We can classify the different types of assessment rules. A first-order assessment rule simply takes into account whether help is given or not. A second-order assessment rule takes moreover into account whether the recipient is good or bad. A third-order assessment rule takes additionally into account whether the donor is good or bad. This leads to 256 value systems. Ohtsuki and Iwasa [22] have shown that only eight of them are stable in the following sense. For a homogeneous population adopting such a value system, there exists a uniquely specified rule of action (prescribing when to help, depending on the donor's and the recipient's image) which leads to cooperation and which cannot be invaded by any other rule of action, and in particular not by **AllC** or **AllD**. Two of the stable rules are of second order, none of first order (Table 14.2).

So far, there exist only very rudimentary theoretical results on the competition between several coexisting assessment rules [23]. Empirically, however, it seems clear that in many societies, several assessment rules coexist. In particular, experiments have consistently shown that SCORING, despite its drawbacks, is adopted by

**Table 14.2** The assessment modules of the leading eight strategies. These strategies obtain the highest payoff values, and are uninvadable by defectors. Strategy L3 corresponds to STANDING and strategy L8 to JUDGING. No SCORING strategy occurs in the list. The assessment only differs on the issues good → bad, bad → bad, and bad ↛ bad

| Situation/strategy | L1 | L2 | L3 | L4 | L5 | L6 | L7 | L8 |
| --- | --- | --- | --- | --- | --- | --- | --- | --- |
| Good → good | Good | Good | Good | Good | Good | Good | Good | Good |
| Good → bad | Good | Bad | Good | Good | Bad | Bad | Good | Bad |
| Bad → good | Good | Good | Good | Good | Good | Good | Good | Good |
| Bad → bad | Good | Good | Good | Bad | Good | Bad | Bad | Bad |
| Good ↛ good | Bad | Bad | Bad | Bad | Bad | Bad | Bad | Bad |
| Good ↛ bad | Good | Good | Good | Good | Good | Good | Good | Good |
| Bad ↛ good | Bad | Bad | Bad | Bad | Bad | Bad | Bad | Bad |
| Bad ↛ bad | Bad | Bad | Good | Good | Good | Good | Bad | Bad |

a substantial part of the players [24–26]. It seems to be cognitively very demanding to adopt higher-order assessment rules, since such rules require information not only on the recipient's past behavior but also on that of the recipient's recipients, etc. It can be argued that a population adopting a higher-order assessment rule could continuously update the images of all the players in a consensual process, but this seems to require an extraordinary amount of information exchange.

Indirect reciprocity was an essential factor in human evolution, because it provided a selective pressure for social intelligence, human language, and moral faculties. This does not imply, of course, that moral rules are innate. Just as we do not inherit a particular language, but have an innate faculty to quickly acquire a language, so we are not born with a ready-made moral system, but have the faculty to adopt one at an early age.

## 14.10 Exceptionalism

Many people, especially among those with a background in the humanities, are uneasy with the application of Darwin's theory to the evolution of moral norms. In their eyes, moral is a taboo topic for natural science, since it has to do with values, rather than with empirical facts. The following quote stems from Pope John Paul II: "Consequently, theories of evolution which [...] consider the mind as emerging from the forces of living matter, are incompatible with the truth about man." It is not only American creationists, but many European intellectuals who would essentially agree. They may accept Darwinism in all other aspects, but shrink from applying it to the so-called higher faculties of humans. The most distinguished "exceptionalist" was Alfred Russell Wallace, the man who almost scooped Darwin. Wallace wrote: "Man's intellectual and moral faculties [...] must have another origin [...] in the unseen universe of Spirit."

Darwin himself never shrank from investigating the evolution of our moral sense. One of his folders, which he later entitled "Old and useless notes on the moral sense" dates from 1837, when he was still in his twenties. Darwin certainly grasped the importance of reciprocation, as is clear from the quote: "We are led by the hope of receiving good in return to perform acts of sympathetic kindness to others", and when he wrote: "[Man's] motive to give aid [...] no longer consists solely of a blind instinctive impulse, but is largely influenced by the praise and blame of his fellow men" he had obviously understood that, in contrast to social insects, human cooperation is to a large extent based on reputation. He may have anticipated a kin selection approach to the evolution of moral faculties when he wrote: "The foundation of moral instincts lies in the social instincts, including in this term the family ties."

Darwin even went so far as to adopt in this context Lamarckist tendencies. This is shown by the following quote: "It is not improbable that virtuous tendencies may through long practice be inherited." Today, this is viewed as extremely improbable. But part of what Darwin may have had in mind can be couched in terms of "gene–culture coevolution" [27]. We know that a culturally shaped environment can cause a selective pressure for the spread of well-adapted genes. To give an example, cattle-raising is certainly a cultural phenomenon, but it provided the appropriate conditions for a genetic disposition that allows most Caucasians to digest lactase products even as adults, long after weaning. In societies without the cattle-raising tradition, for instance in Japan, this genetic predisposition is rare. We can similarly conceive that the culturally shaped social environment of our ancestors led to the spreading of genetic predispositions for what Darwin termed "virtuous tendencies" (such as fairness or solidarity).

## 14.11  Team Efforts

So far, we have only considered interactions between two players. These can, in general, be understood by a cost-to-benefit analysis. The ratio $c/b$ has to be smaller than something – for instance, the coefficient of relatedness $\rho$, or the probability of another round $w$, or the degree of information $q$ about other players. Many interactions occur in larger teams, however, and this raises additional difficulties. The mere concept of reciprocation, for instance, becomes more difficult. Whom do you reciprocate with, if your group contains both cooperators and defectors?

A typical model for such a situation is given by the so-called public goods game. All $N$ players in a group are asked to contribute some amount, knowing that this will be multiplied by a certain factor $r > 1$ and then divided equally among all players. If all contribute the same amount, their return will be the $r$-fold of that amount. But each individual receives only the $(r/N)$th part of his or her investment in return: if $r < N$, it is more profitable to invest nothing, and exploit the contributions of the co-players. However, if the other players follow the same line of action, no one will contribute anything. In actual experiments, players often contribute a substantial amount, but then, from round to round, reduce their contributions gradually. They feel exploited by those who contribute less than they did, and try to retaliate by contributing even less. But this hurts the cooperative players too, who then reduce their contributions in turn, etc.

Obviously, the snag in this game is that exploiters cannot be treated differently from cooperators. If the game is modified so that between the rounds of the public goods contributions players can punish or reward specific individuals, depending on the size of their contributions, then cooperation can often be stabilized at a high level [28–31]. This targeted form of providing positive or negatives incentives – the carrot and the stick – again relies on reciprocation.

Trivers described this in his paper on reciprocal altruism [12]. He wrote: "Altruistic acts are dispensed freely among more than two individuals, an individual being perceived to cheat if he dispenses less than others, punishment coming from the others in the system." And in an essay on "Innate Social Aptitudes of Man", Hamilton has a section on "Reciprocity and Enforcement", in which he describes, first, the problem of reciprocation in larger groups, and then points out that, in a many-person game, cooperators can gang together to punish cheaters. He noted that "There may be reasons to be glad that human life is a many-person game and not just a disjoined collection of two-person games".

## 14.12 Group Selection

The public goods game is another example of a social dilemma. In all such dilemmas, individual advantage and group benefit are at odds. The same tension occurs not only in economics, but also in biology. The controversy between individual selection and group selection has a long history, and is marked by confusion. Many of the earlier arguments involving "the benefit of the group" (or "the good of the species") were plain wrong, but this led to a backlash. For some time, group selection arguments were simply out. It took the patient efforts of several theoreticians to show that, in certain situations, group selection arguments can indeed work: for instance, if the time scales for competition within groups and competition between groups are not too different, if migration between groups falls into the right range, etc. [32–34].

Many evolutionary biologists argue today that group selection arguments can be couched in terms of kin selection, rightly understood (namely via relatedness through assortment, rather than relatedness through descent). Nevertheless, it seems that it is legitimate to phrase an argument in terms of group selection if the relevant evolutionary process can be described in terms of one group eliminating another. Such scenarios, mostly based on murderous conflicts and forms of raiding warfare, seem to have been frequent in the human past. Most of the run-of-the-mill behavior displayed in economic games can be interpreted in terms of self-interest, properly understood: for instance, as attempts to acquire and keep valuable partners. But extreme traits, such as the willingness to sacrifice oneself for others, or the readiness to view other groups as inferior, possibly require different explanations, based on a past history of violent encounters between groups [35].

Darwin stressed the importance of group survival on several occasions, for instance when he wrote: "There can be no doubt that a tribe including many members who [...] were always ready to give aid to each other and to sacrifice themselves for the common good, would be victorious over most other tribes; and this would be natural selection." He did not say: "...and this is group selection", but he obviously was aware of the tension between individual and group selection when he wrote: "He who was ready to sacrifice his life [...] would often leave no offspring to inherit his noble nature. Therefore it seems scarcely possible (bearing in mind that we are not here speaking of one tribe being victorious over another) that the number of men gifted with such virtues could be increased through natural selection." The term in parentheses clearly indicates that Darwin saw no way of explaining the evolution of such traits other than by violent inter-group conflict. In another passage, he stressed that "extinction follows chiefly from the competition of tribe with tribe, and race with race." To many ears, today, this sounds politically incorrect. But Darwin was very conscious of the dark side of human nature. Some of the most remarkable examples of human cooperation occur in war, and other forms of lethal conflict between groups.

# References


1. J. Maynard Smith, E. Szathmáry, *The Major Transitions in Evolution* (Oxford University Press, Oxford, 1995)
2. J. Maynard Smith, *Evolution and the Theory of Games* (Cambridge University Press, Cambridge, 1982)
3. J.W. Weibull, *Evolutionary Game Theory* (MIT Press, Cambridge, MA, 1997)
4. J. Hofbauer, K. Sigmund, *Evolutionary Games and Population Dynamics* (Cambridge University Press, Cambridge, 1998)
5. M.A. Nowak, *Evolutionary Dynamics* (Harvard University Press, Cambridge, MA, 2006)
6. W.D. Hamilton, *Narrow Roads of Geneland: Collected Papers I* (Freeman, New York, NY, 1996)
7. R. Dawkins, *The Selfish Gene*, 2nd edn. (Oxford University Press, Oxford, 1989)
8. A. Grafen, in: *Behavioral Ecology*, 2nd edn., ed. by J.R. Krebs, N.B. Davies (Blackwell Scientific, Oxford, 1984)
9. M. van Veelen, J. Theor. Biol. **259**, 589 (2009)
10. L. Lehmann, L. Keller, J. Evol. Biol. **19**, 1365 (2006)
11. J.A. Fletcher, M. Döbeli, Proc. R. Soc. B **276**, 13 (2009)
12. R. Trivers, Q. Rev. Biol. **46**, 35 (1971)
13. R. Axelrod, *The Evolution of Cooperation* (Basic Books, New York, NY, 1984)
14. K. Sigmund, *The Calculus of Selfishness* (Princeton University Press, Princeton, NJ, 2010)
15. H. Brandt, K. Sigmund, J. Theor. Biol. **239**, 183 (2006)
16. D. Fudenberg, E. Maskin, Am. Econ. Rev. **80**, 274 (1990)
17. M.A. Nowak, K. Sigmund, Nature **364**, 56 (1993)
18. C. Hilbe, IJBC **19**, 3877 (2009)
19. R.D. Alexander *The Biology of Moral Systems* (Aldine de Gruyter, New York, NY, 1987)
20. M.A. Nowak, K. Sigmund, Nature **393**, 573 (1998)
21. M.A. Nowak, K. Sigmund, Nature **437**, 1292 (2005)
22. H. Ohtsuki, Y. Iwasa, J. Theor. Biol. **239**, 435 (2006)
23. S. Uchida, K. Sigmund, J. Theor. Biol. **263**, 13 (2010)
24. C. Wedekind, M. Milinski, Science **288**, 850 (2000)
25. M. Milinski, D. Semmann, T.C.M. Bakker, H.J. Krambeck, Proc. R. Soc. Lond. B **268**, 2495 (2001)
26. M. Milinski, D. Semmann, H.J. Krambeck, Nature **415**, 424 (2002)
27. P.J. Richerson, R. Boyd, *Not by Genes Alone* (University of Chicago Press, Chicago, IL, 2005)
28. T. Yamagishi, J. Pers. Soc. Psychol. **51**, 110 (1986)
29. E. Fehr, S. Gächter, Nature **415**, 137 (2002)
30. K. Sigmund, Trends Ecol. Evol. **22**, 593 (2007)
31. B. Herrmann, U. Thöni, S. Gächter, Science **319**, 1362 (2008)
32. D. Cohen, I. Eshel, Theor. Popul. Biol. **10**, 276 (1976)
33. A. Traulsen, M.A. Nowak, Proc. Natl. Acad. Sci. USA **103**, 10952 (2006)
34. D.S. Wilson, in: *The Innate Mind: Culture and Cognition*, ed. by P. Carruthers, S. Laurence, S. Stich (Oxford University Press, Oxford, 2006)
35. J.K. Choi, S. Bowles, Science **318**, 636 (2007)

# Chapter 15
# Similarities Between Biological and Social Networks in Their Structural Organization

Byungnam Kahng, Deokjae Lee, and Pureun Kim

**Abstract** A branching tree is a tree that is generated through a multiplicative branching process starting from a root. A critical branching tree is a branching tree in which the mean branching number of each node is 1, so that the number of offspring neither decays to zero nor flourishes as the branching process goes on. Moreover, a scale-free branching tree is a branching tree in which the number of offspring is heterogeneous, and its distribution follows a power law. Here we examine three structures, two from biology (a phylogenetic tree and the skeletons of a yeast protein interaction network) and one from social science (a coauthorship network), and find that all these structures are scale-free critical branching trees. This suggests that evolutionary processes in such systems take place in bursts and in a self-organized manner.

## 15.1 Introduction

Darwin's theory of biological evolution is one of the most revolutionary contributions to progress of recent centuries. His renowned book [1], *On the Origin of Species*, published over 150 years ago, contains two major themes, descent with modification and natural selection acting on hereditary variation. Since then, evolutionary theory has developed into diverse areas such as ecology, animal behavior, and reproductive biology. Recently, as research tools on the molecular level have become more sophisticated and available, new directions of evolutionary study have been developed to confirm Darwin's theory. Molecular evolutionary studies that analyze evolutionary processes at the genetic level have drawn considerable attention from the scientific community. Large-scale databases that accumulate various types of biological data have been compiled, so that researchers have easy access and can discover new facts.

B. Kahng (✉)
Department of Physics and Astronomy, Seoul National University, Seoul, Republic of Korea
e-mail: bkahng@snu.ac.kr; kahng@phya.snu.ac.kr

H. Meyer-Ortmanns, S. Thurner (eds.), *Principles of Evolution*,
The Frontiers Collection, DOI 10.1007/978-3-642-18137-5_15,
© Springer-Verlag Berlin Heidelberg 2011

Here we download taxonomy data of evolution from the GenBank database [2], and construct a phylogenetic tree, which provides a cornerstone to understanding evolution at the gene level. Structural properties of the taxonomy tree are examined. We also analyze the skeleton of protein interaction networks, and the evolution of a coauthorship network (taken from social networks). From these examples, we find that there exists a common feature, that there exist scale-free critical branching trees underneath such complex networks. The presence of such trees suggests that evolution takes place in bursts, but in a self-organized manner.

## 15.2 Branching Tree and Fractal Structure

### 15.2.1 Scale-Free and Critical Branching Structure

A tree structure can be generated through a multiplicative branching process starting from a root [3]. A node (ancestor) generates $n$ offspring, in which the number $n$ is not fixed, but distributed with probability $b_n$,

$$
b_n = \begin{cases} \frac{n^{-\gamma}}{\zeta(\gamma-1)} & \text{for } n \geq 1 \,, \\ 1 - \sum_{n=1}^{\infty} b_n & \text{for } n = 0 \,. \end{cases}
\tag{15.1}
$$

This tree structure is heterogeneous in the number of offspring, called the branching number, and thus is a scale-free branching tree. The root can be taken at any vertex in the system. For convenience, here we take as the root the hub, the vertex with the largest degree. When each vertex born in the previous step generates $n$ offspring with probability $b_n$, the criticality condition means that the mean branching number is 1:

$$
\langle n \rangle \equiv \sum_{n=0}^{\infty} n b_n = 1 \,.
\tag{15.2}
$$

Thus, this critical branching tree grows perpetually with offspring neither flourishing nor dying out on the average. In this case, the number of vertices $M$ in the tree within a region with linear size $\ell$ scales in a power law as $M \sim \ell^z$ [3–5] with

$$
z = \begin{cases} (\gamma - 1)/(\gamma - 2) & \text{for } 2 < \gamma < 3 \,, \\ 2 & \text{for } \gamma > 3 \,. \end{cases}
\tag{15.3}
$$

Thus, the branching tree structure is fractal with the fractal dimension $d_B = z$. Such a critical branching tree is similar in topological characteristics to the so-called homogeneous scale-free tree network [6]. On the other hand, when the mean branching number fulfills $\langle n \rangle > 1$ ($\langle n \rangle < 1$), called the supercritical (subcritical) case, $M$ grows (reduces) exponentially with respect to linear size $\ell$ or the distance from the root, that is, $M \sim e^{\ell/\ell_0}$ ($M \sim e^{-\ell/\ell_0}$) with constant $\ell_0$. Thus, the branching tree in the supercritical state exhibits the small-world property.

For later discussions, we recall another property of the branching tree structure briefly [4]. Since the branching tree is stochastically generated, it can die out after just a few branching steps or survive for a large number of branching steps. The distribution of the number of offspring generated from a single ancestor is nontrivial, which is often referred to as the cluster size distribution. When the branching probability is given as $b_n \sim n^{-\gamma}$, the distribution of $s$-size clusters follows a power law $P_b(s) \sim s^{-\tau}$, where the exponent $\tau$ is given as

$$
\tau = \begin{cases} \gamma/(\gamma - 1) & \text{for } 2 < \gamma < 3 , \\ 3/2 & \text{for } \gamma > 3 , \end{cases}
\tag{15.4}
$$

for the critical branching tree.

### 15.2.2 Fractality

Fractal scaling refers to a power-law relationship between the minimum number of boxes $N_B(\ell_B)$ needed to tile an entire network and the lateral size of the boxes $\ell_B$, that is,

$$
N_B(\ell_B) \sim \ell_B^{-d_B} ,
\tag{15.5}
$$

where $d_B$ is the fractal dimension [7]. This method is called the box-covering method.

One may define the fractal dimension in another manner through the mass–radius relation [7]. The average number of vertices $\langle M_C(\ell_C) \rangle$ within a box of lateral size $\ell_C$, called average box mass, scales in a power-law form,

$$
\langle M_C(\ell_C) \rangle \sim \ell_C^{d_C} ,
\tag{15.6}
$$

with the fractal dimension $d_C$. This method is called the cluster-growing method. The formulae (15.5) and (15.6) are equivalent when the relation $N \sim N_B(\ell_B)\langle M_C(\ell_C) \rangle$ holds for $\ell_B = \ell_C$. Such is the case for the conventional fractal objects embedded in the Euclidean space [7] for which $d_B = d_C$. However, for complex networks, the relation (15.6) is replaced by the small-world behavior,

$$
\langle M_C(\ell_C) \rangle \sim e^{\ell_C/\ell_0},
\tag{15.7}
$$

where $\ell_0$ is a constant.

Thus, fractal scaling can be found in the box-covering method, but not in the cluster-growing method for scale-free fractal networks. This contradiction can be resolved by the fact that a vertex is counted into only a single box in the box-covering method, whereas in the cluster-growing method it can be counted into multiple ones. The number of distinct boxes a vertex belongs to in the cluster-growing method follows a broad distribution for scale-free networks. This is in contrast to a Poisson-type distribution obtained from conventional fractal objects [8].

The box-covering method runs as follows:

(i) Label all vertices as "not burned" (NB).
(ii) Select a vertex randomly at each step; this vertex serves as a seed.
(iii) Search the network by distance $\ell_B$ from the seed and burn all NB vertices. Assign *newly burned vertices* to the new box. If no NB vertex is found, the box is discarded.
(iv) Repeat (ii) and (iii) until all vertices are burned and assigned to a box.

This procedure is schematically illustrated in Fig. 15.1. Different Monte Carlo realizations of the procedure may yield a different number of boxes to cover the network. Here, for simplicity, we choose the smallest number of boxes among all the trials. The power-law behavior of the fractal scaling is obtained by at most $\mathcal{O}(10)$ Monte Carlo trials for all fractal networks we studied. It should be noted that the box number $N_B$ we employ is not the minimum number among *all* the possible tiling configurations. Finding the actual minimum number over all configurations is a challenging task by itself.

The origin of the fractal scaling has been studied [9–11]. We demonstrated recently [11] that the fractal property can be understood from the existence of the underlying skeleton [12]. When the skeleton is a critical branching tree, the structure is fractal, because the number of boxes needed to cover the original network is almost identical to that needed to cover the skeleton. Thus, the fractal dimensions of the original network and its skeleton are the same.

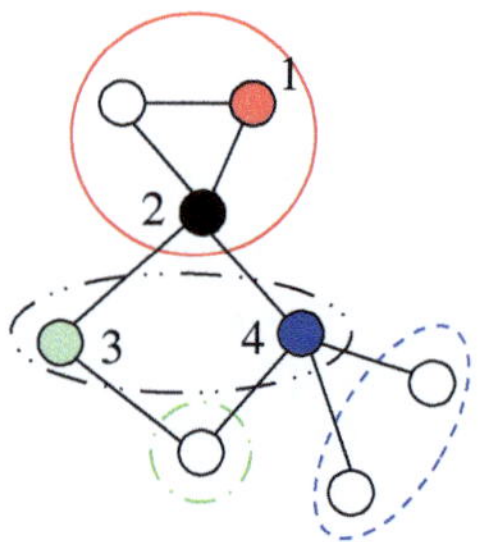

**Fig. 15.1** Schematic illustration of the box-covering algorithm. Vertices are selected randomly, for example, from vertex 1 to 4 successively. Vertices within distance $\ell_B = 1$ from vertex 1 are assigned to a *box* represented by the *continuous (red) circle*. Vertices from vertex 2, not yet assigned to their respective *box* are represented by the dashed–double dotted (*black*) closed curve, vertices from vertex 3 are represented by the dashed–dotted (*green*) *circle*, and vertices from vertex 4 are represented by the *dashed (blue) ellipse*

## 15.3 The Phylogenetic Tree

We have downloaded the taxonomy data from the GenBank database [2], which contains the names of all organisms with at least one nucleotide or protein sequence, and then we examined the structural properties of the taxonomic tree. The obtained taxonomic tree is composed of more than 500,000 nodes, allowing us to investigate the evolutionary pattern of biological organisms. The taxonomic tree indicates that the evolution of organisms proceeds in a branching process manner. The number of branches corresponds to the number of mutants in the next generation. We find that the distribution of the number of branches follows a power law. That is, the probability $P_b(k)$ that $k$ different gene types of descendants are generated obeys a power law, $P_b(k) \sim k^{-\gamma}$ with $\gamma \approx 2.2$. This fact suggests that the evolution takes place in bursts. Moreover, we find that the mean branching rate is 1, suggesting that the biological system maintains itself in self-organized way such that the diversity of organisms maintains its steady state. Besides, the self-similarity and the correlation between branching numbers turn out to be nontrivial.

### 15.3.1 Database

The taxonomy tree starts from a root that has five branches: viruses, viroids, cellular organisms, other sequences, and unclassified. The total number of nodes was about 510,000 when we collected the data in June 2009. As the database is updated daily, that number will have increased since then. The number of species on the tree is about 400,000. The maximum lineage length is 41. For example, *Homo sapiens* is located at the 31st step from the root via the path root $\rightarrow$ cellular organisms $\rightarrow$ Eukaryota (superkingdom) $\rightarrow$ Fungi/Metazoa group (kingdom) $\rightarrow$ Metazoa (kingdom) $\rightarrow$ Eumetazoa $\rightarrow$ Bilateria $\rightarrow$ Coelomata $\rightarrow$ Deuterostomia $\rightarrow$ Chordata (phylum) $\rightarrow$ Craniata (subphylum) $\rightarrow$ Vertebrata $\rightarrow$ Gnathostomata (superclass) $\rightarrow$ Teleostomi $\rightarrow$ Euteleostomi $\rightarrow$ Sarcopterygii $\rightarrow$ Tetrapoda $\rightarrow$ Amniota $\rightarrow$ Mammalia(class) $\rightarrow$ Theria $\rightarrow$ Eutheria $\rightarrow$ Euarchontoglires (superorder) $\rightarrow$ Primates (order) $\rightarrow$ Haplorrhini (suborder) $\rightarrow$ Simiiformers (infraorder) $\rightarrow$ Catarrhini (parvorder) $\rightarrow$ Hominoidea (primates) $\rightarrow$ Hominidae (family) $\rightarrow$ Homininae (subfamily) $\rightarrow$ Homo(genus) $\rightarrow$ Homo sapiens(species) $\rightarrow$ Homo sapiens $\rightarrow$ neanderthalensis (subspecies).

### 15.3.2 Structural Features

The degree distribution of the taxonomy tree follows a power law with the exponent $-2.2$. The mean branching number is $\langle n_b \rangle = 1$, indicating that the taxonomy tree is a critical and scale-free branching tree. However, the degree–degree correlation between two neighboring nodes is nontrivial as $\langle k_{nn} \rangle (k) \sim k^{-0.5}$, which implies that the branching processes between two vertices has nontrivial correlation. Further investigation associated with this numerical result has to be carried out.

## 15.4 Evolution of Protein Interaction Networks

Protein interaction networks (PINs) have been studied in a variety of organisms, including viruses [13], yeast [14], and *C. elegans* [15], in which nodes represent proteins and links are connections between two proteins that physically interact each other. Previous studies have focused on dynamical or computational aspects of interacting proteins as well as their potential links. Construction of evolution models of the PIN with biologically relevant ingredients has been also attractive. One of the successful models was introduced by Solé et al. [16], in which three dynamic processes of edges, such as duplication, mutation, and divergence, were used as key ingredients of the evolution of protein interactions. Similar models that follow this model are listed in the references [17–22]. Here we briefly introduce structural features of the Solé model, and show that the protein interaction network contains a generic critical branching tree underneath it. This feature may reflect that each module in the protein interaction network was generated through evolutionary processes such as mutations and divergences as we observed in the phylogenetic tree.

### 15.4.1 The Solé Model

The model by Solé et al. is a growing network model in which the number of nodes (proteins) increases by one at each time step, which is achieved in the form of duplication: A new node duplicates a randomly chosen pre-existing protein and its links are also endowed by its ancestor. Each link of the new protein is removed with probability $\delta$ (divergence) and the new protein can also form links to any pre-existing node with probability $\beta/N$ (mutation), where $N$ is the total number of nodes existing at each time step. The two parameters $\delta$ and $\beta$ control the densities of short-ranged and long-ranged edges, respectively. This model has been solved analytically [23]. Here, we review some important analytic results.

Let $n_s(N)$ be the density of $s$-size clusters at time $N$, and $g(z) = \sum_s s n_s z^s$ the generation function for $s n_s$, where the sum excludes the giant percolating cluster. $g(1)$ is the density of nodes belonging to finite clusters and $g'(1)$ is the average cluster size, that is, $\langle s \rangle = \sum s^2 n_s$. The model exhibits an unconventional percolation transition in which the parameter $\delta$ turns out to be irrelevant, and thus it is ignored for the time being. Within this scheme, the analytic solution yields that

$$\langle s \rangle = \begin{cases} \dfrac{1 - 2\beta - \sqrt{1 - 4\beta}}{2\beta^2} & \text{for } \beta \le \beta_c \, , \\[3ex] \dfrac{e^{-\beta G} + G - 1}{\beta(1 - e^{-\beta G})} & \text{for } \beta > \beta_c \, , \end{cases} \tag{15.8}$$

where $G$ is the density of the giant cluster $G = 1 - g(1) = 1 - \sum_s s n_s$. $\beta_c$ is the percolation threshold and obtained as $\beta_c = 1/4$. The cluster-size distribution follows a power law,

$$n_s \sim s^{-\tau}, \tag{15.9}$$

where the exponent is solved as

$$\tau = 1 + \frac{2}{1 - \sqrt{1 - 4\beta}}. \tag{15.10}$$

This power-law behavior holds in the entire range $\beta < \beta_c$, in contrast to the behavior of the conventional percolation transition. At the transition point $\beta = \beta_c$, the cluster-size distribution decays as $n_s \sim 1/[s^3(\ln s)^3]$.

The order parameter of the percolation transition is written as

$$G(\beta) \propto \exp\left(-\frac{\pi}{\sqrt{4\beta - 1}}\right). \tag{15.11}$$

Thus, all derivatives of $G(\beta)$ vanish as $\beta \to \beta_c$, and the transition is of infinite order.

The degree distribution was also studied in [23] for general $\delta$. When $\delta > 1/2$ and $\beta > 0$, the degree distribution follows a power law $P_d(k) \sim k^{-\gamma}$, where the degree exponent is determined from the relation

$$\gamma(\delta) = 1 + \frac{1}{1 - \delta} - (1 - \delta)^{\gamma - 2}. \tag{15.12}$$

### 15.4.2 Numerical Results

To examine a percolation transition in parameter space, we measure the mean cluster size $\langle s \rangle$ as a function of $\beta$ at a fixed parameter value $\delta = 0.95$, corresponding to the case with little duplication of links. We find that the mean cluster size exhibits a peak at transition point $\beta_c$, which is estimated to be $\beta_c \approx 0.29$ (Fig. 15.2) . This critical point is regarded as a percolation threshold. The percolation threshold $\beta_c$ varies as

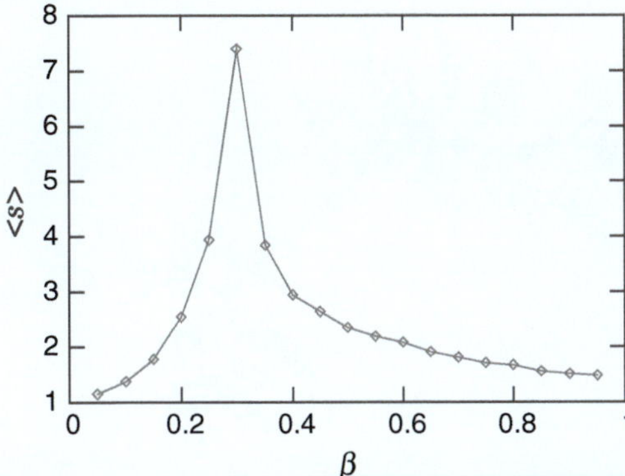

**Fig. 15.2** Mean cluster size $\langle s \rangle$ versus $\beta$ at $\delta = 0.95$. The data, obtained from system size $N = 10^5$ ($\diamond$), display a peak at the percolation transition estimated to be $\beta \approx 0.29$. All data points are averaged over 100 configurations

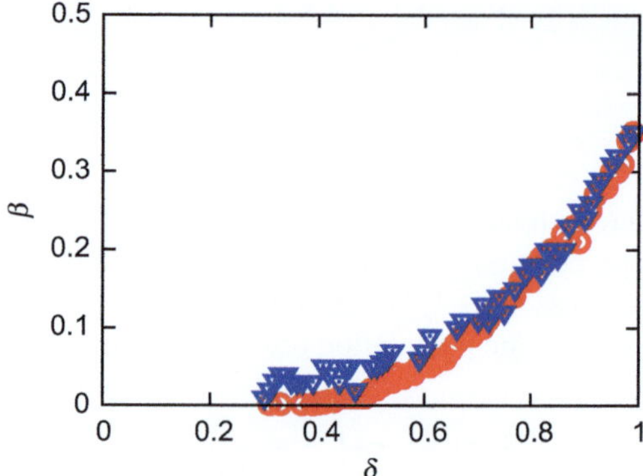

**Fig. 15.3** Plot of phase boundary between percolating and nonpercolating phases in the parameter space of $\beta$ and $\delta$. Data are obtained from system sizes $N = 10^3\,(\triangledown)$ and $10^4\,(\circ)$

a function of $\delta$. Thus, we plot in Fig. 15.3 the mean cluster size as a function of $\beta$ and $\delta$. The peak locus lies in the small $\beta$ region, indicating that a small fraction of long-range edges is sufficient to develop the giant cluster.

We show a giant cluster of the model network with small system size $N = 10^4$ at the percolation threshold in Fig. 15.4. This network is constructed with parameter values $\beta = 0.29$ and $\delta = 0.95$. The network topology is effectively a tree but with small-size loops within it.

We examine the degree distributions of the giant component at evolution steps $N = 10^3$ and $N = 10^4$ with parameter values $\delta = 0.58$ and $\beta = 0.16$ and show them in Fig. 15.5. It shows a power-law behavior. The dashed line in Fig. 15.5 has a slope of $-2.94$, the theoretical value obtained from (15.12).

In order to see the fractality of the model network, we measure the number of boxes $N_B$ as a function of box size $\ell_B$ using the box-covering method. In Fig. 15.6,

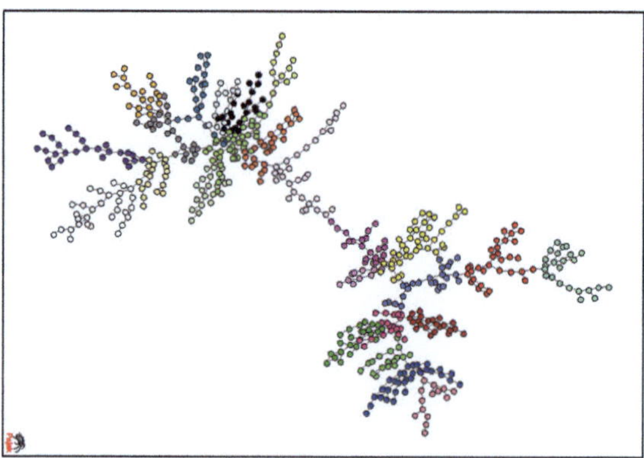

**Fig. 15.4** Snapshot of the giant component near the percolation threshold $\delta = 0.95$ and $\beta = 0.29$ with size $N = 568$

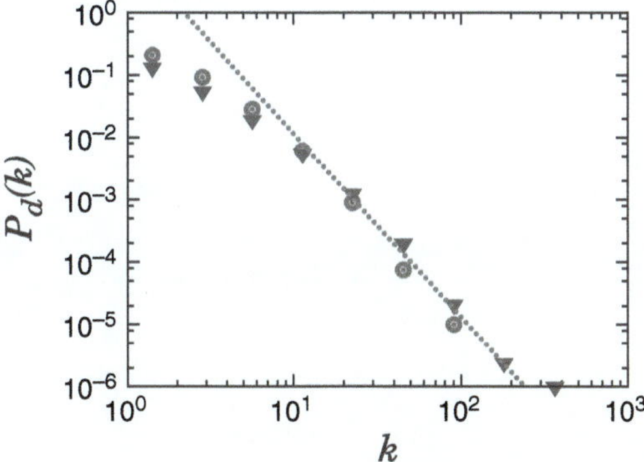

Fig. 15.5 Degree distribution $P_d(k)$ versus $k$ for the giant component of the Solé model with $\delta = 0.58$ and $\beta = 0.16$. Shown are the distributions for $N = 10^3$ (○) and $10^4$ (▽). The *dotted line* is the predicted line with slope $-2.94$

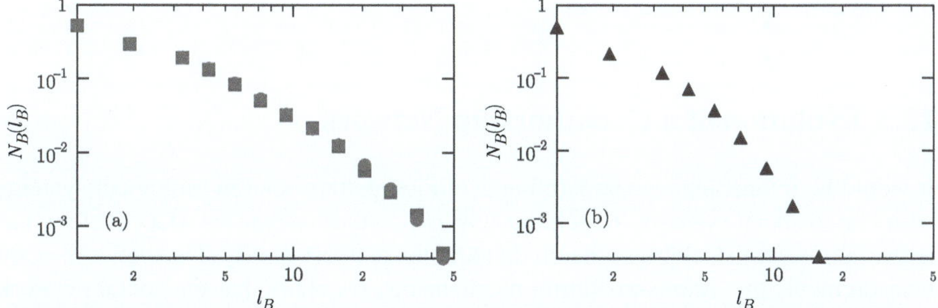

Fig. 15.6  **a** Fractal scaling analysis of the giant component (□) and its skeleton (●) near the percolation transition point $\delta = 0.95$ and $\beta = 0.29$ (□, ●) of the Solé model. The number of boxes follows a heavy-tailed distribution. **b** However, data obtained at $\delta = 0.6$ and $\beta = 0.3$ (△), located far away from the percolation threshold, decay faster than the previous one. All data points are log-binned and averaged over 100 configurations

$N_\mathrm{B}(\ell_\mathrm{B})$ exhibits a heavy-tailed distribution with respect to $\ell_\mathrm{B}$ when the data are obtained near the percolation threshold ($\delta = 0.95$ and $\beta = 0.29$). The numbers of boxes covering the skeleton for each box size are also shown in Fig. 15.6a: They overlap with those covering the entire network. Since power-law behavior is not manifest, one may wonder if this model network is indeed a fractal. Thus, we present $N_\mathrm{B}(\ell_\mathrm{B})$ for the network obtained from a different parameter set, particularly, located far away from the percolation threshold. Indeed, $N_\mathrm{B}(\ell_\mathrm{B})$ for this case decays fast compared with that obtained near the percolation threshold.

The criticality of the skeleton is checked. We measure the mean branching number function $\bar{n}(d)$ as a function of distance from a root. Indeed, it fluctuates around 1, implying that the skeleton can be regarded as a critical branching tree and thus manifestly a fractal (Fig. 15.7). Since the box numbers required to cover the

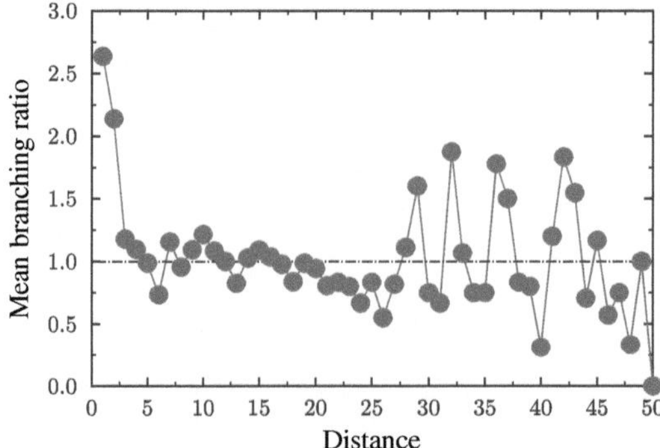

**Fig. 15.7** Mean branching number as a function of distance from a root for the skeleton of the giant component of the Solé model produced by the parameter values $\delta = 0.95$ and $\beta = 0.29$

entire network with each box size and the skeleton only are the same, we can say that the entire network is a fractal.

## 15.5 Evolution of a Coauthorship Network

It would be interesting to compare the evolution pattern seen in biological systems with that in social systems. To achieve this goal, we use a data set of a coauthorship network of scientists doing research on complex network theory. Through fine-scale measurements on various evolution mechanisms, we show that the social network follows a couple of distinct evolutionary steps: a tree-like structure forms at an early stage and large-scale loop structure develops at a later stage. We also show that a genuine skeleton structure underneath the coauthorship network is also a critical and scale-free branching tree.

### 15.5.1 Data Collection

To track time evolution of the coauthorship network, we first identify a set of ground-breaking papers, which will serve as the root papers, which all subsequent papers should cite. For complex network theory, we chose two papers, one by Watts and Strogatz about the small-world network [24] and the other by Barabási and Albert on the scale-free network [25], as the root papers. We further chose three early review papers [26–28]. Then we considered that the authors of papers that cited any of these five papers form the complex network coauthorship network. According to Web of Science, by December 2008 there were 5,008 such papers in total, with the information on the list of authors and the publication time in units of months, written by 6,816 nonredundant authors (in terms of their last name and initials) for the period spanning 127 months (from June 1998 to December 2008).

In the coauthorship network, a link is made between two nodes (researchers) if they are coauthors of at least one paper. The weight of the link is given by the number of papers they have coauthored. To track the time evolution of the network, we generated the coauthorship network for each month, from the papers published up to that month.

## 15.5.2 Evolution of a Large-Scale Structure

Temporal analysis reveals a global structural transition of the network through three major regimes (Fig. 15.8c): (I) nucleation of small isolated components, (II) formation of a tree-like giant component by cluster aggregation, and (III) network entanglement by long-range links. Subsequently, the network reaches the steady state, in which the mean separation between two nodes stabilizes around a finite value. The locality constraint, that is, new links are formed much more locally than globally, played an important role in sustaining the network's tree-like structure in regime II. Here, by tree-like structure we mean that the network is dominated by short-range loops and devoid of long-range connections, thus becoming a tree when coarse-grained into the network of supernodes, corresponding in this case to groups led by each principal investigator. This implies that most papers are the result of in-group collaborations. If the locality effect were weak, the inter-

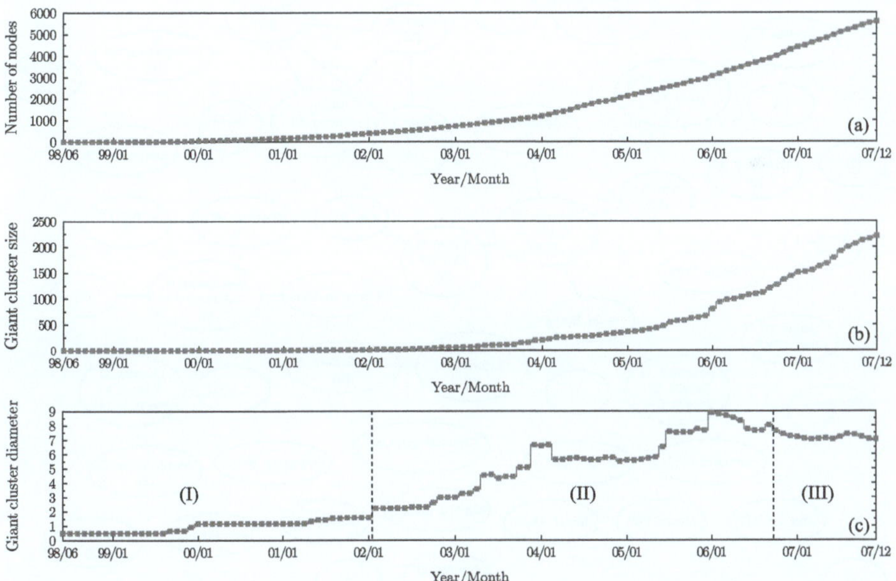

**Fig. 15.8** Evolution of large-scale structure of the complex network coauthorship network. **a** The number of nodes as a function of time. **b** The size of the largest component as a function of time. **c** The mean separation between two nodes of the largest component as a function of time

mediate stage II would not appear. Moreover, such a tree-like structure is a fractal [29, 30] and is sustained even underneath the entangled network even in the late regime III. This structure is unveiled upon the removal of inactivated edges, and has the same fractal dimension as in the tree-like structure. This implies that a hidden ordered structure with the same fractal dimension underlies the evolution process.

To see whether the general pattern of evolution conforms to that of conventional network growth models [25–27], we visualized snapshots of the network in time, as shown in Figs. 15.9, 15.10, 15.11 and 15.12. The network grows both by the expansion and merging of the existing components and by a continuous introduction of new components, as seen in the sea of small components surrounding the largest component at all stages. In the early stage, the largest component grows in a tree-like manner, in that it branches more and deeper as time goes on, but rarely establishes links between branches. It leads to the gradual increase of the largest component size

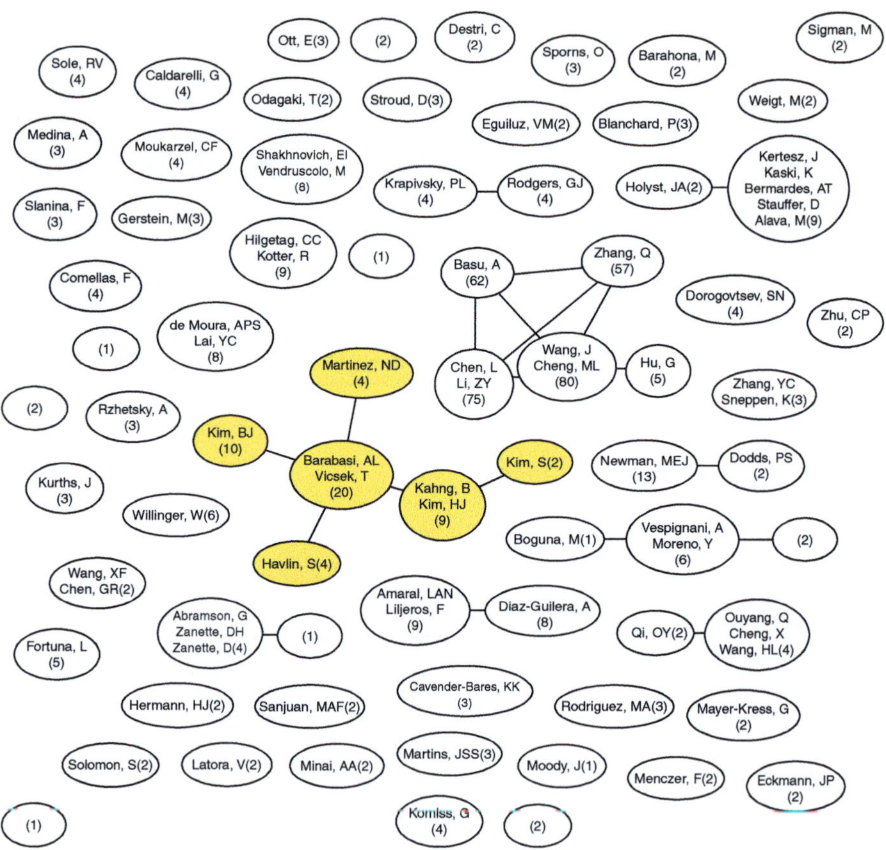

**Fig. 15.9** Snapshot of the coauthorship network in October 2002 in regime I

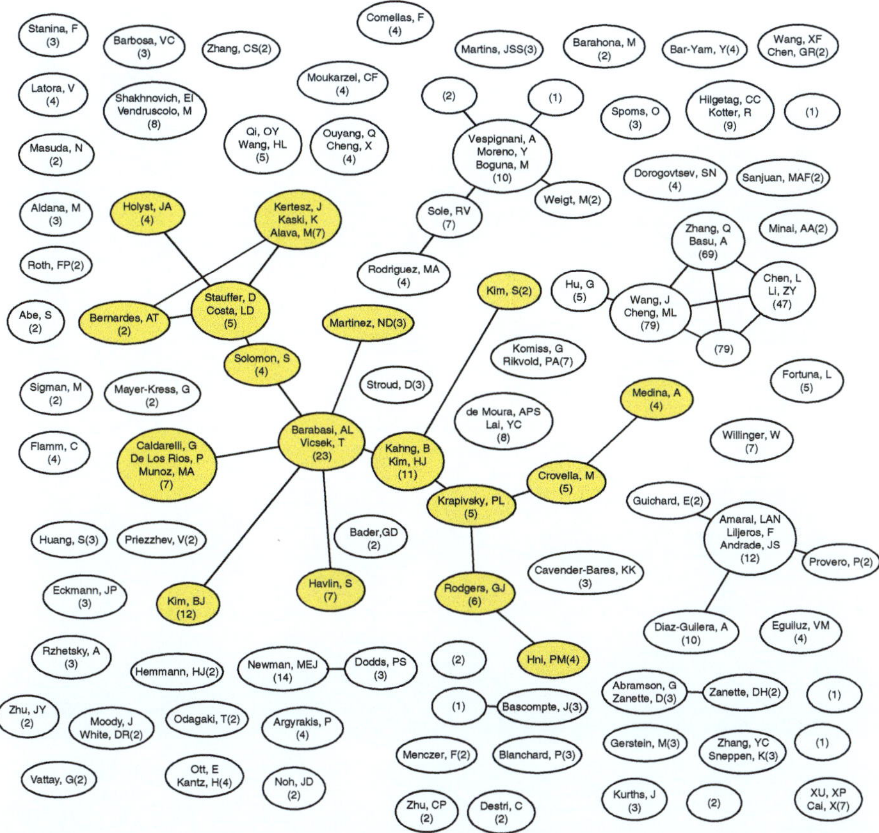

**Fig. 15.10** Snapshot of the coauthorship network in August 2003 in early regime II. The network is effectively a tree

and diameter, with a few intermittent jumps resulting from the merging of the small but macroscopic component to the largest one. Not until 2004 does a large-scale loop appear formed by a long-distance interbranch link (Figs. 15.11 and 15.12). Such a long-range loop formation can be monitored by the sudden drop in the largest component diameter. From then on more and more large-scale loops are formed, resulting in more entangled giant component structure.

### 15.5.3 Fractal Structure and Critical Branching Tree

The network has grown both by the expansion and merging of existing components and by the continuous introduction of new components. In intermediate time regime II, the giant component grows in a tree-like manner, it branches

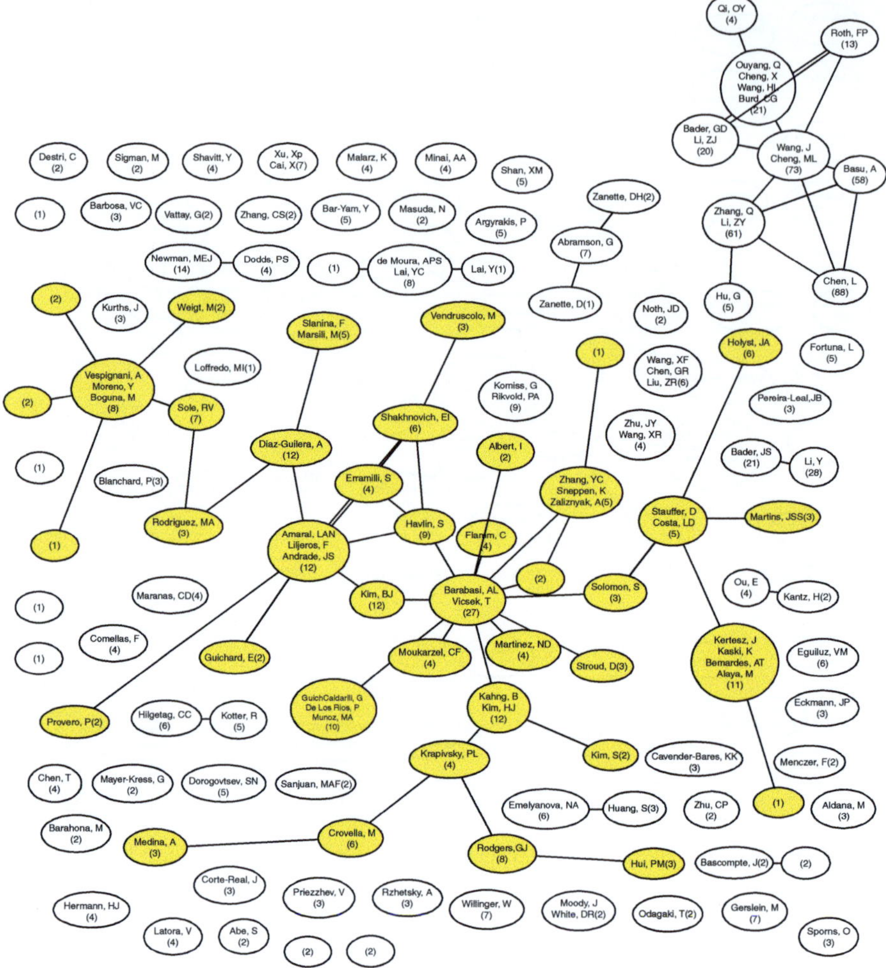

**Fig. 15.11** Snapshot of the coauthorship network in February 2004 in regime II

out more, and more deeply, with the passage of time, but rarely establishes links between branches. Component sizes become inhomogeneous in the growth process. The ramified network topology is reminiscent of the diffusion-limited aggregation (DLA) cluster. We find indeed that the giant component is a fractal with the fractal dimension $d_{\mathrm{B}} \approx 1.7$, close to that of the DLA cluster in two dimensions. We also find that the skeleton of the fractal network is a critical branching tree (Fig. 15.13).

Coauthorship links may be no longer active if the collaboration has ceased long ago. Thus, to ensure that the generic structural features remain robust, we examine how the overall network structure is affected in the presence of a link degradation process, that is, removal of redundant edges. The central issue would be whether the

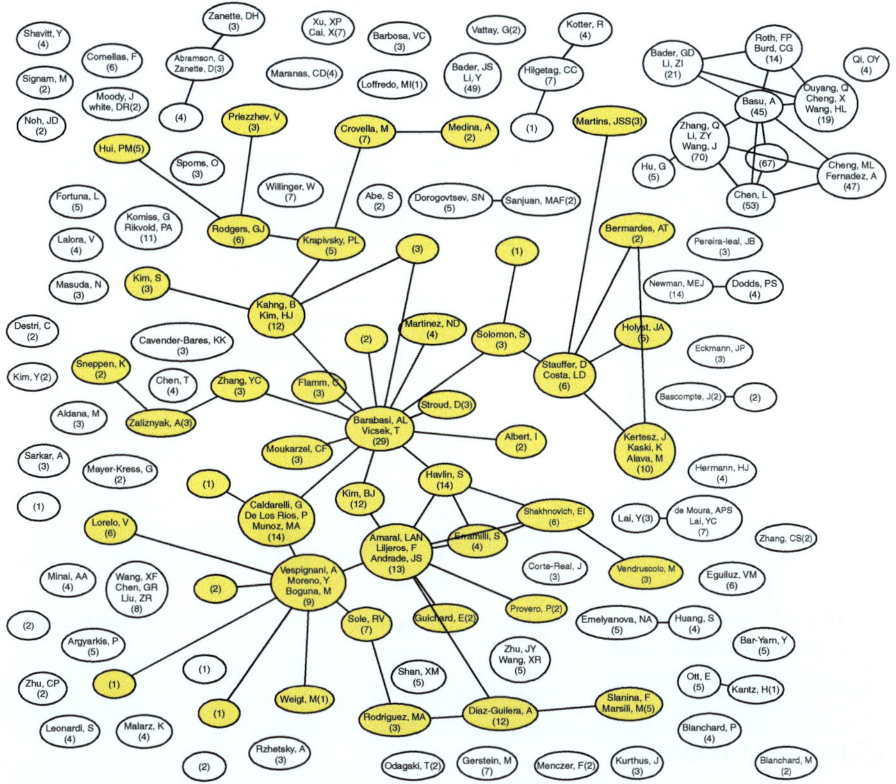

**Fig. 15.12** Snapshot of the coauthorship network in March 2004 in regime II

giant component persists in the system. To this end, for each month, we removed all the links that had not been re-activated during the previous 2 years – a typical postdoctoral contract period. We find that 86% of the links formed up to the year 2006 eventually disappeared before the end of 2008 according to the 2-year inactivation rule. However, the giant component not only persists upon degradation, but is also more stable, in the sense that its relative size $S$ has been stable at $\approx 10\%$ of the total network since 2000. At the same time, the link-degraded giant component (LDGC) is highly dynamic, in that its members constantly change over time. At the end of 2008, 1,195 nodes formed the LDGC, among which only 272 were the LDGC members in December 2006, when it was composed of 727. This indicates that complex network research is still a vigorous field [31]. The LDGC exhibits a tree-like structure throughout the observation period, implying that such a tree-like spanning component structure exists to provide a dynamic backbone underlying the complex original interwoven network. Furthermore, the LDGC resembles the original giant component in regime II, and their fractal dimensions are the same as $d_{\mathrm{B}} \approx 1.7(1)$. The skeleton of the remaining fractal structure is also a critical branching tree (Fig. 15.13).

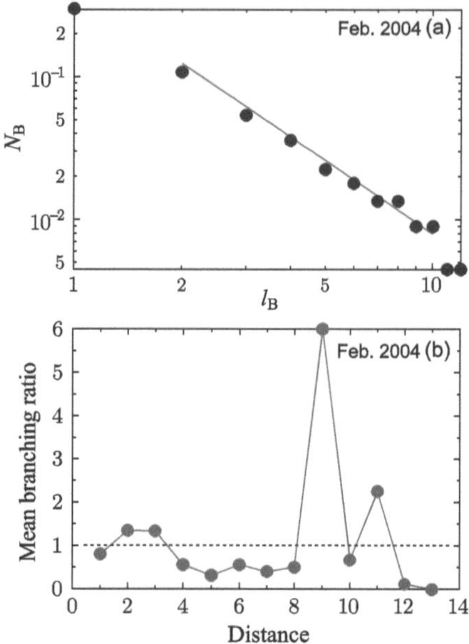

**Fig. 15.13** **a** Fractal scaling and **b** mean branching number for the coauthorship network. Data are for the giant component in February 2004

## 15.6 Conclusions

In this chapter, we have studied the structural properties of the phylogenetic tree, the skeletons of the yeast protein interaction network, and the coauthorship network (whose authors do research on complex networks). Through this study, we found the evolution patterns of these systems. Interestingly, there exists a common feature arising in all these systems, which is the presence of critical and scale-free branching trees. This suggests common evolutionary processes behind these structures: that evolution takes place in bursts, but in a self-organized manner in biological and social systems.

## References


1. C. Darwin, *The Origin of Species by Means of Natural Selection, or the Preservation of Favored Races in the Struggle for Life* (John Murray, London, 1859)
2. E.W. Sayers, T. Barrett, D.A. Benson, S.H. Bryant, K. Canese, V. Chetvernin, D.M. Church, M. DiCuccio, R. Edgar, S. Federhen, M. Feolo, L.Y. Geer, W. Helmberg, Y. Kapustin, D. Landsman, D.J. Lipman, T.L. Madden, D.R. Maglott, V. Miller, I. Mizrachi, J. Ostell, K.D. Pruitt, G.D. Schuler, E. Sequeira, S.T. Sherry, M. Shumway, K. Sirotkin, A. Souvorov, G. Starchenko, T.A. Tatusova, L. Wagner, E. Yaschenko, J. Ye, *Database Resources of the National Center for Biotechnology Information*, Nucleic Acids Res. **37**(Suppl. 1) (Database issue), D5 (2009); Epub Oct 21 (2008)

 


3. T.E. Harris, *Theory of Branching Processes* (Springer, Berlin, 1963)
4. K.-I. Goh, D.-S. Lee, B. Kahng, D. Kim, Phys. Rev. Lett. **91**, 148701 (2003)
5. A. Saichev, A. Helmstetter, D. Sornette, Pure Appl. Geophys. **162**, 1113 (2005)
6. Z. Burda, J.D. Correira, A. Krzywicki, Phys. Rev. E **64**, 046118 (2001)
7. J. Feder, *Fractals* (Plenum, New York, NY, 1988)
8. J.S. Kim, K.-I. Goh, B. Kahng, D. Kim, Chaos **17**, 026116 (2007)
9. S.-H. Yook, F. Radicchi, H. Meyer-Ortmanns, Phys. Rev. E **72**, 045105(R) (2005)
10. C. Song, S. Havlin, H.A. Makse, Nat. Phys. **2**, 275 (2006)
11. K.-I. Goh, G. Salvi, B. Kahng, D. Kim, Phys. Rev. Lett. **96**, 018701 (2006)
12. D.-H. Kim, J.D. Noh, H. Jeong, Phys. Rev. E **70**, 046126 (2004)
13. M. Flajolet, G. Rotondo, L. Daviet, F. Bergametti, G. Inchauspe, P. Tiollais, C. Transy, P. Legrain, Gene **242**, 369 (2000)
14. T. Ito, K. Tashiro, S. Muta, R. Ozawa, T. Chiba, M. Nishizawa, K. Yamamoto, S. Kuhara, Y. Sakaki, Proc. Natl. Acad. Sci. USA **97**, 1143 (2000)
15. A.J.M. Walhout, R. Sordella, X.W. Lu, J.L. Hartley, G.F. Temple, M.A. Brasch, N. Thierry-Mieg, M. Vidal, Science **287**, 116 (2000)
16. R.V. Sole, R. Pastor-Satorras, E.D. Smith, T. Kepler, Adv. Complex Syst. **5**, 43 (2002)
17. J.C. Nacher, M. Hayashida, T. Akutsu, Physica A **367**, 538 (2006)
18. G.P. Karev, Y.I. Wolf, A.Y. Rzhetsky, F.S. Berezovskaya, E.V. Koonin, BMC Evol. Biol. **2**, 18 (2002)
19. S. Wuchty, Mol. Biol. Evol. **18**, 1694 (2001)
20. G.P. Karev, F.S. Berezovska, E.V. Koonin, Bioinformatics **21**, 12 (2005)
21. A. Beyer, T. Wilhelm, Bioinformatics **21**, 1620 (2005)
22. K.-I. Goh, B. Kahng, D. Kim, J. Kor. Phys. Soc. **46**, 565 (2005)
23. J. Kim, P.L. Krapivsky, B. Kahng, S. Redner, Phys. Rev. E **66**, 055101 (2002)
24. D.J. Watts, S.H. Strogatz, Nature **393**, 440 (1998)
25. A.-L. Barabási, R. Albert, Science **286**, 509 (1999)
26. R. Albert, A.-L. Barabási, Rev. Mod. Phys. **74**, 47 (2001)
27. S.N. Dorogovtsev, J.F.F. Mendes, Adv. Phys. **51**, 1079 (2002)
28. M.E.J. Newman, SIAM Rev. **45**, 167 (2003)
29. C. Song, S. Havlin, H.A. Makse, Nature **433**, 392 (2005)
30. K.-I. Goh, G. Salvi, B. Kahng, D. Kim, Phys. Rev. Lett. **96**, 018701 (2006)
31. G. Palla, A.-L. Barabási, T. Vicsek, Nature **446**, 664 (2007)

# Chapter 16
# From Swarms to Societies: Origins
# of Social Organization

Alexander S. Mikhailov

**Abstract** What are the distinguishing features of socially organized systems, as contrasted to the other known forms of self-organization? Can one define a society in abstract terms, without referring to the specific nature of its elements and thus making the definition applicable to a broad class of systems of various origins? Is social evolution different from biological evolution? This chapter attempts to approach such questions in a general perspective, without technical details and mathematical equations.

## 16.1 What Is a Society?

In 1944, Erwin Schrödinger gave a series of lectures in Dublin and later published a book [1] with the title *What is Life?* This small book had a great impact and continues to fascinate researchers. Analyzing experimental data, Schrödinger conjectured that the material basis of biological evolution should be at the molecular level, because only molecules can guarantee safe storage of genetic information and its precise transfer to the next generations. His ideas opened the way to the discovery of the DNA code.

In the same book, Schrödinger also touched on a different aspect of the problem. Why are biological processes so different from what is observed in physical systems? How can one explain that biological systems tend to maintain and increase their order, in apparent contradiction to the second law of thermodynamics, which states that entropy – the measure of disorder – must increase with time? According to him, this behavior is possible because living biological systems are open, receiving energy and/or matter from external sources and dissipating energy or releasing the products into the environment. Together with the flows of mass or energy, entropy can be exported, so that its content within a system remains constant or even decreases with time, despite persistent entropy production.

A.S. Mikhailov (✉)
Department of Physical Chemistry, Fritz Haber Institute of the Max Planck Society, D-14195 Berlin, Germany
e-mail: mikhailov@fhi-berlin.mpg.de

H. Meyer-Ortmanns, S. Thurner (eds.), *Principles of Evolution*,
The Frontiers Collection, DOI 10.1007/978-3-642-18137-5_16,
© Springer-Verlag Berlin Heidelberg 2011

Explicit examples of self-organization processes were subsequently given by Turing [2] and Prigogine [3], initiating many publications on nonlinear pattern formation far from thermal equilibrium. In such studies, reaction–diffusion systems are usually considered. They can be viewed as formed by active nonlinear elements with local interactions through diffusional flows caused by concentration gradients. It has been found that such systems can support a variety of dissipative structures maintained under energy supply. Not only uniform synchronous oscillations, stationary patterns, or traveling waves, but also chaotic regimes and turbulence are possible [4, 5].

Investigations revealed that essentially the same models are repeated again at different levels of biological or ecological organization. For example, actively traveling waves can be observed inside single biological cells and are found in cell populations. They are of principal importance for operation of the heart and are also observed in the brain. Similar wave structures exist in spatially distributed ecological populations. The lesson learned from such investigations was that in mathematical biology attention should be focused on the search for and analysis of generic models describing self-organization processes [6].

With this experience, it seems natural to ask whether there are also some characteristic models that should be used to describe social phenomena. Do social self-organization phenomena represent a variant of behavior already seen in more primitive chemical or biological systems? Or are we dealing here with a distinctive kind of models that become essential only at the next, and higher, level of self-organization?

## 16.2 Swarms and Active Fluids

While particles in a reaction–diffusion system are passively transported by diffusion, agents of a social system are able to move themselves, determining the direction and the magnitude of their motion. This ability is already characteristic for simple, single-cell organisms. In the behavior known as chemotaxis, bacteria can control their active motion, steering according to gradients of chemical substances. The direction and the velocity of active motion of an agent can moreover be influenced by its perception of other active agents in its neighborhood. Again, this is possible already for primitive microorganisms where chemical communication is typical (a cell releases chemical substances into the growth medium, whose presence can be sensed by the other cells, affecting their motions). Such behavior is characteristic for animals such as fish, birds, or sheep that are permanently sensing the presence of other individuals around them and adjusting their motion accordingly [7, 8].

If interactions are attractive, compact groups of actively moving agents become formed. Such groups are generally known as swarms, although they may also bear different specific names (schools of fish, flocks of birds, or herds of sheep). In the swarms, coherent collective motion of a population takes place. Usually, a swarm moves towards areas with higher nutritional value or to avoid predators. Swarms can be seen as possessing a certain degree of collective intelligence [9]. An optimal size of a swarm is typically maintained. Running into an obstacle, a swarm becomes split, but later rejoins. Separation into a number of smaller swarms can take place under a predator attack.

Investigations of swarm models have become popular within the last decade. The motivation is to understand the collective behavior of particular biological species. Many projects dealing with swarm models are, however, stimulated by military applications. The vision is to develop flocks of miniature flying (or otherwise moving) robots that collectively attack and destroy targets.

Under spatial confinement, a high density of agents can be maintained even when attractive interactions between them are absent (short-range repulsive interactions are obviously always present, e.g., as an effect of excluded volume). A population behaves then as an active fluid. Bacterial films provide a good example of such systems. Active fluids are described by models that are an extension of the Navier–Stokes equations for classical fluids (see, e.g., [10]).

Under certain conditions, human populations behave as swarms or active fluids. This is clear when such behavior as evacuation under panic is considered. Pilgrims circling around the sacred stone in Mecca provide an impressive example. A closely related behavior is exhibited by traffic flows. Cars on highways, controlled by intelligent drivers, behave not very differently from primitive bacteria forming biofluids [11, 12]. It is, however, obvious that swarm and active-fluid behavior cannot be considered as a distinguishing property of social self-organization.

## 16.3  Internal Dynamics and Communication

Even molecules may have different internal states, transitions between which follow dynamical laws. The level of internal organization increases sharply when biological cells are considered. Animals and humans are agents with great internal complexity. To specify a state of such an agent, it is not enough to indicate only its spatial location.

In addition to spatial coordinates, an active agent is characterized by a number of internal coordinates, so that the dynamics of a population of agents proceeds both in the coordinate and the internal space. Note that the dimension of the internal space of a population is much larger than that of the coordinate space.

Communicating, agents can exchange information about their internal states. Single cells can communicate using chemical signals or elastic strains. Animals use optical and acoustic signals. Humans have developed sophisticated communication tools, such as languages, and electronic communication is currently employed. In a modern society, local human contacts are complemented by communication through mass media, enabling persons or groups to address a large audience.

When discussing communication aspects, emphasis is often placed on the exchange of signals immediately influencing the dynamics of involved agents, but information can also be laid aside and stored, allowing its communication at a later time and even information transfer to the next generation. Stored information accounts for the emergence of culture, which has played a fundamental role in the transition from biological to social evolution. Note that culture does not necessarily involve language and other advanced means. Material culture is already expressed, for example, in the modifications of environment resulting from the activity of agents. When ants chemically mark their tracks, a primitive form of material culture is already involved [13].

Communication is typically nonlocal in the internal space, in terms of internal variables of communicating agents. There is no general reason to expect that a signal, reflecting an internal state of an individual, will only affect other individuals in close internal states.

## 16.4 Synchronization

Through interactions, coherent dynamics of a group of agents can emerge. Such coherence is already characteristic for swarms which represent groups of agents compactly located in the coordinate space and traveling there as single entities. Coherence can also develop in the internal space of agents. Then, swarms are formed with respect to the internal coordinates of their members. Generally, coherence is revealed in synchronization of individual processes and implies the presence of correlations between dynamical variables of the agents.

When the internal dynamics of elements is cyclic, so that the same internal motions are repeatedly performed, phase variables can be introduced. The phase of an oscillator specifies its current position within the cycle. If complete phase synchronization takes place, dynamical states of all oscillators become identical and the entire population oscillates as a single element. It can also be that not phases but velocities of internal motions become synchronized.

In heterogeneous populations, it may happen that only a subset of agents undergoes synchronization, starting to move as a swarm, while other elements are not entrained. Synchronization can persist in the presence of fluctuations and noise. Not only periodic oscillators, but even elements with intrinsically chaotic dynamics can synchronize [14–16].

Coherence must play an important role in any society. Indeed, the ultimate reason for the existence of a society is that, collectively, a group of agents is able to accomplish tasks beyond the reach of individual members. But this is possible only if actions of the agents are coherent.

However, social organization is certainly much more complex than simple synchronous dynamics. When the states of all members are nearly identical, the whole population behaves just like a single element. Such collective dynamics, while being highly ordered, would obviously lack the complexity and richness expected for a society.

## 16.5 Clustering

Investigations show that, in a population of identical agents, clusters can spontaneously develop as a result of interactions. Effects of dynamical clustering have been extensively discussed by Kaneko in his studies of globally coupled logistic maps [17]. They were also considered for chaotic Rössler oscillators (see [16]).

A good illustration of clustering was provided by a study [18] in which a population of identical agents, each possessing its own neural network, was investigated. The neural network of an individual agent was chosen to generate persistent chaotic oscillations. The agents could communicate, exchanging limited information about their mental states (for details, see [18]). The communication was global, in the sense that each agent responded only to an average of the signals received by it from all other population members and each agent communicated with all of them. A subset of neurons in each network was sensitive to communication, so that the dynamics was determined by a combination of the internal signal and the population-averaged external signals. The relative weight of the external signals in such a combination determined the interaction intensity.

When interactions were absent or very weak, the "mental" dynamics of different agents was independent and noncorrelated. As the interaction strength was gradually increased, a remarkable transition took place, however. The internal states (i.e., instantaneous network activity patterns) of some agents became identical. These coherently operating agents formed several groups or clusters, with the internal states being the same within any of them. The rest of the population remained nonentrained and their internal states were random.

The increase of interaction intensity led to the growth of coherent clusters, until eventually they included all population members. The fully clustered state persisted within an interval of interaction strength. As the interactions were further increased, the clusters disappeared and became finally replaced by a synchronous regime where the internal states of all agents were identical [18].

This sequence of transitions is typical for various globally coupled populations of chaotic elements, including logistic maps and Rössler oscillators (see [16]). In the above example, however, individual oscillators have complex internal organization, each representing a certain neural network. As the strength of global coupling is increased, asynchronous dynamics of individual oscillators become transformed into the synchronous dynamics of the entire ensemble. The final full synchronization is, however, preceded by the regimes with dynamical clustering, where coherent oscillator groups are formed.

The properties of clustered dynamics are interesting. While the dynamics of all elements within a particular cluster are identical, such dynamics is different for different clusters, depending on their relative sizes. Identical agents become spontaneously distributed into a set of coherently operating groups and their internal dynamics gets differentiated, depending on the particular coherent group to which they belong. Thus, symmetry is broken and an ordered state of a population emerges.

Such phenomena can be viewed as a paradigm of primary social self-organization, where a homogeneous population undergoes spontaneous structuring. The seeds of coherently operating groups emerge and grow as the intensity of communication and its efficiency are increased. Each group has its own dynamics, affected, however, by the interactions with other coherent groups and with nonentrained agents, if they are still present.

For populations of logistic maps, special numerical investigations have been performed, revealing that they behave as dynamical glasses [19, 20]. This means that such systems have a large number of different coexisting dynamical attractors, each corresponding to a particular partition of elements into the clusters. Depending on initial conditions, the same population may disaggregate into a different number of clusters of variable sizes. The collective dynamics of a population is different for each of the cluster partitions. Moreover, so-called "replica symmetry" can become violated, implying that, among various possible kinds of collective dynamics in different clustered states, one would always be able to find dynamical behavior of any degree of similarity. It seems feasible that other populations of chaotic elements, including the above example of communicating neural-network agents, also represent dynamical glasses in their clustered states.

It should be noted that clustering is possible even for agents that represent simpler, periodic oscillators. In this case, however, the number of developing clusters and their sizes are usually uniquely determined by interactions between them [16].

Moreover, clusters can develop as a result of a Turing instability in globally coupled systems. In this case, oscillations are absent. All oscillators belonging to a cluster are in the same stationary internal state, but these states vary for different clusters [21].

Generally, clustering means that functional structure develops in an initially uniform population, with different functional roles played by self-organized groups. The presence of such structure distinguishes clustered populations from homogeneous swarms.

## 16.6 Hierarchies

Hierarchies are ubiquitous in nature. Quarks and gluons, as elementary particles, form protons and neutrons, which make up the atomic nuclei, that, together with electrons, combine to form atoms. Atoms give rise to molecules, which in turn constitute solid bodies, fluids, or gases. Interactions between them produce all the systems of the macroscopic world. This physical hierarchy extends further to planets, stars, and galaxies.

The lowest level of the biological hierarchy is formed by elementary biochemical reactions that combine to produce complex chains. A sophisticated system of such reaction chains builds up a living cell. Interacting cells constitute a biological organism. Animals and plants form populations whose interactions determine all processes in the biosphere.

The above examples refer to gross hierarchies whose existence is obvious. A detailed analysis would, however, also often reveal the hierarchical structure of various particular processes in biological organisms, ecological systems, and human societies. Apparently, the hierarchical organization is not accidental. It must be essential for functioning of complex living systems.

To understand the origins and the role of natural dynamic hierarchies, let us return to physics. Subsequent structural levels correspond here to the operation of different forces responsible for interactions between elements. It can be noted that forces acting at lower levels are significantly stronger. The hierarchy of structures has its parallel in a hierarchy of interactions – from extremely strong nuclear forces to relatively weak electromagnetic interactions and further to very weak gravitational effects.

The hierarchy of interactions makes physical systems nearly decomposable. If we could switch off electromagnetic forces, we would see that atomic nuclei still exist, but atoms and molecules already do not form. If gravitational forces were eliminated, planets, stars, and galaxies would be absent, but all lower levels of the structural physical hierarchy would be left intact.

The decomposability implies that, to describe structure formation at a certain hierarchical level, one needs to take into account only the forces operating at this particular level. For example, atomic nuclei are produced by strong nuclear interactions between protons and neutrons. The influence of the electromagnetic force can here be neglected. If we move one step higher and consider atoms and molecules, the dominant role is played by electromagnetic interactions. Such interactions operate in a system made of nuclei and electrons. Though the nuclei actually have a complex internal structure, they can be viewed as simple particles at such a higher structural level. This becomes possible because electromagnetic forces are too weak to interfere considerably with the internal organization of nuclei.

Note that the same forces can give rise to interactions of varying strength. The interactions between atoms and molecules are also essentially electrostatic, though they are much weaker than the interactions between particles inside an atom. This is explained by the fact that the total electric charge of such composite particles is zero, and hence the principal electrostatic forces are shielded to a large extent. The residual forces give rise to weak dipole–dipole or van der Waals interactions, which are responsible for the formation of liquids or solids.

The decomposability of physical systems plays a fundamental role. If all interactions had the same strength, the separation into different structural levels would have not been possible. Then the whole Universe would have been represented just by a single huge nondifferentiated structure.

Apparently, a similar decomposability underlies various biological and social hierarchies. The problem is how to compare and define the "strength" of biological or social interactions. Indeed, all physical forces are quantitatively well defined by the respective laws. Their intensities are measured in the same units and can easily be compared. In contrast to this, no universal dynamical laws are available for biological and social systems. At most, we have here various phenomenological models that describe particular aspects of their behavior.

The strength of interactions in a physical system determines the time scale of the processes resulting from such interactions. The stronger the interactions, the shorter the characteristic time scale of the respective process. Indeed, if an elastic string is stiffer and a stronger force is needed to expand it, the oscillation period of this string is smaller. The structural hierarchy of physical systems corresponds to a hierarchy of their time scales.

This suggests that we can estimate and compare the strength of chemical, biological or social interactions by looking at their characteristic time scales. Viewed from this perspective, interactions between biochemical reactions in a living cell are strong, since they lead to characteristic times of a millisecond. Physiological processes in a human body correspond to weaker interactions, since their characteristic times would typically lie in the range of seconds or minutes. Social dynamics of small human groups proceeds on the scale of hours or days, whereas large social groups evolve only on the scale of months or years.

The separation of time scales is important. Because of it, slow processes of a higher structural level cannot directly interfere with the dynamics at lower levels. However, they can effectively guide this dynamics by setting conditions under which it takes place. Variables of a higher structural level can play the role of control parameters for the subordinated systems of a lower level. Their slow evolution induces instabilities and bifurcations in such systems, switching their dynamical behavior. On the other hand, low-level processes also cannot interfere with the high-level dynamics. These processes are so rapid that their influence on the slow dynamics of a high-level structure is simply averaged out.

Thus, biological and social systems are also nearly decomposable. Self-organization of structural units at any hierarchical level is determined by interactions with the time scale corresponding to this level. When higher levels are considered, elements of a previous level can be viewed as simple objects described by a small number of relevant properties. Consequently, theoretical understanding and mathematical modeling of emerging hierarchically organized patterns becomes possible.

From a general perspective, hierarchical organization provides a solution of an apparent contradiction between the complexity of a dynamics and its stability and predictability. At a first glance, the collective dynamics of more complex systems, consisting of a larger number of various interacting components, would generally be expected to be more complicated and less predictable. If, however, a complex system is appropriately hierarchically designed, its behavior can still be quite regular.

Emergence of hierarchies must represent an important aspect of self-organization behavior. At present, however, this kind of self-organization remains very poorly understood. The difficulty is the formulation of simple mathematical models that would allow spontaneous development of a hierarchy characterized by variation of time scales of the processes taking place at different levels. Although some examples are available, much remains to be done in this direction.

## 16.7 Networks

As we have noted above, interactions between agents are usually not local in terms of their internal states, that is, even agents with strongly different internal conditions can communicate and affect the internal dynamics of each other. Moreover, spatial locality of interactions, requiring that only immediate spatial neighbors communicate, is also gradually being eliminated in biological and social evolution. Acoustical and optical communication has long ranges and the range of modern electronic communication is practically infinite.

Nonlocality of interactions allows an agent to broadcast information to the entire population. On the other hand, it is also obvious that, even in relatively small populations, the amount of information received by an agent from other population members under such global communication would be too large. Only some selected pieces of information, essential for all population members (e.g., alerting of potential common dangers), need to be broadly broadcast.

Most of the information is communicated in an addressed manner, so that an agent effectively interacts only with a relatively small number of other agents. Thus, networks are established. The nodes of a network are individual agents; two nodes are connected by a link if there is communication between them. These networks are distributed dynamical systems. Each agent, occupying a network node, has its own internal dynamics (and may also actively move in the coordinate space). Signals, received by an agent along the links from other agents, affect its internal dynamics. In heterogeneous populations with variation in the properties of agents, the presence of a connection between two nodes can be determined by particular properties of these agents. Then, the connections are fixed and independent of the dynamical processes taking place inside the network. However, connections can also be established or broken depending on the current (or past) internal states of the agents. Already available links can be used more or less intensively for communication, depending on the internal dynamics of agents. Generally, networks are flexible and their architecture can change, reflecting changes in the activity patterns of the population (and also strongly influencing such activity patterns).

The presence of self-organizing dynamical networks can be considered as a general property that distinguishes societies from more primitive forms of population organization, such as swarms.

## 16.8 Coherent Patterns and Turbulence

Interactions between dynamical elements – nodes of a network – may lead to the development of coherent patterns of network activity. The simplest kind of coherence is synchronization, that is, a regime where internal states of all network elements become identical or sufficiently close. Synchronization is already possible for globally coupled systems and it is also known for spatially extended reaction–diffusion systems, where it corresponds to stable uniform oscillations. More complex forms of coherent dynamics, which would correspond to clustering or traveling waves in reaction–diffusion models, should, however, be also possible in networks.

Coherence develops through interactions between elements, and only if these interactions are strong enough. Decreasing the intensity of interaction, destruction of coherent activity patterns, and emergence of chaos (or network turbulence) can usually take place.

Generally, the difference between coherent regimes and turbulence lies in the presence of correlations between internal dynamical states of network elements. Coherence can thus be revealed through the analysis of correlations in the behavior of individual agents. Another way to characterize such differences is to consider the number of effective degrees of freedom involved in the generation of a particular dynamics (in the theory of dynamical systems, this corresponds to the "embedding dimension" of a system).

Coherent network dynamics is governed by a relatively small number of underlying collective dynamical variables, known as order parameters. The dynamics of individual agents are enslaved by order parameters and many internal variables of the agents are controlled by them. In the state of extensive chaos or turbulence, the order parameters disappear and the embedding dimension of the dynamics becomes proportional to the overall size of the system.

Coherent patterns in spatially extended reaction–diffusion systems represent some spatial structures, stationary or time-dependent. In networks where spatial separations and coordinates are irrelevant, coherent patterns are some network structures.

The transition from synchronization to turbulence has been numerically investigated for networks of coupled logistic maps [22]. Similar to globally coupled populations of logistic maps, clustered dynamical states were found within an interval of interaction intensities in the transition region. Several clusters of different sizes were usually observed. Within a cluster, the states of elements were close (but not identical).

Each cluster represents a coherent group and can be viewed as a new, effective dynamical unit. Such new units continue to interact and their mutual dynamics determines the collective behavior of the entire population. While internal states of agents in a group are nearly identical, their temporal evolution depends on the current activity states of other coherent groups.

Considering interactions between the groups, one can notice that they are generally different and their presence and intensity are determined by the architecture of network connections between individual agents. Obviously, two groups of agents can only interact if there are some interactions between their members. Therefore, a system of interacting coherent groups will itself represent a certain dynamical network.

For logistic maps, different cluster partitions are usually possible for the same parameter values. This means that, in the clustering regime, such systems are characterized by multistability. Depending on initial conditions, the same population can build up various coherent states with different cluster organization. The number of coherent groups, their sizes, and composition can be different in each of these states. Since interacting coherent groups form a network, the same network system can give rise to various socially organized networks with different collective dynamics, depending on a particular cluster partition [22].

The transition from synchronization to turbulence, proceeding through self-organized coherent patterns, has also been investigated for networks of periodic oscillators. On the other hand, self-organized stationary patterns, representing an analog of Turing patterns in reaction–diffusion systems, are also possible in networks [21].

## 16.9 Feedback and Control

A society can operate in an ordered and predictable way only if it finds itself in a coherent state and is not degenerated to the state of chaos. On the other hand, its organization should also be flexible, not hindering social evolution and adaptation to environmental changes. To maintain ordered functioning of a society, some control mechanisms should usually be employed.

One mechanism consists in imposing rigid centralized control, where each society member receives commands dictating what its actions should be. In an absolutist state, the sovereign directly controls all actions of his subordinates. Remarkably, centralized control re-emerged in communist states, motivated by the ideology that can be traced back to the mechanistic concepts of the nineteenth century. If the laws of a society are all known, should it not be possible to determine, based on such laws, what are the correct actions of every society member and enforce them?

However, such rigid controls are only possible in systems with primitive organization. When each element of a society is involved in a variety of subtle interactions with other members, centrally issued commands will often destructively interfere with intrinsic social interactions and their effects may be unpredictable. Instead of imposing order, rigid central control may well bring a society to chaos.

A society is a self-organized system. Forces and perturbations used for its control should not interfere destructively with the processes responsible for self-organization. Instead, steering of social processes should be performed by creating conditions and biases that favor the development of the required self-organized structures. The control is particularly efficient near critical points and bifurcations, where even weak perturbations may be enough to select a particular kind of coherent collective behavior.

In principle, control perturbations needed to maintain a system in a desired self-organized state can be computed if the laws governing the system dynamics are known. However, this is rarely the case and, moreover, environmental conditions can vary, making such a method impracticable. Instead, control methods employing certain feedbacks may be employed.

In the feedback control, signals and perturbations acting on a system are collectively generated by all population members or a subset of them. Thus, control perturbations are automatically adjusted to the current activity state of a system and, moreover, they also accommodate changes in the environmental conditions. In this manner, particular self-organized states can be induced and stabilized.

As an example, a recent study [23] can be mentioned. There, a network formed by interacting phase oscillators was considered. Interactions were chosen in such a way that network turbulence was intrinsically established in the absence of control. Global feedback was used, so that each oscillator additionally experienced a force collectively generated by the whole population. The feedback was chosen in such a way that, acting alone, it would have induced synchronization. When the feedback was introduced and its intensity varied, the transition from network turbulence to complete synchronization could be observed. Furthermore, it was also possible to choose the feedback intensity so that the system was kept inside the transition region, that is, in the regime with partial synchronization characterized by certain coherent patterns.

The laws imposed by parliaments and governments in modern societies can be viewed as providing feedback controls, aimed at maintaining a society in a desired state. They should not be too strong and restrictive, disrupting important self-organization processes. Ideally, they must be designed in such a way that only certain biases and preferences are created.

## 16.10 Social Evolution

The transition from biological to social evolution took place when, instead of genetically transferring acquired changes, cultural transfer of innovations from one generation to another became possible. Once social evolution had been initiated, it led, within an unprecedentedly short historical time, to explosive growth of culturally transferred information.

In a changing world, society must evolve in response to variable challenges. At the same time, it should also exhibit robustness, so that it is not destroyed by perturbations.

Robustness can be achieved through negative feedbacks, as already suggested by Wiener in his cybernetics approach. Recent investigations additionally show that robustness can also be enhanced at the structural level, by using networks with special self-correction properties [24]. Such networks, resulting from an evolution process, can maintain their functions despite structural perturbations or action of noise.

On the other hand, a certain degree of fluctuations and variability must be retained to ensure efficient evolution of a society. Similar to biological evolution, a persistent flow of "mutations" is needed to provide the material for subsequent selection. In social systems, mutations represent innovations and seeds of new social structures. Such fluctuations should not be suppressed, hindering social progress.

The presence of fluctuations can also be of vital importance to guarantee that a society recovers, by rearranging itself, after large-scale perturbations. Recovery is facilitated if the society had in advance the seeds of new structures that are required for responding to a perturbation. Otherwise, adaptation may need to go through the process of chaos development, with a danger of complete disruption of social organization.

Thus, the optimal control strategy should be aimed at balancing within the interval separating rigid organization from social chaos. It has been already proposed that balancing on the edge of chaos might be a characteristic property of all complex living systems. There are good reasons to believe that this can be essential for complexly organized social systems, too.

## 16.11 Open Questions and Perspectives

Further progress in mathematical modeling of societies would require extensive studies of self-organization phenomena in dynamical networks. What are the instabilities and bifurcations possible in such systems? What kinds of self-organized coherent patterns, beyond full synchronization and clustering, can spontaneously develop in the networks? What are their characteristic features and how can the presence of self-organized patterns be detected in the data? What are the properties of network turbulence? What coherent patterns can exist in the regimes of intermittent turbulence in networks?

Control of network dynamics is a field where very little research has so far been carried out. How can feedback be introduced that would not impose complete synchronization, but instead stabilize certain desired coherent patterns of network activity? What kinds of feedback are needed if one only wants to suppress hazardous fluctuations, leaving intact basic social self-organization processes?

In addition to dynamical processes that take place in given networks, dynamics of the networks must also be considered. The architecture of a network can change depending on its activity pattern. Growth, disintegration, and fusion of the networks are possible. Feedback may act on the network architecture, so that network architectures that generate needed activity patterns become stabilized or established.

How can networks interact? What are the evolution processes in network populations? How can "culture" be defined in abstract terms and how can cultural evolution be modeled mathematically?

How can networks with specific dynamical properties be designed and engineered? Can functional networks with desired coherent activity patterns, which are robust against random damage and noise, be constructed?

Providing answers to these questions would be important not only for social research. The constructed generic models and approaches can be used in the design and control of robot populations with the elements of social organization. They will be essential in the introduction of mechanisms governing interactions between robots and human communities.

In this chapter, I have covered a broad spectrum of problems and could only outline them briefly. My aim was not to provide a review of the considered topics and I have not provided here a systematic list of references to specific publications. Further details of the discussed models and related references can be found in the monograph [25].

## References


1. E. Schrödinger, *What Is Life? The Physical Aspect of a Living Cell* (Cambridge University Press, Cambridge, 1944)
2. A. Turing, Philos. Trans. R. Soc. Lond. B **237**, 37 (1952)
3. I. Prigogine, R. Lefever, J. Chem. Phys. **48**, 1695 (1968)
4. G. Nicolis, I. Prigogine, *Self-Organization in Nonequilibrium Systems* (Wiley, New York, NY, 1977)
5. S. Mikhailov, *Foundations of Synergetics I. Distributed Active Systems*, 2nd edn. (Springer, Berlin, Heidelberg, 1994)
6. J.D. Murray, *Mathematical Biology* (Springer, Berlin, Heidelberg, 1989)
7. S. Camazine, J.L. Denebourg, N.R. Franks, J. Sneyd, G. Theraulaz, E. Bonabeau, *Self-Organization in Biological Systems* (Princeton University Press, Princeton, NJ, 2001)
8. J.D. Farmer, F. Schweitzer, *Brownian Agents and Active Particles* (Springer, Berlin, Heidelberg, 2003)

 9. E. Bonabeau, M. Dorigo, G. Theraulaz, *Swarm Intelligence: From Natural to Artificial Systems* (Oxford University Press, Oxford, 1999)
10. J. Toner, Y. Tu, Phys. Rev. E **58**, 4828 (1998)
11. I. Prigogine, R. Herman, *Kinetic Theory of Vehicular Traffic* (Elsevier, New York, NY, 1971)
12. D. Helbing, *Verkehrsdynamik* (Springer, Berlin, Heidelberg, 1997)
13. E.O. Wilson, *The Social Insects* (Belknap, Cambridge, MA 1971)
14. Y. Kuramoto, *Chemical Oscillations, Waves, and Turbulence* (Springer, Berlin, Heidelberg, 1984)
15. A. Pikovsky, M. Rosenblum, J. Kurths, *Synchronization: A Universal Concept in Nonlinear Science* (Cambridge University Press, Cambridge, 2002)
16. S.C. Manrubia, A.S. Mikhailov, D.H. Zanette, *Emergence of Dynamical Order: Synchronization Phenomena in Complex Systems* (World Scientific, Singapore, 2004)
17. K. Kaneko, I. Tsuda, *Complex Systems: Chaos and Beyond* (Springer, Berlin, Heidelberg, 2001)
18. D.H. Zanette, A.S. Mikhailov, Phys. Rev. E **58**, 872 (1998)
19. S.C. Manrubia, A.S. Mikhailov, Europhys. Lett. **53**, 451 (2001)
20. S.C. Manrubia, U. Bastolla, A.S. Mikhailov, Eur. Phys. J. B **23**, 497 (2001)
21. H. Nakao, A.S. Mikhailov, Nat. Phys. **6**, 544 (2010)
22. S.C. Manrubia, A.S. Mikhailov, Phys. Rev. E **60**, 1579 (1999)
23. S. Gil, A.S. Mikhailov, Phys. Rev. E **79**, 026219 (2009)
24. P. Kaluza, M. Vingron, A.S. Mikhailov, Chaos **18**, 026113 (2008)
25. A.S. Mikhailov, V. Calenbuhr, *From Cells to Societies: Models of Complex Coherent Action* (Springer, Berlin, Heidelberg, 2002)

# Index



H. Meyer-Ortmanns, S. Thurner (eds.), *Principles of Evolution*,
The Frontiers Collection, DOI 10.1007/978-3-642-18137-5,
© Springer-Verlag Berlin Heidelberg 2011

# THE FRONTIERS COLLECTION

*Series Editors:*

A.C. Elitzur L. Mersini-Houghton M.A. Schlosshauer M.P. Silverman
J.A. Tuszynski R. Vaas H.D. Zeh